Fluid Mechanics and Its Applications

Volume 112

Series editor

André Thess, German Aerospace Center, Institute of Engineering
Thermodynamics, Stuttgart, Germany

Founding Editor

René Moreau, Ecole Nationale Supérieure d'Hydraulique de Grenoble,
Saint Martin d'Hères Cedex, France

Aims and Scope of the Series

The purpose of this series is to focus on subjects in which fluid mechanics plays a fundamental role.

As well as the more traditional applications of aeronautics, hydraulics, heat and mass transfer etc., books will be published dealing with topics which are currently in a state of rapid development, such as turbulence, suspensions and multiphase fluids, super and hypersonic flows and numerical modeling techniques.

It is a widely held view that it is the interdisciplinary subjects that will receive intense scientific attention, bringing them to the forefront of technological advancement. Fluids have the ability to transport matter and its properties as well as to transmit force, therefore fluid mechanics is a subject that is particularly open to cross fertilization with other sciences and disciplines of engineering. The subject of fluid mechanics will be highly relevant in domains such as chemical, metallurgical, biological and ecological engineering. This series is particularly open to such new multidisciplinary domains.

The median level of presentation is the first year graduate student. Some texts are monographs defining the current state of a field; others are accessible to final year undergraduates; but essentially the emphasis is on readability and clarity.

More information about this series at http://www.springer.com/series/5980

Roberto Mauri

Transport Phenomena
in Multiphase Flows

 Springer

Roberto Mauri
DICI
Università degli Studi di Pisa
Pisa
Italy

ISSN 0926-5112   ISSN 2215-0056   (electronic)
Fluid Mechanics and Its Applications
ISBN 978-3-319-38370-5   ISBN 978-3-319-15793-1   (eBook)
DOI 10.1007/978-3-319-15793-1

Springer Cham Heidelberg New York Dordrecht London
© Springer International Publishing Switzerland 2015
Softcover reprint of the hardcover 1st edition 2015

Translation from the Italian language edition: *Elementi di fenomeni di trasporto* by Roberto Mauri,
© Pisa University Press 2005. All rights reserved

Printed on acid-free paper

Springer International Publishing AG Switzerland is part of Springer Science+Business Media
(www.springer.com)

Preface

This textbook is an introduction to the transport of the three quantities that are conserved in nature, namely mass, momentum, and energy (the transport of electric charge can be seen as a particular case of the transport of mass). It is the outgrowth of 30 years of teaching this material to students of chemical, mechanical, nuclear, and biomedical engineering, both in the U.S. and in Italy.

No transport phenomena textbook could be written without referring to the first modern book on this subject, namely *Transport Phenomena*, published by R.B. Bird, W.E. Stewart, and E.N. Lightfoot in 1960, which has constituted the gold standard for all textbooks that have been written after it. In that book, the authors intended to answer the "current demand in engineering education, to put more emphasis on understanding basic physical principles than on the blind use of empiricism." To understand the meaning of this statement, one has only to look at the typical heat and mass transfer book of the 1940s: no mathematics is required beyond elementary algebra and no partial differential equations can be found in those texts. In fact, Bird, Stewart, and Lightfoot went the opposite way, as they first applied the mathematical framework of continuum mechanics to derive the fundamental governing equations at the microscopic level and only afterward, by integrating them, they obtain the macroscopic equations of mass, momentum, and energy balance. Therefore, microscopic balances precede any coarse-grained analysis: for example, the Navier-Stokes equation is derived in Chap. 3, while for the Bernoulli equation, one has to wait until Chap. 7.

This very rigorous approach is coupled to an extremely powerful idea: that of unifying all types of transport phenomena, describing them within a common framework, which is in terms of cause and effect, respectively, represented by the *driving force* and the *flux* of the transported quantity.

In this textbook we retain this basic idea. However, I decided to reverse the way this material is presented to students. Fifty-five years have passed since the appearance of Bird, Stewart, and Lightfoot's textbook, and engineering students are now much more proficient in advanced mathematics. On the other hand, I feel, they have not been exposed enough to "common sense" physics. In fact, often I find myself presenting this subject to students who had no previous exposure to fluid dynamics or heat and mass transfer and have no idea of what a pressure drop or a heat flux are. Therefore, I try to strike a balance between a rigorous explanation of the fundamental laws that govern these subjects and an intuitive approach that stresses where these laws come from. Presenting the Navier-Stokes equation before any basic macroscopic balance would be like explaining electromagnetism by deriving Maxwell's equation before showing the basic experimental results by Faraday and Ampere. This is why I think that, from a didactical point of view and in light of the type of students we are dealing with today, it is better to describe phenomena first from a macroscopic point of view, even at the expense of mathematical rigor, and only afterwards, more advanced treatments can be carried over.

This textbook has been written as a teaching tool for a two-semester course (or two one-semester courses) on transport phenomena. It is a modular teaching tool, though. So, for example, if transport phenomena are taught in a single one-semester course, one should follow only the macroscopic treatments, described in Chaps. 1–5, 9–11, 14–16, postponing the study of the material covered in the other chapters to a subsequent, more advanced course. In any case, the book offers an abundant resource in the sense that the material covered is much more than what one can hope to cover in two semesters. So, it is up to the instructor to choose which subjects should be favored, also based on the students' needs: mechanical engineering students will be more interested in turbulence, while biomedical engineering students will tend to prefer surface phenomena.

Special mention should be made of the problems that are proposed at the end of each chapter. First of all, their importance cannot be underestimated: a student cannot claim to understand a subject until she/he can solve problems. However, while in Anglo-Saxon universities, problem solving is part of the students' homework assignments, elsewhere, they are solved in class, and therefore, they become an integral part of the course. This is why here the solution of the problems are provided in the Appendix.

Finally, I would like to answer a common complaint that I heard from my students, namely that in this textbook, they cannot find all the physical properties that are required to solve some of the proposed problems. This deficiency is deliberate, being motivated by the fact that in this way the students are forced to look outside and find the missing data.

Pisa, November 2014   Roberto Mauri

Acknowledgments

I would like to thank the many generations of students whose questions and complaints have contributed to shape this textbook. I would also like to thank a few colleagues who have enlightened me on how transport phenomena should be taught. In particular, I would like to mention John F. Brady, who showed me how to find the correct scaling of different problems in very intuitive ways, and Andrea Lamorgese, for his line-by-line comments of the text.

Pisa, November 2014

Roberto Mauri

Contents

Chapter 1
Thermodynamics and Evolution

Abstract This is an introductory, yet fundamental chapter, where all the relevant concepts that are encountered in transport phenomena are defined. After a brief introduction (Sect. 1.1) on what transport phenomena consist of, in Sect. 1.2 we describe the relation between thermodynamics and transport phenomena, by defining the condition of local equilibrium. We explain that, under very general conditions, although far from thermal and mechanical equilibrium, we can speak of thermodynamic quantities such as temperature and pressure. This idea is further explored in Sect. 1.3, where the basic concepts of continuum mechanics are briefly sketched. Then, in Sect. 1.4, we show that mass, momentum, and energy can be transported through two fundamentally different modalities, namely convection and diffusion. The former is a time reversible process due to a net movement of the fluid, and the related convective fluxes admit exact analytical expressions. On the other hand, diffusion is intrinsically irreversible, and diffusive fluxes are expressed through so called constitutive relations, that characterize the fluid at the molecular level. In the case of ideal gases, as shown in Sects. 1.5–1.7, diffusion of momentum, energy and mass can be modeled rigorously, leading to Newton's, Fourier's and Fick's constitutive relations, respectively. The analogy between different transport phenomena is further explored in Sect. 1.8, showing that diffusion can be modeled through a random walk process, so that the mean square displacement of the appropriate tracer of momentum, energy or mass grows linearly with time. Finally, in Sect. 1.9, a few examples of diffusion are presented.

1.1 Introduction

Transport phenomena, as their name indicates, consist of the flows of conserved physical quantities, such as energy, momentum and matter, from one region to another within systems that are constituted of a very large number of elementary particles, like molecules or atoms. As in thermodynamics, these systems are

© Springer International Publishing Switzerland 2015
R. Mauri, *Transport Phenomena in Multiphase Flows*,
Fluid Mechanics and Its Applications 112, DOI 10.1007/978-3-319-15793-1_1

assumed to be continuous, although sometimes, for sake of clarity, we will refer to their molecular structures.

The main distinctive characteristics of this discipline is to unify the physical and mathematical approach of all types of transport. To clarify this point, let us consider the heat transport, when a difference of temperature between two reservoirs induces a heat flux from hot to cold regions. Now, if we consider other types of transport, we realize that they can also be interpreted in the same way, that is in terms of a cause, i.e., a *driving force*, and an effect, that is the *flux* of the transported quantity. Let us see some examples.

- A pot of water on a hot plate. The temperature difference between the two sides of the pot wall causes a heat flux from the hot side to the cold side that heats the water.
- A drop of ink in a water glass. The concentration difference between the inside and the outside of the drop induces a flux of ink from regions of high concentration to low that continues until the ink is uniformly distributed.
- A pressure difference within a pipe induces a mass flux of the fluid that is contained within the pipe from regions of high pressure to low that continues until the pressure is uniform.
- The velocity difference between two regions of a fluid induces a momentum flux from regions at high velocity to low that continues until the momentum is uniform.
- An electric potential difference induces an electric current, that is a flux of electric charge, from regions at high potential to low.

Let us consider the first example: because of the heat transported from warm to cold regions, that is from the hot plate to the water, the water temperature will increase, tending to become equal to that of the hot plate, when the heat will stop flowing. Then, the heat flux remains stationary (i.e., constant in time) only when the temperature of the hot plate and that of the water are maintained constant, for example by continuously cooling the water, i.e., an energy consuming operation. Now, since at steady state the internal energy of the pot wall is constant, this energy removal must balance the energy entering through the heat flux.

In general, an isolated physical system tends to move towards a state of stable equilibrium, where temperature, composition, pressure, momentum and electric field are uniform (i.e., constant in space). On the other hand, if the differences (or, we should say, the gradients) of these quantities, i.e., the driving forces, are nonzero, then a constant flux of energy, mass, momentum and electric charge is induced. This state of non-equilibrium can be maintained only through the introduction of an energy into the system, which is then *dissipated*[1] during the transport process. In other words, during the transport of a physical quantity some amount of

---

[1]The term *energy dissipation* here means *heat conversion*. In fact, as the energy is conserved, it cannot be consumed.

energy is dissipated that, at steady state, must be continuously reintegrated, by introducing it into the system from the outside.

Going back to the example of the pot on the hot plate, the transport of heat does not occur with the same ease in any material, as it is easier when the pot is made of a heat conducting material. This is why we will use a pot made of aluminum (i.e., an excellent conductor) when we want to boil some water as fast as possible, while we will use a pot made of terracotta (i.e., a poor conductor) when we want to prepare a sauce, requiring small and regular heat fluxes for a long cooking time. This sort of friction, opposing heat flux, is a characteristic of our system, denoted *thermal resistance*, defined as the proportionality term between temperature gradient and heat flux. In general, in transport phenomena we assume that the driving force is proportional to the flux through a coefficient, called *resistance*, measuring the opposition of the system against transport. This resistance is the cause of the energy dissipation.

In the following, the concepts of driving force, flux and resistance will be applied to fluid mechanics, heat transport and mass transport, by deriving balance equations of three of the four physical quantities that are conserved in nature: momentum, energy and mass. The transport of the fourth quantity, namely the electric charge, will not be considered here although it can be considered as a particular case of the transport of mass and thus it can be studied following the same procedure.

1.1.1 Statics and Dynamics

If we pour hot water into a container whose walls are impermeable to the transfer of mass, momentum and heat (i.e., the container is closed, rigid and insulated), the water, that is initially subjected to internal movements, has a non-uniform temperature and is partially evaporating, eventually will stop moving, will reach a uniform temperature and will stop evaporating,[2] meaning that it will reach, respectively, mechanical, thermal and chemical equilibrium. In general, if we consider a complex physical system, i.e., composed of a very large number of elementary particles, and we isolate it, the system will evolve until it reaches a state, called of equilibrium, after which it will stop evolving in time. These rules are the basis of thermodynamics and constitute its two laws: the first law states that the energy of an isolated system remains constant so that, although energy can be converted from one form to another, it cannot be created nor destroyed. The second law states that, during its evolving from one state to another, an isolated system tends to assume its state of stable equilibrium.

---

[2]In other words, microscopically, the quantity of water evaporating will be balanced by that of the vapor condensing.

So, thermodynamics tells us that the evolution of an isolated system progresses towards its stable equilibrium state, always conserving its total energy. However, if we wish to know how such evolution takes place, thermodynamics offers no help. In fact, despite its misleading name, thermodynamics is the science that studies complex systems at equilibrium, so it is of little help in studying *transformations*, that are all processes taking place in non- equilibrium conditions. For example, in the previous instance with hot water, thermodynamics predicts that the water will reach a final equilibrium temperature and it even indicates what this temperature will be, but it cannot say anything about how long this process will take, nor how it will happen.

At this point, it would be tempting to conclude that thermodynamics is useless: after all, when we look around, all surrounding systems are far from being at equilibrium. Our body, for example, is constantly out of thermal equilibrium when it is alive, because its 37 °C temperature rarely coincides with the ambient temperature, and equilibrium is reached only after its death. Our conclusion, however, would be wrong because thermodynamics indicates also the limits and the constraints of any process. For example, our body succeeds in maintaining its state of thermal non-equilibrium at the expense of a given energy input contained within the air and the food that we inject. Thermodynamics predicts the minimum input of energy that we must inject in order to produce the minimum work that is necessary to remain alive. However, if we want to determine in which way and how fast this energy input will be dissipated, thermodynamics is of no help: to do that, we must know how energy (and mass and momentum, as well) moves from one region to another. This is the subject of *transport phenomena*.

1.2 Local Equilibrium

In this section, we want to determine when a system can be regarded to be, locally, at thermodynamic equilibrium, so that all thermodynamic variables (temperature, pressure, entropy, internal energy, chemical potential, fugacity, etc.) can be still defined and we can apply, locally, all the thermodynamic relations. This is the so-called condition of *local equilibrium*.

The importance of the condition of local equilibrium is obvious: while some ideal processes (including reversible processes) can be described as if they were composed of a series of equilibrium states and therefore can be successfully modeled using thermodynamics, real systems typically evolve through states that are far from equilibrium (even when their initial and final states are at equilibrium) so that thermodynamics cannot, strictly speaking, be applied. Yet, when we study, for example, a heat exchanger, in which heat passes from a hot to a cold source and a fluid flows from high pressure to low, we encounter terms such as temperature and pressure distribution, i.e., we use terms, such as temperature and pressure, which were introduced in thermodynamics to characterize equilibrium systems. This would seem to indicate that even in conditions of apparent non-equilibrium, in which the temperature and pressure are not uniform, these variables can be sometimes defined *locally* and in this section we intend to study when this is possible.

We know that in a system composed of a finite number N of particles and maintained at thermodynamic equilibrium, any thermodynamic intensive variables A fluctuates around its equilibrium value, $\bar{A}$, which is constant in time and uniform in space, so that the relative value of these fluctuations, $\delta A/\bar{A}$, is proportional to $1/\sqrt{N}$. Thus, we can say that a system is in a condition of equilibrium when these local fluctuations are greater than the variations due to spatial or temporal inhomogeneity of $\bar{A}$. More precisely, a system is in a condition of local equilibrium when the following two conditions are met:

(a) You can divide the system into elementary volumes (which constitute what are generally referred to as "material points") that are large enough to contain a large number N of particles, so that the fluctuations δA of each physical quantity A are small, i.e., $\delta A/\bar{A} \ll 1$.

(b) The variation of A due to the macroscopic gradient $\nabla \bar{A}$ is smaller than the fluctuations δA. Similar considerations apply to the temporal variations of $\bar{A}$.

Therefore, denoting by λ the linear dimension of these elementary volumes, the condition of local equilibrium requires that:

$$\frac{\lambda |\nabla \bar{A}|}{\bar{A}} \le \frac{\delta A}{\bar{A}} \ll 1. \tag{1.2.1}$$

For example, in a gaseous system, assuming we want to define any quantity within a 0.1 % accuracy (i.e., $\delta A/\bar{A} = 10^{-3}$) and since $\delta A/\bar{A} \propto 1/\sqrt{N}$, where N is the number of particles contained in the elementary volumes, the volume will contain $N \approx 10^6$ elementary particles, occupying a volume $\lambda^3 \approx N/n \approx 10^{-14}\,\text{cm}^3$, corresponding to a linear dimension $\lambda \approx 0.1$ µm. Note that the number density n of an (ideal) gas is:

$$n = \frac{\rho N_A}{M_w} = \frac{10^{-3}\,\text{g/cm}^3\; 6 \times 10^{23} part./\text{moli}}{10\,\text{g/moli}} \approx 10^{20}\frac{part}{\text{cm}^3}. \tag{1.2.2}$$

Hence we see, for example, that the maximum gradient of temperature that we can impose while satisfying the condition of local equilibrium is $\nabla \bar{T} \approx 10^{-3}\bar{T}/\lambda$, and therefore for ordinary temperatures the condition of local equilibrium requires: $\nabla \bar{T} < 10^4$ K/cm, which is clearly satisfied in all reasonable cases. For liquid or solid systems, the condition of local equilibrium is applicable even more easily.

At this point, we need to think about how to change the fundamental thermodynamic concepts of pressure, temperature and density when they are applied to non-homogeneous and non-stationary systems, in particular in the case of fluids in motion. We realize now that the last two variables do not require any fundamental "revision": as they are related to the energy and mass of the system, they are in fact inherently scalar quantities and, therefore, can be defined as the temperature $T(\mathbf{r}, t)$ and the density $\rho(\mathbf{r}, t)$ of the system at a point $\mathbf{r}$ and at a certain time t. In contrast, when applied to a continuous system, the concept of pressure, being related to the momentum of the system, must be reconsidered in the context of continuum mechanics.

1.3 Introduction to Continuum Mechanics

In a macroscopically continuous system, the particles that compose it microscopically are subjected to external forces, such as gravity or electric fields, as well as to forces due to intermolecular interactions. Depending on the magnitude of such forces, the system is defined as solid, liquid or gaseous. If we imagine to divide the system into two parts through a separation surface, each part will exert a force on the other, equal to the sum of the intermolecular forces that the particles located on one side exert on those located on the other side of the surface. At the macroscopic level, this force per unit area is called stress, σ, and can be decomposed into its two components, normal and tangential with respect to the separation surface (see Fig. 1.1). The normal component is called pressure, p, while the tangential is the shear stress τ.

In solid bodies, the intermolecular forces are so strong that elementary particles have their own fixed position (think of the atoms in a crystal lattice), so that, macroscopically, solids tend to maintain their shape. In contrast, fluids deform freely, and when they are sheared they change their shape, as layers of fluid slide relative to one another. During this sliding, microscopic friction forces develop within the fluid, that translate macroscopically into a shear stress. When the fluid is

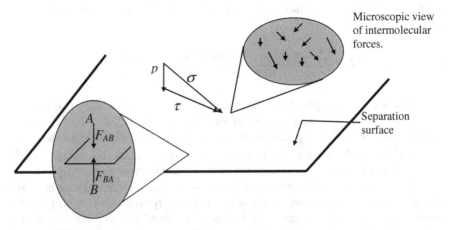

Fig. 1.1 Normal and tangential components of a surface force with respect to the separation surface

at rest, however, these dissipative forces are obviously absent, and therefore, as we shall see, only normal stresses, i.e., pressure, are present.

As is well known from thermodynamics, the state of a pure substance at equilibrium, such as a fluid at rest, is completely determined as long as two independent variables are fixed.[3] Therefore, for a given temperature T and pressure p, the fluid density ρ is fixed.[4] If the density changes are very small compared to variations in T and p, the fluid can be approximated as incompressible, while when such changes are significant, the fluid is compressible (just think of the ideal gas, where ρ is proportional p/T). Typically, liquids are assumed to be incompressible and gases are compressible, even if, obviously, there are cases in which large temperature excursions cause relevant density changes in liquids as well while, conversely, in other processes the changes of pressure and temperature are so small that the gas density can be considered constant.

1.3.1 Pressure

As we have seen in the previous section, the pressure is the force per unit area applied in a direction perpendicular to the surface of an object. More precisely, it is the limit of the force/area ratio when the area tends to zero, with $[p] = M L^{-1} T^{-2}$. The SI unit of pressure is the Newton per square meter, which is called the *Pascal* (Pa), i.e., 1 Pa = 1 N m^{-2}. As 1 Pa corresponds to the pressure exerted by a 1 g weight on a 10 cm × 10 cm area, it is very small, thus pressures are often expressed in kilopascals.

Now, we can imagine to draw a surface about any point of a fluid at rest, separating the fluid into two parts. Each of these two parts will exert an equal and opposite force on the other, because if it were not so, the net force resulting from the action of the fluid on the surface will cause the movement of the fluid.[5] Indeed, a simple force balance on the two faces of the element of volume of infinitesimal thickness of Fig. 1.2 shows that $p_{up} = p_{down}$, that is the pressure of the fluid located on one side of the separation surface is equal to the pressure of the fluid on the other side. As for the tangential stresses, although the force balance of Fig. 1.2 shows that $\tau_{up} = \tau_{down}$, these tangential forces would cause a relative sliding of the fluid layers, so that we may conclude that the shear stresses in a fluid at rest are null.

At this point, the fundamental question that we ask is whether the pressure exerted at a point depends on the orientation of the surface, that is whether the ratios,

---

[3]This is a consequence of Gibbs' phase rule.

[4]The density is the inverse of the specific volume and is defined as a mass per unit volume, $[\rho] = M L^{-3}$. Within the SI system, it is measured in Kg m^{-3}.

[5]Microscopically, the stress exerted by the fluid located on one side of the separating surface on the fluid located on the other side is equal to the sum of all the forces F_{AB} describing the interaction between all pairs of molecules A and B located on opposite sides of the separating surface (see Fig. 1.1). Clearly, as $F_{AB} = -F_{BA}$, we may conclude that $\sigma_{up} = \sigma_{down}$.

Fig. 1.2 Pressure and shear forces exerted by a fluid on both sides of a virtual surface

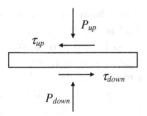

Fig. 1.3 Forces on a fluid element at rest

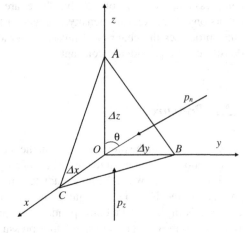

$$p_x = \lim_{\Delta S \to 0} \frac{\Delta F_x}{\Delta S_x}; \quad p_y = \lim_{\Delta S \to 0} \frac{\Delta F_y}{\Delta S_y}; \quad p_z = \lim_{\Delta S \to 0} \frac{\Delta F_z}{\Delta S_z},$$

are equal to each other. After all, although in thermodynamics we take for granted that the pressure is a scalar,[6] so that $p_z = p_y = p_z$, this is by no mean obvious. To answer this question, consider the tetrahedron of Fig. 1.3, built around the Cartesian axes xyz, and write the forces exerted on its faces by the surrounding fluid along, for example, the z direction. Since those are pressure forces (in fact, with no fluid movement, the forces are perpendicular to the surfaces), the force balance will include the force $F_z = p_z S_{OBC}$ exerted on the OBC face, the z component of the force $F_n = p_n S_{ABC}$ exerted on the ABC face, and the gravity force, i.e.,

$$p_z S_{OBC} - p_n S_{ABC} \cos \theta - \rho g V_{OABC} = 0,$$

where ρ is the fluid density, g the acceleration of gravity, $V_{OABC} = \Delta x \Delta y \Delta z / 6$ is the volume of the tetrahedron, θ is the angle indicated in figure, while p_z and p_n are the

---

[6]This is due to the fact that the pressure, p, is thermodynamically conjugated with the volume, V: as V is a scalar, so is p.

mean pressures exerted by the fluid on the corresponding faces. Now, considering that $S_{OBC} = \Delta x \Delta y/2$, $S_{OBC} = \Delta x \Delta y/(2\cos\theta)$, dividing by $\Delta x \Delta y/2$ we obtain:

$$p_z = p_n + \rho g\, \Delta z/3.$$

At this point, squeezing the tetrahedron towards the origin, keeping the angle θ constant, obviously Δz tends to zero, while the mean pressures will be replaced by the local pressures at the origin, indicating that $p_z = p_n$. Similar consideration can be drawn along the other directions, so that we may conclude that

$$p_x = p_y = p_z = p, \tag{1.3.1}$$

indicating that the pressure at a point in a static fluid is independent of the orientation of the surface.

1.3.2 Shear Stresses

In the previous sections we have repeatedly stated that the shear stress is related to the internal friction of the fluid. The easiest way to see it is to note that, given two adjacent layers of fluid that are moving with different speeds, due to the mutual attraction the slower molecules will tend to accelerate, retarding those that move more rapidly, which is tending, at the end, to cancel any difference in speed. In the case of a gas, the same result is obtained as an effect of molecular diffusion: faster molecules diffuse in regions populated by slower particles and, through the consequent collisions, transmit momentum to the slower molecules. Obviously, if no force were applied from the outside, the velocity difference between the two layers could not be maintained and the fluid would end up moving with a uniform velocity. At the macroscopic level, these molecular interactions can be seen (see Fig. 1.4) as a net flux of momentum, J_Q, directed along the direction normal to the fluid motion. The momentum flux, being the momentum that crosses the unit area per unit time, is a force per unit area, i.e., it has the same units as a stress.

This same mechanism, by which two adjacent fluid threads tend to move at the same speed, can be described through the shear stresses τ, which are "friction" forces along the direction of motion of the fluid. Now, considering that the shear stress has the same units as the momentum flux, the forces acting on the faces of a

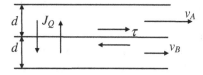

Fig. 1.4 Momentum fluxes and shear stresses

Fig. 1.5 Forces acting on the
faces of a fluid element

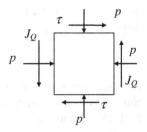

fluid element are indicated in Fig. 1.5, where we have taken into account that: (1) as already noted, in a force balance at a microscopic level the volumetric forces, such as gravity, are negligible compared to the surface forces, and therefore can be omitted; (2) the pressure is independent of the direction of the surface on which it acts; (3) the two shear stresses acting on the horizontal surfaces must be equal to each other, as shown by a simple balance of forces; the same goes for the two momentum fluxes J_Q acting on the vertical surfaces. Hence, imposing that the moment of the forces indicated in Fig. 1.5 be equal to zero yields:

$$\tau = J_Q, \tag{1.3.2}$$

indicating that, as a consequence of the angular momentum balance, shear stress and momentum flux have equal magnitude, though they are perpendicular to each other.

1.4 Convection and Diffusion

Physical quantities such as mass, momentum, and energy can be transported in two fundamentally different ways. The first mode, called convection, is related to the overall, macroscopic motion of the medium in which these quantities are defined. Thus, for example, the convective transport of energy (heat) consists of the net motion of a warmer fluid towards a colder region: through their coherent motion, the fluid molecules carry a thermal energy, related to the fluid temperature. The same can be said of the convective transport of mass and momentum, where the fluid molecules carry a mass or a momentum.

The second mode of transport is diffusion, in which mass, momentum and energy move from one point to another without any net displacement of matter; as we will see in Sects. 1.5 and 1.6, diffusion is related to the incoherent motion of the fundamental constituents of the medium. In the case of energy transport, this mechanism refers to the process in which heat moves from the warmer to the colder regions of a body without the particles making up the body having any overall, coherent motion (think of a metal spoon that heats up when immersed in a hot liquid). In the case of momentum transport, diffusion is the process by which faster fluid threads can accelerate adjacent slower ones; therefore, J_Q in Eq. (1.3.2) is of a diffusive type.

In the following, we describe transport phenomena in terms of fluxes of mass, momentum, and energy, defined as the amount of, respectively, mass, momentum, and energy that cross a surface of unit area per unit time.

1.4.1 Convective Fluxes

As previously noted, during convection all fluid particles move with the same, ensemble velocity v. Therefore, within a fluid moving with a uniform velocity v, consider a cylinder of section S and length $v\Delta t$ (see Fig. 1.6); in a time interval Δt, the fluid mass $M_c = \rho V$, with $V = Sv\Delta t$ the volume of the cylinder, will cross the section S. Therefore, based on the definition of mass flux as the mass M_c divided by S and by Δt, we obtain:

$$J_{Mc} = \frac{M_c}{S\Delta t} = \rho v. \qquad (1.4.1)$$

In the same way, the convective flux of momentum, J_{Qc}, equals the ratio of the momentum Qc contained in this cylinder and the product of S by Δt, where Q_c is the product of the momentum per unit volume, $Q_c/V = \rho v$, by the volume, $V = Sv\Delta t$, i.e.,

$$J_{Qc} = \frac{Q_c}{S\Delta t} = \rho v^2. \qquad (1.4.2)$$

Finally, the convective flux of internal energy J_{Uc}, equals the ratio of the internal energy U_c contained in this cylinder and the product of S by Δt, where U_c is the product of the internal energy per unit volume, $U_c/V = \rho c_v(T - T_0)$, by the volume, $V = Sv\Delta t$, i.e.,

$$J_{Uc} = \frac{E_c}{S\Delta t} = \rho c_v v(T - T_0). \qquad (1.4.3)$$

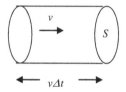

Fig. 1.6 Fluid moving with a uniform velocity

Here c_v is the specific heat per unit mass, while $T - T_0$ is the temperature change relative to an arbitrary temperature reference, T_0.

In general, the convective flux of any quantity A is $J_A = av$, where $a = A/V$ is the density of A.

1.4.2 Diffusive Fluxes and Constitutive Relations

In the previous Section we studied the interaction between adjacent fluid threads and saw that: (1) the flux of momentum in a direction perpendicular to the motion, which is diffusive in nature (as there is no convection in that direction) is equal to the shear stress in the longitudinal direction; (2) shear stress and shear rate are linked to each other. In fact, the relationship between shear stress and shear rate, called constitutive relation of the shear stress, is one of the fundamental dynamic properties of a fluid.

Typically, as the velocity gradient increases, the shear stress also increases, i.e., the curve $\tau = \tau(dv/dz)$ is monotonically increasing. The study of the relationship between shear stress and shear rate is the main subject of rheology.

The simplest rheological behavior of fluids is that of the so-called Newtonian fluids, in which the shear stress is proportional to the velocity gradient. All gases, most of pure liquids and solutions with low molecular weight (e.g., oils and dilute solutions of polymers) behave as Newtonian fluids and therefore follow *Newton's constitutive relation*:

$$\tau = -\mu \frac{dv}{dz}. \tag{1.4.4}$$

where μ is the fluid *viscosity*, sometimes also called *dynamic viscosity*, which then expresses the internal friction, or the internal resistance of the fluid to flow. Dimensionally, the viscosity has units $[\mu] = M\,L^{-1}T^{-1}$ and in the SI system it is expressed in $N\,m^{-2}\,s$, i.e., $kg\,s^{-1}\,m^{-1}$, commonly called *Poiseuille*.[7] In CGS units, instead, the *Poise* (P) is used, which is defined as $1\,P = 1\,g\,cm^{-1}\,s^{-1} = 0.1\,kg\,m^{-1}\,s^{-1}$ or, more conveniently, in *centipoise* $(1\,cP = 0.01\,P)$. The viscosity of water at room temperature is about $1\,cP = 10^{-3}\,kg/ms$. In summary: $1\,Poiseuille = 10\,P = 1000\,cP$.

The constitutive equation of the shear stress can be rewritten as:

$$J_{Qd} = -v \frac{d(\rho v)}{dz}, \tag{1.4.5}$$

establishing that the diffusive momentum flux, J_{Qd}, is proportional to the gradient of the momentum density $q = \rho v$, through a proportionality coefficient $v = \mu/\rho$.

---

[7] Both poiseuille and poise are named after Jean Léonard Marie Poiseuille (1797–1869), a French physicist and physiologist.

The quantity v is called *kinematic viscosity* has units L^2T^{-1} and is expressed in m²/s in the SI system.

This equation is similar to *Fourier's constitutive relation*[8] for the diffusive transport of heat (commonly called *conduction*),

$$J_{Ud} = -k\frac{dT}{dz},\qquad(1.4.6)$$

where J_{Ud} is the diffusive heat flux, i.e., a flux of internal energy, while k is the coefficient of thermal conductivity. Denoting by c_v the specific heat per unit mass, whereas $u = c_v (T - T_0)$ is the specific internal energy (per unit mass) and ρu is the energy density (i.e., the internal energy per unit volume), Fourier's law states that the diffusive flux of internal energy is proportional to the energy density gradient via the coefficient of thermal diffusivity $\alpha = k/\rho c_v$, that is,

$$J_{Ud} = -\alpha\frac{d(\rho u)}{dz}.\qquad(1.4.7)$$

Thermal diffusivity α has the same units, L^2T^{-1}, as kinematic viscosity v and therefore in the SI system it is expressed in m²/s.

The Eqs. (1.4.5) and (1.4.7) are similar to *Fick's constitutive relations*,[9] describing the specific (i.e., per unit mole or mass) diffusive transport of a species A in a mixture,

$$J_{Ad} = -D\frac{dc_A}{dz},\qquad(1.4.8)$$

where the diffusive flux of A is proportional to the gradient of concentration (that is the molar density, or the number of moles per unit volume) through a diffusion coefficient, D, having again units of L^2T^{-1}.

Therefore, kinematic viscosity, v, thermal diffusivity, α, and molar diffusivity, D, all have the same L^2T^{-1} units. For sake of convenience, these quantities are expressed in the CGS system, defining the *Stokes* (1 S = 1 cm²/s) and the *centistokes* (1 cS = 0.01 S), where 1 m²/s = 10^4 S = $10^6 cS$. Physically, v is a measure of the "rate" of momentum diffusion, i.e., how rapidly it can propagate in the absence of convection, exclusively due to molecular interactions. In the same way, α e D indicate the "rate" at which heat or mass can diffuse in the absence of any coherent motion due to convection.

---

[8]Jean Baptiste Joseph Fourier (1768–1830) was a French mathematician and physicist.
[9]Adolf Eugen Fick (1829–1901) was a German physician and physiologist.

1.5 Viscosity

Consider a gas of uniform density, ρ, and temperature, T, that moves macroscopically along the x direction with an average velocity, $u = u(z)$, depending on the transverse coordinate, z. As the gas molecules are subjected to thermal fluctuations, in addition to the this macroscopic (i.e., average) movement there is a chaotic motion as well, characterized by an average velocity v, with $v \approx \sqrt{3kT/m} \ll u$, where m is the mass of a single molecule. Now, of the n molecules per unit volume composing the gas, about one-third will have a velocity directed along the z-axis and half of these, or $n/6$ molecules, will have an average velocity v in the $+z$ direction and $n/6$ molecules in the $-z$ direction. Therefore, on average, there will be $nv/6$ molecules per unit time passing through a unit area of the $z = constant$ plane from the bottom to the top (that is, $nv/6$ is the upwards average molecular flux), and the same number of molecules will flow downward (see Fig. 1.7). On average, these molecules have had their last collision with another molecule at a distance from the z plane equal to the mean free path λ. Therefore, since the mean longitudinal velocity, u, is a function of z, each molecule that cross the z plane moving downwards will be carrying a momentum (i.e., its x-component) equal to $mu(z + \lambda)$, while those moving upward will be carrying a momentum $u(z - \lambda)$. This suggests that the net transport of the x-momentum along the $+z$-direction, i.e., the diffusive momentum flux (because there is no convective transport along the z-direction) is given by:

$$J_{Qd} = \tau = \left(\frac{1}{6}nv\right)[mu(z - \lambda)] - \left(\frac{1}{6}nv\right)[mu(z + \lambda)],\qquad(1.5.1)$$

where we have considered that momentum flux is equal to the shear stress. Expanding $u(z)$ in Taylor series, and neglecting higher-order terms, we obtain:

$$u(z + \lambda) = u(z) + \lambda\frac{du}{dz} + \cdots \quad \text{and} \quad u(z - \lambda) = u(z) - \lambda\frac{du}{dz} + \cdots$$

Fig. 1.7 Flow field at a molecular scale

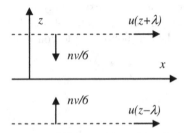

so that,

$$J_{Qd} = \tau = -\mu \frac{du}{dz}. \tag{1.5.2}$$

where μ is the viscosity,

$$\mu = \frac{1}{3} nm\lambda v = \frac{1}{3} \rho\lambda v, \tag{1.5.3}$$

with $\rho = mn$ denoting the fluid density. This result, obtained by Maxwell in 1860, allows us to make some interesting predictions. In fact, the mean free path can be easily estimated by considering that the number of molecules, moving with a mean velocity v, hitting a given "target" molecule during the time interval Δt is equal to the number of molecules contained in a cylinder having as a base the cross section $\sigma_0 = \pi d^2$ (where d is the molecular diameter) and $v\Delta t$ as its height (trivially, during the time interval Δt, the target molecule will be hit solely by the molecules contained within that cylinder). This implies that the number of collisions per unit time is $\pi n v d^2$ and the mean time interval between two successive collisions is

$$\tau = \frac{1}{\pi n v d^2}.$$

Therefore, we see that the mean free path is:

$$\lambda = v\tau = \frac{1}{\pi n d^2} \left(= \frac{1}{n\sigma_0} \right). \tag{1.5.4}$$

Finally, considering that $v \approx \sqrt{kT/m}$, we may conclude that the viscosity of a gas is[10]

$$\mu \approx \frac{\sqrt{mkT}}{d^2} \left(\approx \frac{1}{\sigma_0} \sqrt{mkT} \right). \tag{1.5.5}$$

The most interesting feature of this result is that μ does not depend on n, and therefore is independent of the pressure. In fact, as we increase pressure, and therefore density, the number of intermolecular collisions will increase, as there will be more molecules carrying momentum along the transversal direction, but, as the mean free path decreases, the contribution of each of these collisions will decrease in the same proportion. It should be noted, however, that this analysis applies only to low density fluids, where $d \ll \lambda$, while as λ decreases it loses its validity. In particular, in liquids, where $d \approx \lambda$, it must be radically modified.

---

[10]A more rigorous analysis shows that $\lambda = (\sqrt{2} n\sigma_0)^{-1}$ and $v = [(8kT)/(\pi m)]^{1/2}$.

The other conclusion we can draw from this analysis is that in gases $\mu \propto \sqrt{T}$, i.e., viscosity increases with temperature. In fact, considering that collisions between molecules are partially inelastic, we find that the cross section tends to decrease with temperature, obtaining at the end $\mu \propto T^{0.7}$. This behavior is very different from that of liquids, where viscosity decreases as temperature increases, due to a decrease of the strength of the intermolecular interaction.

1.6 Thermal Conductivity

Here, we want to model the transport of heat induced by a temperature gradient. Consider that each molecule carries an internal energy $\varepsilon = mc\,(T - T_0)$, where c is the specific heat per unit mass, m is the mass of a molecule and T_0 is a reference temperature. Proceeding as in the previous section (see Fig. 1.8), we find that the net transport of internal energy (i.e., the heat flux) along the $+z$-direction is given by Eq. (1.5.1), with $\varepsilon = mc\,(T - T_0)$ replacing mu,

$$J_{Ud} = \left(\frac{1}{6}nv\right)[\varepsilon(z - \lambda)] - \left(\frac{1}{6}nv\right)[\varepsilon(z + \lambda)]. \tag{1.6.1}$$

Expanding $\varepsilon(z) = mc\,[T(z) - T_0]$ in Taylor series, and neglecting higher-order terms, we obtain:

$$T(z + \lambda) = T(\lambda) + \lambda\frac{dT}{dz} + \cdots \quad T(z - \lambda) = T(\lambda) - \lambda\frac{dT}{dz} + \cdots$$

so that,

$$J_{Ud} = -k\frac{dT}{dz}, \tag{1.6.2}$$

where k is the thermal conductivity,

$$k = \frac{1}{3}\rho c\lambda v, \tag{1.6.3}$$

Fig. 1.8 Temperature field at a molecular scale

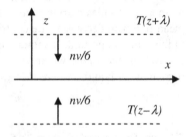

with $\rho = mn$. Finally, substituting for the mean free path, λ, from Eq. (1.5.4), together with the mean fluctuating velocity, $v \approx \sqrt{kT/m}$, we obtain:

$$k \approx \frac{c}{d^2}\sqrt{mkT}\ \left(\approx \frac{c}{\sigma_0}\sqrt{mkT}\right). \tag{1.6.4}$$

Thus, in gases, thermal conductivity, like viscosity, does not depend on pressure; variations of k with temperature, instead, depend on how the specific heat depends on temperature. Comparing (1.6.3) with (1.5.3) we see that:

$$\alpha = \frac{k}{\rho c} = \frac{\mu}{\rho} = v = \frac{1}{3}\lambda v \quad \Rightarrow \quad Pr = \frac{v}{\alpha} = 1, \tag{1.6.5}$$

where Pr indicates the *Prandtl* number. Therefore, we see that in ideal gases thermal diffusivity is equal to momentum diffusivity (that is, kinematic viscosity). In reality, we find experimentally that for real gases the Prandtl number is approximately 0.7; considering the drastic approximation in the above analysis, this discrepancy is extraordinarily small. Finally, note that in liquids the two diffusivities can be very different from each other, as it can happen that $Pr \gg 1$ (think of glycerin or motor oils) or $Pr \ll 1$, as in quicksilver (see Appendix 1).

1.7 Molecular Diffusivity

Here, we want to model the molecular transport of a chemical species A within a binary mixture, induced by a concentration gradient (see Fig. 1.9). In this case, consider that each molecule of species A carries a mass m from regions with concentration $c_A(z + \lambda)$ to regions with concentration $c_A(z - \lambda)$, where $c_A = mn$. Again, proceeding as in the previous two sections, we obtain the following expression for the mass flux of species A:

$$J_{Ad} = \left(\frac{1}{6}mv\right)[n(z - \lambda)] - \left(\frac{1}{6}mv\right)[n(z + \lambda)]. \tag{1.7.1}$$

Fig. 1.9 Concentration field at a molecular scale

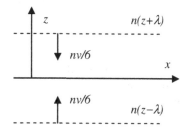

Expanding $n(z)$ in Taylor series, and neglecting higher-order terms, we obtain:

$$n(z + \lambda) = n(z) + \lambda\frac{dn}{dz} + \cdots; \quad n(z - \lambda) = n(\lambda) - \lambda\frac{dn}{dz} + \cdots,$$

so that,

$$J_{Ad} = -D\frac{dc_A}{dz}, \tag{1.7.2}$$

where D is the coefficient of molecular diffusion (or diffusivity),

$$D = \frac{1}{3}\lambda v. \tag{1.7.3}$$

Finally, substituting the expressions for the mean fluctuating velocity, v, and the mean free path, λ, we obtain:

$$D \approx \frac{1}{nd^2}\sqrt{\frac{kT}{m}}\left(\approx \frac{1}{n\sigma_0}\sqrt{\frac{kT}{m}}\right). \tag{1.7.4}$$

Thus, we conclude that $D \propto 1/P$ and $D \propto \sqrt{T}$.

Comparing (1.7.3) with (1.6.5), we see that in ideal gases all three diffusivities are equal to each other, i.e.,

$$D = \alpha = v \quad \Rightarrow \quad Sc = \frac{v}{D} = 1, \tag{1.7.5}$$

where Sc indicates the *Schmidt* number. Experimentally, for real gases we find that the Schmidt number is approximately 0.7, that is very close to the Prandtl number. In liquids (and solids, where applicable), however, the results are very different. In fact, while in gases D, α e v are close to one another, with magnitudes of, approximately, 10^{-1} cm^2/s, in liquids D is of $O(10^{-5}$ cm^2/s), while α and v are of $O(10^{-2}$ cm^2/s) in many cases, so that Sc turns out to be of $O(10^3)$. A qualitative explanation of this fact can be found in Sect. 14.4.

1.8 Molecular Diffusion as an Example of Random Walk

Molecular diffusion can also be modeled as the random movement of the gas molecules, assuming that their mutual collisions are statistically independent from one another. The displacement vector **R** of each molecule after N collisions is the sum of N elementary displacements r_i, each describing the distance traveled by the molecule between two successive collisions,

$$\mathbf{R} = \sum_{i=1}^{N} \mathbf{r}_i. \qquad (1.8.1)$$

Obviously, since each elementary displacement is random, that is $\langle \mathbf{r}_i \rangle = \mathbf{0}$, then we have $\langle \mathbf{R} \rangle = \mathbf{0}$. This result agrees with the fact that all diffusive processes do not involve any net displacement of matter. Now assuming, for sake of simplicity, that each elementary displacement has the same length, with $|\mathbf{r}_i| = \ell$, we obtain:

$$\langle R^2 \rangle = \langle \mathbf{R} \cdot \mathbf{R} \rangle = \sum_{i=1}^{N} \langle r_i^2 \rangle + \sum_{i=1}^{N} \sum_{j \neq i=1}^{N} \langle \mathbf{r}_i \cdot \mathbf{r}_j \rangle = N\ell^2. \qquad (1.8.2)$$

This result can be justified by considering the case $N = 2$: the product $\langle \mathbf{r}_1 \cdot \mathbf{r}_2 \rangle$ equals $\ell^2 \langle \cos \theta \rangle$, where θ is the angle between the two vectors. As this angle is random, the product is zero.

At this point, it can be proved[11] that ℓ is proportional to the mean free path λ that was used in the previous sections through the following relation:

$$\ell = \sqrt{2}\lambda = \sqrt{2}v\tau, \qquad (1.8.3)$$

where τ is the mean time interval between two successive collisions and v is the mean velocity along any direction. Finally, considering that

$$N = (\Delta t)/\tau,$$

we obtain:

$$\langle R^2 \rangle = 6D(\Delta t), \qquad (1.8.4)$$

where we have defined the diffusivity coefficient as in Eq. (1.7.3), i.e.,

$$D = \frac{1}{3}\lambda v. \qquad (1.8.5)$$

Note that by decomposing the molecular motion along the three Cartesian directions x, y and z, and defining $R^2 = X^2 + Y^2 + Z^2$, we obtain:

$$\langle X^2 \rangle = \langle Y^2 \rangle = \langle Z^2 \rangle = 2D(\Delta t). \qquad (1.8.6)$$

So, the diffusion process can been described in two different ways. On one hand, we assume that the system is non-homogeneous and consider the molecular flux resulting from the applied concentration gradient: the ratio between these quantities,

---

[11]F. Reif, "Statistical Thermal Physics", McGraw Hill, p. 486.

which are sometimes referred to as *thermodynamic flux* and *thermodynamic force*, defines the coefficient of gradient diffusion. On the other hand, consider a homogeneous system and follow the random motion of a single molecule, due to its thermal fluctuations: (one half of) the time growth of its mean square displacement defines the coefficient of self-diffusion. The fact that the coefficients of gradient diffusion and of self-diffusion are equal to each other is by no means obvious and constitutes perhaps the simplest example of application of the fluctuation-dissipation theorem, proved by H.B. Callen and T.A. Welton in 1951.

1.9 Examples of Diffusive Processes

Suppose we have a uniform solid slab at room temperature, and imagine that at time $t = 0$ the wall temperature changes abruptly. At this point, the temperature inside the slab begins to change, with larger delay with respect to $t = 0$ as we move farther away from the walls. Indeed, at a distance l from the wall, the abrupt change in wall temperature will be felt only after a time of order $t = l^2/\alpha$, where α denotes thermal diffusivity. Thus, if the body under examination has a linear dimension L, it will take a time of order L^2/α for the body to reach roughly the new equilibrium state, with the temperature everywhere equal to the given wall temperature (see Fig. 1.10).

In the same way, if in a glass containing, for example, water at rest we introduce a small drop of ink, we see that after an interval of time t the size of the droplet will be of order $l = \sqrt{(Dt)}$; then, assuming that the glass has a linear dimension L, after a time of order L^2/D a steady state will be reached, where the ink concentration is uniform and the water in the glass has a uniform color.

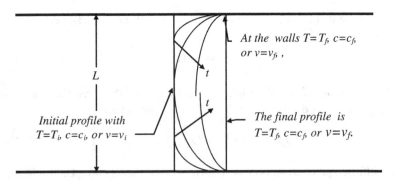

Fig. 1.10 Profiles of temperature T, concentration c and velocity v as a function of time in a system (a solid or a fluid) located between two plates at a distance L, when T, c and v change abruptly at the walls. The characteristic time to reach steady state equals the ratio between L^2 and the diffusivity of energy (i.e., α), mass (i.e., D), or momentum (i.e., v)

Now, as we understand the diffusion of energy as almost the same as the diffusion of mass, we can move on to the study of the diffusion of momentum, showing that this process obeys the same rules. Suppose we have a fluid between two plates, initially at rest (see Fig. 1.10), and assume that at time $t = 0$ the two plates are set in motion.[12] Again the fluid velocity, and therefore its momentum, within the system will begin to change, but not immediately. Indeed, the abrupt change in wall velocity will be felt at a distance l from the wall after a time $t = l^2/v$, where v is the kinematic viscosity. Thus, if the distance between the plates is equal to L, it will take a time of order L^2/v for the system to reach, roughly, its steady state, with a uniform velocity profile.

Note that, although the units of α, D and v are identical to one another, their numerical values are very different. For example, typical values for water are: $\alpha \approx 10^{-3}$ cm^2/s, $D \approx 10^{-5}$ cm^2/s and $v \approx 10^{-2}$ cm^2/s. Therefore, the rate for mass diffusion is much slower than that for both thermal and momentum diffusion.

1.10 Problems

1.1 On Thanksgiving in the U.S., turkeys are sold in supermarkets with the following table attached, indicating the cooking time as a function of turkey mass. Based on what we saw in Sect. 1.8, we expect that the cooking time should be proportional to the square of the typical size of the turkey, so that:

$$(\text{cooking time}) \times (\text{mass of turkey})^{-2/3} = C = \text{constant}$$

(a) Compare this prediction with the experimental data of the table.
(b) Determine the thermal diffusivity of the meat (obviously, its mean value) and compare it with that of water.

Mass (kg)	Cooking time per unit mass (min/kg)
2.5–4.5	45–55
4.5–7.5	40–45
8.0–11.5	35–40

---

[12]Another set of examples is when one of the two walls is set in motion, or changes its temperature or composition, while the other is fixed. In this case, the steady state corresponds to a linear velocity, temperature or concentration profile, and, again, in each case the time required to reach these final conditions is of the order of L^2/α, L^2/D and L^2/v, respectively.

Chapter 2
Statics of Fluids

Abstract We begin this chapter with a discussion on hydrostatic (Sect. 2.1), presenting a few applications (Sect. 2.2). Then, in Sects. 2.3–2.5, a few interfacial effects in fluids are described, first by defining surface tension, and then by showing a few examples of capillary hydrostatics, such as the Young-Laplace theory and the contact angle.

2.1 Hydrostatic Equilibrium

From experience we know that in a fluid at rest the pressure is constant along any horizontal section, but varies with height. In this paragraph we intend to study such variation.

Consider the vertical fluid column shown in Fig. 2.1. A simple balance of all vertical forces acting on an elementary volume $S dz$ yields the following result:

$$pS - (p + dp)S - \rho g\, S dz = 0$$

where ρ is the fluid density, g the gravity field and we have assumed that the vertical, z-axis is directed upward, i.e. in the direction opposite to gravity. At the end we obtain:

$$dp = -\rho g\, dz \qquad (2.1.1)$$

Note that Eq. (2.1.1), as well as all the other following results, is independent of the area S, as one could easily predict, based on symmetry considerations. Now, in order to integrate this expression, we must know how ρ depends on p. The two most important cases are those of incompressible fluids and ideal gases.

Fig. 2.1 Hydrostatic
equilibrium

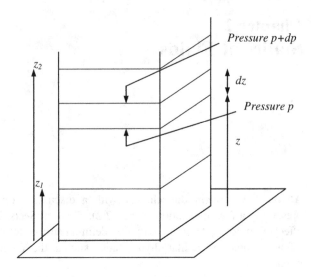

2.1.1 Incompressible Fluids

When the density can be considered constant, Eq. (2.1.1) is can be easily integrated
between $z = z_1$ and $z = z_2$, obtaining the so-called *Stevin*'s law,[1]

$$p_1 - p_2 = \rho g(z_2 - z_1), \tag{2.1.2}$$

showing that the pressure increases linearly with height as we move downward.
Note that $\rho g(z_2 - z_1)$ is the weight of a fluid column of unit area and height
$(z_2 - z_1)$; as such, it has the units of a pressure,

$$[\rho g \Delta z] = ML^{-3} \times LT^{-2} \times L = ML^{-1}T^{-2} = \left[\frac{\text{force}}{\text{surface}}\right] = [\text{pressure}],$$

i.e., Pascal, in the SI system, with $1\text{Pa} = 1\text{Nm}^{-2}$.

Applying Stevin's law, it comes natural to measure pressure in terms of the
height of a column of a given fluid. The most common fluid that is used is mercury,
as its density, $13{,}600 \text{ kg m}^{-3}$, is almost independent of pressure and temperature.
So, the pressure of 1 atm, equal to 101,325 Pa, corresponds to that exerted by a
column of mercury 760 mm high, and it is indicated as 760 mm Hg (1 mm Hg is
also denoted as 1 Torr [2]).

---

[1]Named after Simon Stevin (1548–1620) a Flemish mathematician and engineer.
[2]Named after Evangelista Torricelli (1608–1647), an Italian mathematician and physicist.

2.1.2 Ideal Gases

In an ideal gas, density ρ is connected to pressure through the well-known equation,

$$\rho = \frac{M_W p}{RT},$$ (2.1.3)

where $R = 8314.47$ J/Kmol, K is the gas constant, T the absolute temperature, and M_W is the molecular mass, that is the mass, expressed in kilograms, of 1000 mol of fluid, expressed in kilograms per kilomoles, i.e., $[M_W] = $ Kg/Kmol.

Substituting (2.1.3) into (2.1.1), we obtain:

$$dp/p + (gM_W/RT)dz = 0.$$

Assuming T constant, this equation can be integrated between sections $z = z_1$ and $z = z_2$, obtaining:

$$\ln(p_2/p_1) = (gM_W/RT)(z_1 - z_2) \quad \text{i.e.,} \quad \frac{p_2}{p_1} = \exp\left\{-\frac{gM_W}{RT}(z_2 - z_1)\right\}, \quad (2.1.4)$$

Equation (2.1.4) is called the *isothermal barometric equation*.

2.2 Manometers

A manometer is a pressure measuring instrument, in particular one that uses a column of liquid by applying Stevin's law. The simplest manometer is the U-tube sketched in Fig. 2.2, whose lower part is filled with the heavier A fluid, while the upper parts are filled with the lighter B and C fluids, which are immiscible with A.

Fig. 2.2 U-tube manometer

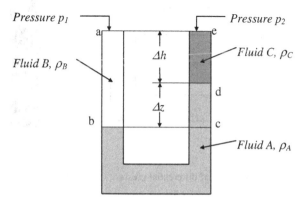

In general, the difference $p_1 - p_2$ between the pressures at the two sides of the device is proportional to the difference in fluid heights. In fact, consider the following equality:

$$p_1 - p_2 = (p_a - p_b) + (p_b - p_c) + (p_c - p_d) + (p_d - p_e).$$

Now, applying Stevin's law (2.1.3) to each homogeneous column, we obtain:

$$p_1 - p_2 = -\rho_B g(\Delta h + \Delta z) + 0 + \rho_A g \Delta z + \rho_C g \Delta h$$

i.e.,

$$p_1 - p_2 = (\rho_A - \rho_B)g\Delta z + (\rho_C - \rho_B)g\Delta h. \qquad (2.2.1)$$

Basically, the U-tube can be used in two ways, (see Fig. 2.3), measuring either the pressure drop within a conduit or the relative pressure in a reservoir. In the first case, we measure the difference between the pressures at two points, e.g. before and after a valve, when there is only one fluid flowing. Therefore, as here fluids B ad C are identical, i.e., $\rho_C = \rho_B$, Eq. (2.2.1) becomes,

$$p_1 - p_2 = (\rho_A - \rho_B)g\Delta z, \qquad (2.2.2)$$

showing that the differential pressure is independent of Δh.

In the second type of U-tube application, one arm is in contact with the fluid whose pressure we want to measure, while the other arm is left open to ambient conditions. Therefore, applying Eq. (2.2.1) with the air density ρ_C neglected, we obtain the relative pressure, that is the fluid pressure p_1 referred to the atmospheric pressure.

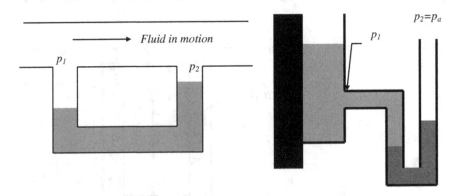

Fig. 2.3 Measures of differential pressure (*left*) and relative pressure (*right*)

2.3 Surface Tension

As the size L of a system increases, the impact of all effects related to the presence of surfaces separating two or more phases decreases as $1/L$ and therefore, in most applications, it is negligible. However, there are systems where this is not true, in which surface phenomena have an important, often decisive, role. In particular, consider the following three examples of such systems:

1. The surface of separation between a liquid and its vapor, both at equilibrium.
2. A soap bubble, composed of a liquid film surrounded on both sides by a gas.
3. A film of a lubricant (oil, for example) on the free surface of a liquid (e.g., water).

The separation region between two phases, known as interface, is not really a two-dimensional surface (in the mathematical sense of the term). Instead, it is a region of small but non-zero thickness, in which the properties of the system vary continuously from those of one phase to those of the other phase. However, if we are not too close to the critical point, when the distinction between the phases vanishes, the thickness of this region is of the order of a few molecular radii, i.e., of the order of one hundredth of a micron. Therefore, for convenience, we will assume that the interface is a zero-thickness surface, where the properties of the system, such as density and composition, may jump discontinuously.

Indicating by A_s the area of separation between two fluids, consider the work δW that is necessary to compress it reversibly by an infinitesimal amount, dA_s:

$$\delta W = \sigma dA_s \qquad\qquad (2.3.1)$$

The quantity σ defined in (2.3.1) is called *surface tension*; its units are force per unit length, or energy per unit area, i.e., $[\sigma] = \mathrm{Nm}^{-1} = \mathrm{Jm}^{-2}$.

To better understand the meaning of surface tension, consider a film with a rectangular boundary, so constructed that one of the sides is mobile (see Fig. 2.4). Here, the surface tension σ is the force per unit length directed along the normal to the inner contour and tending to decrease the surface of the film. The expression (2.1.4) has the same formal expression as the work $\delta W = -pdV$ required to

Fig. 2.4 Film with a rectangular boundary

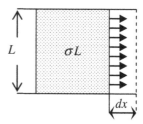

reversibly change the volume of a system. So, we can say that σ has the same role for surfaces that $-p$ has for volumes. Note the sign of the surface tension: if it were $\sigma > 0$, the force on the contour delimiting the surface would be directed inward, and therefore the interface between two phases would tend to increase indefinitely, with the result that, in general, there might not exist two coexisting phases. Conversely, since $\sigma > 0$, surface tension tends to contract the interfacial surface as much as possible, albeit satisfying all the thermodynamic constraints (for example, when one phase is incompressible, its volume must remain constant). That is why a water droplet in air, in the absence of gravity, assumes a perfectly spherical shape. So, we may conclude that, while pressure tends to maximize volume, surface tension tends to minimize surface. From a different perspective, if we "break" the continuity of a liquid by placing a gas bubble within the liquid bulk, the liquid will tend to fill it back, that is, to recover its continuity; on the contrary, if we "break" the continuity of a stretched membrane (for example, by cutting a balloon), the cut will widen.

The simplest example explaining the importance of surface tension is the formation of a liquid drop at the tip of a leaky faucet (or a capillary conduit). We know that the drop will grow until its weight overcomes the surface forces that keep it attached to the edge of the faucet, at which point the drop falls. We saw that surface tension acts along the contact line between the liquid and the capillary, which is a circle of diameter D_c; therefore, it induces a force F_σ directed along the tangent to the liquid-gas interface and pointing upward, since it tends to "close" the droplet (see Fig. 2.5). Just before falling, a force balance allows us to calculate the maximum size reached by the droplet,

$$(\Delta\rho)g\left(\pi D_d^3/6\right) = \sigma\pi D_c \cos\theta. \tag{2.3.2}$$

Now, the angle θ is unknown. We observe that as the drop grows, it assumes an elongated, pear-like shape due to its weight, until, just before the detachment, we see that $\theta = 0$, approximately Therefore, we obtain:

Fig. 2.5 Pendant liquid drop

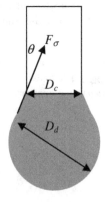

$$D_d = \left(\frac{6\sigma D_c}{(\Delta\rho)g}\right)^{1/3} \tag{2.3.3}$$

or, in dimensionless form,

$$\frac{D_d}{D_c} = \left(\frac{6\sigma}{D_c^2(\Delta\rho)g}\right)^{1/3} = \left(\frac{6}{Bo}\right)^{1/3} = 1.82\, Bo^{-1/3}, \tag{2.3.4}$$

where we have defined the following Bond[3] number:

$$Bo = \frac{(\Delta\rho)gD_c^2}{\sigma}, \tag{2.3.5}$$

The Bond number is the ratio between gravity, $(\Delta\rho)gL^3$, and surface forces, σL, where L is a characteristic linear dimension of the system, which, in this case, we have taken to be the capillary diameter (of course, we could take the radius as well).

The expression (2.3.4) is in excellent agreement with the experimental data, confirming that the ratio between the diameter of the drop and that of the capillary is inversely proportional to the cubic root of the Bond number; the proportionality coefficient, though, results to be slightly smaller, namely about 1.6, instead of 1.82. This discrepancy is due to the fact that when the drop falls, it leaves behind some residual liquid that remains attached to the tip of the capillary, so that the volume of the falling drop is less than that of the liquid just before detachment. An important consequence of this result is that we can easily determine the surface tension by measuring the diameter of the falling drop (see Problem 2.7) and applying the relation

$$\sigma = 0.244\frac{D_d^3\rho g}{D_c} \tag{2.3.6}$$

which can be derived from (2.3.4), using the coefficient 1.6 instead of 1.82.

2.4   The Young-Laplace Equation

The equation of Young[4]-Laplace[5] is the basis of capillary hydrostatics, stating that surface tension induces a pressure difference across an interface at equilibrium that is proportional to the surface curvature. This rule applies to all kinds of interfaces,

---

[3]Named after the English physicist Wilfrid Noel Bond (1897–1937). It is also called Eötvös number, after the Hungarian physicist Loránd Eötvös (1848–1919).

[4]Thomas Young (1773–1829). English mathematician, physicist, physiologist and Egyptologist.

[5]Pierre-Simon, marquis de Laplace (1749–1827). French astronomer and mathematician.

in particular to the case of a liquid drop at equilibrium with its vapor phase, as well as the case of a vapor bubble at equilibrium with its liquid phase.

2.4.1 Thermodynamic Approach

Consider a single component system composed of two phases in both thermal and chemical equilibrium, contained in a fixed volume reservoir. The force acting at the interface equals the variation of the appropriate thermodynamic potential (the Helmholtz free energy in this case, as temperature and volume are constant) with respect to a virtual displacement of the interface. Therefore, at equilibrium, the position of the interface must minimize the Helmholtz free energy.

Based on (2.3.1), the isothermal variation of the Helmholtz free energy of a system composed of two coexisting phases separated by an interface will contain an extra work-like term, related to the surface tension, i.e.,

$$(dA)_T = -p_1 dV_1 - p_2 dV_2 + \mu dN_1 + \mu dN_2 + \sigma dA_s \qquad (2.4.1)$$

where the subscripts 1 and 2 indicate phase 1 or 2, while the subscript i refers to the chemical species; so that V_i and N_i denote the volumes and the number of moles of phase i, respectively. Naturally, we have considered that at equilibrium the chemical potential of any species in the two phases are equal to each other, i.e., $\mu_1 = \mu_2 = \mu$. For simplicity, let us assume that one phase lays inside a drop of radius R, while the other phase consists of the continuum phase outside. Then, indicating the inside and the outside of the drop with the subscripts "i" and "o", respectively, the variation (2.4.1) of the Helmholtz free energy can be rewritten as:

$$dA = -p_i dV_i - p_o dV_o + \sigma dA_s + \mu(dN_i + dN_o)$$

Now, since at equilibrium the force acting at the interface must vanish, after noting that the total volume, $V_i + V_o$, and the total number of moles, $N_i + N_o$, are constant, we obtain:

$$F = \frac{dA}{dR} = -(p_i - p_o)\frac{dV}{dR} + \sigma\frac{dA_s}{dR} = 0,$$

where $V = V_i$. Furthermore, as $dV/dR = 4\pi R^2$ and $dA_s/dR = 8\pi R$, we finally obtain:

$$(p_i - p_o) = \frac{2\sigma}{R} \qquad (2.4.2)$$

In the case of a plane interface, the pressures of the two phases are equal to each other, and we obtain the usual condition of mechanical equilibrium. For non-spherical surfaces, this relation can be generalized as:

$$(p_i - p_o) = \sigma\left(\frac{1}{R_1} + \frac{1}{R_2}\right), \qquad (2.4.3)$$

where R_1 and R_2 are the principle radii of curvature. Note that the equation of Young-Laplace states that the pressure inside a drop is always larger than that outside, regardless of whether we have a liquid drop suspended in a vapor phase or a vapor bubble suspended in a liquid phase.

A simpler way to derive the equation of Young-Laplace is to consider a (reversible) transfer of liquid volume dV from the outside to the inside the drop. As the work required to perform such a transfer, $\delta W = (p_i - p_o)dV$, must equal the extra work σdA^s absorbed by the system in order to extend the surface area by an amount dA^s, we find again:

$$(p_i - p_o)4\pi R^2 dR = \sigma 8\pi R dR \Rightarrow (p_i - p_o) = 2\sigma/R.$$

2.4.2 Mechanical Approach

Consider a fluid at pressure p_i contained within a sphere of radius R, at equilibrium with another fluid located outside the sphere with pressure p_o. Now, let us isolate on the interface a differential area of side $ds = Rd\theta$, and impose that the sum of all forces on it be zero (see Fig. 2.6). Therefore, the pressure forces, $(p_i - p_o)ds^2$, directed outward, must balance the surface forces, directed inward, acting on the four sides of the square, given by $4(\sigma ds)\sin(d\theta/2)$, which is equal to $(2\sigma/R)ds^2$. In a generic case, the surface has two radii of curvature, so that we readily obtain the Young-Laplace Eq. (2.4.3).

We should stress that the pressure difference between the inside and the outside of a drop or a bubble is relevant only at the micron scale. In fact, as the surface tension of liquids (with air) is about 0.05 N/m, applying the Young-Laplace equation we find that when the radius of the drop (or the bubble) is $R = 1$ mm, the pressure difference is $\Delta p \approx 100$ N/m$^2 = 1$ mbar, while $\Delta p \approx 1$ bar when $R = 1$ μm.

Fig. 2.6 Pressure difference between the inside and the outside of a droplet/bubble

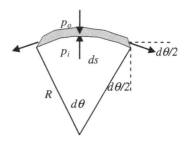

Example: liquid rise in a capillary

When a narrow capillary tube of radius R_c is immersed in a liquid, in most cases the liquid is observed to rise, until it reaches a height h (see Fig. 2.7). The liquid column is subjected to two forces, opposite to one another: a capillary force, $F_\sigma = 2\pi R_c \sigma \cos\theta$, where θ is the contact angle (see Sect. 2.5), and a gravitational force, $F_g = \pi R_c^2 h(\Delta\rho)g$, where $\Delta\rho$ is the density difference between liquid and air. At equilibrium equating the two forces and denoting by $D_c = 2R_c$ the capillary diameter, we obtain:

$$h = \frac{4\sigma}{(\Delta\rho)gD_c}\cos\theta, \tag{2.4.4}$$

that is,

$$\frac{h}{D_c} = \frac{4}{Bo}\cos\theta, \quad \text{where } Bo = \frac{(\Delta\rho)gD_c^2}{\sigma}, \tag{2.4.5}$$

is the Bond number (2.3.5). Note that for obtuse contact angles, i.e. when the liquid does not "want" to wet the solid surface, then $\cos\theta < 0$ and therefore h is negative. The most important application of this phenomenon corresponds to that of a liquid that wets completely the surface of the capillary tube, i.e., when $\theta = 0$. In particular, if the liquid is water, with $\sigma = 0.072$ N/m, we see that, in SI units, $h \cong 2.8 \times 10^{-5}/D_c$; thus, when the capillary has a $D_c = 0.2$ mm diameter, we find $h \cong 14$ cm, while when $D_c = 1\,\mu$m we obtain $h \cong 28$ m. Therefore, this phenomenon, too, can be used to measure surface tension.

Fig. 2.7 Liquid rise in a capillary tube

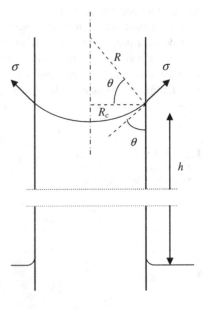

In the analysis above we have assumed that $R_c \ll h$, so that the volume of the liquid column, $\pi R_c^2 h$, does not depend on the exact shape of the meniscus, which would introduce a correction of $O(\pi R_c^3)$. This implies that the surface forces dominate gravity, i.e., $Bo \ll 1$. When this is true, we can assume that the interface is perfectly spherical, with a constant curvature of radius $R = R_c/\cos\theta$ (see Fig. 2.7). Therefore, imposing that the pressure drop across the meniscus equals the hydrostatic losses, we obtain:

$$\Delta p = \frac{2\sigma}{R} = (\Delta\rho)gh, \tag{2.4.6}$$

which coincides with Eq. (2.4.4). Note that the pressure below the meniscus is less than the ambient pressure; then, moving downward, the pressure gradually increases, until at the bottom of the column it reaches again its ambient value.

2.5 Contact Angle

The *contact angle* θ is the angle where a liquid/gas interface meets a solid surface (see Fig. 2.8). Common experience tells us that the smaller the contact angle the more evenly the liquid will spread over the solid surface until, at $\theta = 0$, complete wetting takes place. As the contact angle increases, the liquid does not readily wet the solid surface until, when $\theta > \pi/2$, the liquid tends to form a droplet that may easily run off the surface (think of a drop of water on a Teflon surface).

A given system of solid, liquid, and gas at a given temperature and pressure has a unique equilibrium contact angle. In fact, the variation of the free energy due an increment of the solid area wetted by the liquid is:

$$dA^s = dS(\sigma_{SL} - \sigma_{SG}) + dS\sigma_{LG}\cos\theta,$$

where σ_{SL}, σ_{SG} and σ_{LG} are the surface tensions between solid and liquid, solid and gas, and liquid and gas, respectively. At equilibrium, $dA^s = 0$ and therefore we obtain the so-called Young equation:

$$\cos\theta = \frac{\sigma_{SG} - \sigma_{SL}}{\sigma_{LG}}, \tag{2.5.1}$$

Fig. 2.8 Liquid/gas interface and contact angle

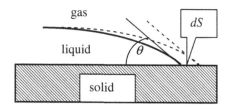

Thus, knowing the liquid-gas surface tension, σ_{LG}, and the contact angle, θ, we can determine the difference $\sigma_{SG} - \sigma_{SL}$ but not their absolute values. This relation can also be obtained from a simple force balance among the three interfacial forces at the 3-phase contact line, directed inward from the corresponding separation surface. As shown in Fig. 2.9, when $\sigma_{SG} > \sigma_{SL}$, that is when the surface tension between solid and gas is larger than that between solid and liquid, then $\cos\theta > 0$ and the contact angle is acute; on the other hand, when $\sigma_{SG} < \sigma_{SL}$, the contact angle is obtuse. The extreme cases of complete wetting and no wetting are not described by Eq. (2.5.1). In fact, when $\sigma_{SG} \geq \sigma_{SL} + \sigma_{LG}$, there is no contact line, as the solid is completely wetted by the liquid, so that $\theta = 0$; also, when $\sigma_{SL} \geq \sigma_{SG} + \sigma_{LG}$, the liquid does not wet the solid at all and the gas displaces the liquid completely, so that $\theta = \pi$.

The contact angle can be measured directly by photographic techniques, such as the static sessile drop method, using a high resolution cameras, or, in alternative, using the capillary rise method of Sect. 2.4.2, or the falling drop method of Sect. 2.3. The contact angle can also be determined by measuring the maximum height h reached by a free liquid on a plane vertical rigid wall, sketched in Fig. 2.10. In this case, the assumption $R \ll h$ that, as we saw, is often satisfied in the liquid rise in a capillary, can never be applied. In fact, the upward capillary force, $F_\sigma = \sigma L \cos\theta$, where L is the slab width, now balances the weight of the liquid that has been dragged upward. Thus, the problem can be solved only by determining the exact shape $y(x)$ of the meniscus, as shown below. An approximate solution (and the exact scaling), though, can be found by assuming that the shape of the meniscus is circular. In that case, applying Eq. (2.4.3), with $R_1 = R$ and $R_2 \to \infty$, we obtain: $\sigma/R = \alpha\Delta\rho g h$, with $R(1 - \sin\theta) = h$ (see Fig. 2.10), while $0 < \alpha < 1$ is an unknown constant. Thus, we may conclude:

Fig. 2.9 Liquid-gas surface tension

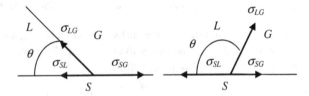

Fig. 2.10 Interface shape

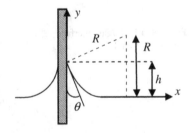

$$\sin \theta = 1 - \alpha \frac{(\Delta \rho) g h^2}{\sigma} = 1 - \alpha \left(\frac{h}{\lambda_c} \right)^2, \quad \text{with } \lambda_c = \sqrt{\frac{\sigma}{g \Delta \rho}}, \qquad (2.5.2)$$

As shown below, the exact solution of the problem yields $\alpha = \frac{1}{2}$. Here, λ_c is the *capillary length*, indicating the lengthscale on which surface tension effects on the shape of an interface are likely to be comparable with gravity effects, i.e., when $Bo_\lambda = \lambda_c^2 \rho g / \sigma = 1$. In particular, for clean water in air at normal temperatures, $\lambda_c = 2.7$ mm, indicating that larger drops are unstable and break down. Soap bubbles in air, instead, can reach a much larger size, $\lambda_c \approx 4$ m. Equation (2.5.2) shows that the capillary length also characterizes the height reached by a meniscus on a plate; for example, when the liquid wets perfectly the solid surface, with $\theta = 0$, we find $h = \sqrt{2} \lambda_c$.

Exact interface shape

Here, we want to determine the height h that is reached by a meniscus wetting a plate (Fig. 2.10). At any gas-liquid interface, since the pressure in the gas phase is constant, the condition for equilibrium at any point of the interface is,

$$(\Delta \rho) g y - \sigma \left(\frac{1}{R_1} + \frac{1}{R_2} \right) = \text{const.} \qquad (2.5.3)$$

In our case, the constant on the RHS is zero, because far from the wall the interface becomes plane. In addition, our problem is two-dimensional, so that $R_1 = R$ and $R_2 \rightarrow \infty$. Then, considering that the curvature of a given curve $y(x)$ is $R^{-1}(x) = y'' / (1 + y')^{3/2}$, where we have denoted $y' = dy/dx$, Eq. (2.5.3) becomes,

$$\frac{1}{\lambda_\sigma^2} y - \frac{y''}{(1 + y')^{3/2}} = 0 \qquad (2.5.4)$$

where λ_c is the capillary length (2.5.2). Integrating once, we obtain:

$$\frac{1}{2} \left(\frac{y}{\lambda_\sigma} \right) + \frac{1}{(1 + y')^{1/2}} = C \qquad (2.5.5)$$

Applying again the boundary conditions far from the walls, we find $C = 1$. Finally, considering that $y'(x = 0) = \tan^{-1} \theta$ so that $[1 + y''(0)]^{-1/2} = \sin \theta$, we obtain Eq. (2.5.2), with $\alpha = \frac{1}{2}$.

2.6 Problems

2.1   Indicate the pressure at the bottom of the three vases shown in Fig. 2.11 and the forces acting on their lateral faces, assuming that all vases have circular sections. Assume that the liquid is water, p_o is the atmospheric pressure, and $L = 10$ cm.

Fig. 2.11 Pressure in three vases

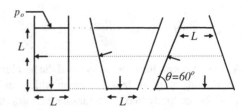

2.2 Industrial centrifuges rotate at a such large angular velocity ω that the centrifugal force greatly exceeds gravitational forces and the liquid free surface is consequently a cylinder, as shown in Fig. 2.12. Considering that the centrifugal force on the indicated liquid element of mass $dm = \rho 2\pi r L dr$ is equal to $\omega^2 r dm$, show that, assuming that the liquid is incompressible, a force balance yields at the end, the following pressure difference between the surfaces $r = r_1$ and $r = r_2$:

$$p_2 - p_1 = \rho\omega^2\left(r_2^2 - r_1^2\right)/2.$$

2.3 Determine the free surface profile of an incompressible liquid contained in a cylinder rotating with angular velocity ω.

2.4 Air temperature in the atmosphere decreases with elevation by about 5 °C every 1000 meters. If at sea level the air temperature is 15 °C and pressure is 760 mmHg, determine the height at which the air pressure is 380 mmHg? Assume that air behaves as an ideal gas.

2.5 Estimate the error that we would make if in the previous problem air were treated as an incompressible fluid, with the density of air at 0 °C temperature and 570 mmHg pressure.

2.6 The U-tube manometer of Fig. 2.2 is used to measure the pressure drop across a valve in a pipe where carbon tetrachloride (with 1.6 specific weight) is flowing. If the U-tube is filled with mercury (with 13.6 specific weight) and indicates an $h = 200$ mm height difference, how much is the pressure drop, expressed in Pascal?

Fig. 2.12 Centrifuge rotating at high speed

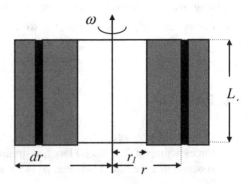

2.7 Water drops at 20 °C leak from a 0.14 cm diameter capillary pipe. If we collect 20 drops, having a total weight of 0.448 g, determine the surface tension of water.

2.8 Recently, two researchers have found that the dripping frequency, f (in s^{-1}), of a leaky capillary depends on the volumetric flow rate, $\dot{V}$ (in cm^3/s), as $X = 1.5\dot{V}$, with $X = f\,\sigma\pi D_c/\rho g$, where σ denotes the surface tension, D_c the capillary diameter and ρ the liquid density. Compare this result with Eq. (2.3.4).

2.9 It is known experimentally that the maximum volumetric flow rate $\dot{V}$ of a dripping capillary (meaning that above that limit the flow becomes continuous) corresponds to a unit Weber number, that is $We = \left(16\rho\dot{V}^2\right)/\left(\sigma\pi^2 D_c^3\right) = 1$, where σ denotes the surface tension, D_c the capillary diameter and ρ the liquid density. (a) Determine the maximum dripping frequency, f; (b) estimate f when the liquid is water and $D_c = 1$ mm.

Chapter 3
General Features of Fluid Mechanics

Abstract In this chapter we introduce some basic concepts of fluid mechanics. After observing in Sect. 3.1 that viscous effects are necessary to justify the existence of friction and drag forces, in Sect. 3.2 we define the Reynolds number, *Re*, as the ratio between convective and diffusive momentum fluxes, deriving a scaling of the drag force in the limits of small and large *Re*. Then, in Sect. 3.3, we observe that within these scaling laws there are some inconsistencies, which can be resolved only by introducing the concept of boundary layer. Finally, after a brief discussion on the boundary conditions, in Sect. 3.5 we present a brief qualitative overview of turbulence. A more complete description of this very complex subject can be found in Chap. 18.

3.1 Introduction

The behavior of a small material element of a moving fluid greatly depends on its distance from the walls. In fact, far from the boundaries its frictions and all dissipative forces are generally negligible, so that the fluid behaves as incompressible and inviscid. As in this case there are no diffusive shear stresses, the characteristics of this motion depend only on the pressure field. This is the so-called *inviscid flow*, which has been studied in depth, starting in the 19th century.

The inviscid flow model is completely described by Newtonian mechanics, following the laws of conservation of mass and momentum, as it does not present any energy dissipation, that is there is no conversion of the mechanical energy into heat. Obviously, although this type of movement can also exist close to the boundaries, the fluid velocity changes very rapidly in that region, thus producing an energy dissipation (proportional to the velocity gradient, as we saw in Sect. 1.4) that cannot be described within the framework of the potential flow. That means that, even if convective fluxes are dominant almost everywhere, at the wall the fluid velocity is zero, so that near the boundaries viscosity-related diffusive effects are dominant. One could object that this is a little region of space, but we can never

© Springer International Publishing Switzerland 2015
R. Mauri, *Transport Phenomena in Multiphase Flows*,
Fluid Mechanics and Its Applications 112, DOI 10.1007/978-3-319-15793-1_3

neglect it, since that's where friction takes place, and in most cases the net interaction between body and fluid is the only thing that matters. Consider, for example,
the motion of a solid body immersed in a stagnant fluid: if we assumed that the
resulting fluid flow were everywhere potential (even at the wall), mechanical energy
would be conserved and we could not explain why, instead, the fluid exerts a force
on the body in the direction opposite to body motion, or, equivalently, why the
body exerts a force on the fluid trying to *resist* it from moving. This resistance
force, which is well known to anybody riding a bicycle on a windy day, is generally
called *drag* force.[1] In fact, the inner product of the velocity of the body by the force
exerted on it equals the power, i.e. the mechanical energy per unit time, required to
maintain a constant velocity difference between body and fluid. Therefore, we see
that at steady state, as this mechanical energy is continuously produced (by the
engine pushing on the body), it must also be consumed, i.e., dissipated. To account
for that, potential forces are of no use, as they are time reversible, and therefore we
must consider viscous forces. The impossibility to account for such fundamental
quantity as the drag force in classical hydrodynamics is called the *d'Alembert
paradox*.[2]

3.2 The Reynolds Number

Denoting by momentum flux the momentum that crosses per unit time a cross
section of unit area, the characteristics of a fluid motion are determined by the
relative importance of the momentum flux due to convection with respect to that
due to viscous diffusion. As shown in Sect. 1.4, the former describes the coherent
motion of a fluid moving with uniform velocity, V, and equals $J_{Qc} = \rho V^2$, while the
latter describes the effect of a non-uniform velocity field, with $J_Q = \mu dv/dz$, and can
be approximated as $J_Q \approx \mu V/L$, where μ is the fluid viscosity, while L is a characteristic distance of the fluid flow. The ratio between these two momentum fluxes
is a non-dimensional quantity, called *Reynolds number*, named after its inventor[3]

---

[1]Note that the potential flow model, instead, can account for the existence of the so-called *lift*
force, perpendicular to the flow direction, since by definition it does not do any work.

[2]Jean-Baptiste le Rond d'Alembert (1717–1783) was a French mathematician, engineer, philoso-
pher, and music theorist, who, in 1752, proved that potential flow theory results in the prediction of
zero drag. It was a fortune that the brothers Orville and Wilbur Wright were not expert fluid
dynamicists, so that they did not hesitate in starting their flight experiments well before Prandtl
introduced the concept of boundary layer, thus explaining why airplanes can fly.

[3]Named after the Irish fluid dynamicist Osborne Reynolds (1842–1912), although it was introduced first by George Gabriel Stokes in 1851. Starting from the consideration that the fluid motion
in a pipe can only depend on the diameter of the tube, and the viscosity, density and mean velocity
of the fluid, Reynolds understood that the number that bears his name is the only way these four
quantities can be grouped together, forming a non-dimensional quantity (see Problem 3.1).

$$Re = \frac{\text{convective momentum flux}}{\text{diffusive momentum flux}} = \frac{J_{Qc}}{J_Q},$$

that is:

$$Re = \frac{\rho VL}{\mu} = \frac{VL}{\nu}. \tag{3.2.1}$$

Once more, we can use the analogy between momentum transport and the transport of energy and mass, to describe the physical characteristics of the latter by defining the *Peclet number*[4] as the ratio between convective fluxes and diffusive fluxes (of energy or of mass); to distinguish the two cases, we will call them Peclet number for heat transfer and for mass transfer, and define them as follows:

$$Pe_t = \frac{VL}{\alpha}; \quad Pe_m = \frac{VL}{D}. \tag{3.2.2}$$

The Reynolds number (and the Peclet numbers as well) gives a rapid indication of the type of transport, i.e., the flow regime, that we are considering. For example, consider a rigid body immersed in a fluid flow. When Re is large, convection prevails on diffusion; therefore, considering that the force is the momentum that the fluid transmits to the body per unit time and that $J_{Qc} = \rho V^2$ is the momentum that crosses the unit area per unit time, we conclude that the resistance force F exerted by the fluid on the body is proportional to $J_{Qc}S$, where $S = L^2$ is the area of the body cross section in a direction perpendicular to the motion and L is a characteristic linear dimension, i.e.,

$$F \approx L^2 \rho V^2. \tag{3.2.3}$$

In this expression and in the following, the symbol "$\approx$" indicates "of the same magnitude as". Therefore, $A \approx B$ is equivalent to $A = O(B)$, meaning not only that A and B are proportional to each other, but also that their ratio is constant and of $O(1)$, that is, $A/B \approx 1$ or $A/B = O(1)$.

Equation (3.2.3) indicates that the resistance force that a fluid exerts on a moving body at large Reynolds number are proportional to the square of the velocity, the cross section of the body and the fluid density, while they are independent of the fluid viscosity. As we will see in the following, though, this expression is correct only for turbulent, that is when Re exceeds a certain critical value (see Sect. 3.5 or, more extensively, Chap. 18). For example, in the case of a solid sphere of radius R immersed in a turbulent flow field, we find: $F \approx 0.22\pi R^2 \rho V^2$.

---

[4]Named after the French physicist Jean Claude Eugène Péclet (1793–1857).

On the other hand, for small *Re*, the fluid flow is completely determined by viscous dissipation, and convection can be neglected altogether. In this case, $F \approx SJ_Q$, that is,

$$F \approx \mu L V, \tag{3.2.4}$$

with $S \approx L^2$. This relation shows that the resistance force exerted on a body by a fluid flowing at low Reynolds number is proportional to the fluid velocity, a linear dimension L of the body, and the fluid viscosity, while it is independent of its density. For example, in the case of a solid sphere of radius R moving at low speed in a quiescent fluid we find the Stokes law, Eq. (22.2.14), i.e., $F = 6\pi\mu R V$.

Similar results are obtained when we study the lift force.

3.3 Boundary Layer and Viscous Resistance

Consider the flow of the wind around a building or that of a river around a pillar. Assuming that L in (3.2.1) is equal to the size of the building or the radius of the pillar (i.e., defining the Reynolds number in terms of those macroscopic lengths), we see that the Reynolds number is very large. This implies that the fluid motion described on the macroscopic scale, L, is inviscid, so that mechanical energy is conserved on that macroscale. However, we must take into account the fact that the fluid at the wall has zero velocity (see Sect. 3.4), so that, near the wall, the velocity profile has the form indicated in Fig. 3.1. There, we define a fluid layer of thickness δ, called the *boundary layer*, within which the influence of the wall is felt, where velocity gradients and shear stresses are particularly important. Thus, far from the wall, at $y > \delta$, convection is dominant and the velocity field is potential (uniform, in this case), while near the wall, at $y < \delta$, viscous forces prevail, so that the ensuing dissipation can account for the existence of a drag force, resolving d'Alembert's paradox described in Sect. 3.1. The concept of boundary layer was introduced by L. Prandtl[5] in 1904; it is based on the fact that, as convection dominates far from the wall, while diffusion prevails (as velocity is zero) at the wall, then there must be a distance δ where the two transport modalities balance each other. Accordingly, at the edge of the boundary layer, when $y \approx \delta$, the forces due to convective momentum fluxes balance the forces due to diffusive momentum fluxes. As a force equals the momentum transferred per unit time, it is the product of the momentum flux by the area crossed by it; therefore, we obtain:

$$J_{Qc}S_C \approx J_Q S_Q. \tag{3.3.1}$$

---

[5]Ludwig Prandtl (1875–1953) was a German engineer. He is considered the founder of modern aeronautical engineering. In 1904 he published the seminar paper "fluid flow in very little friction," where he described the boundary layer and its importance for drag and streamlining.

Fig. 3.1 Velocity profile near
a wall

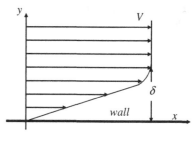

Fig. 3.2 Boundary layer at a
tube entrance

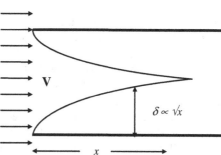

Now, convective momentum fluxes occurs longitudinally, i.e. along the direction of
the mean flow, so that they cross an area $S_C = \delta L$, while diffusive momentum fluxes
operates along the transversal direction, with $S_Q = L^2$. Therefore, from Eq. (3.3.1)
we see that $\rho V^2/L \approx \mu V/\delta^2$, that is,

$$\frac{\delta}{L} \approx \frac{1}{\sqrt{Re}}, \tag{3.3.2}$$

showing that the relative size of the boundary layer decreases as the square root of
the Reynolds number. Note, however, that $\delta \approx \sqrt{L}$, that is the boundary layer
thickness increases as the square root of the macroscopic length. That means that at
the entrance of a tube, as indicated in Fig. 3.2, the boundary layer will thicken as we
move away from the tube edge, until it will occupy the whole cross section. As
shown in Problems 3.2 and 3.3, that will occur at a distance $L \approx DRe$ from the tube
edge, where D is the tube diameter, while $Re = VD/\nu$ is the Reynolds number. At
larger distances from the tube edge, the flow field is fully developed and the
momentum transport will take place, so to say, inside the boundary layer, so that it
will be controlled by viscous dissipation.[6]

   The case of a fluid flowing past a submerged object is completely different. There,
in fact, considering the boundary layer profile of Fig. 3.3, we see that we can always
find a region sufficiently far from the wall where the flow is unperturbed. From these

---

[6]This analysis is valid only when the flow field is laminar (see Sect. 3.5).

Fig. 3.3 Boundary layer
around a sphere

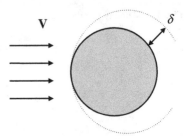

simple observation, we see that the flow in a tube is a very particular case, as we are never too far from the walls and therefore their influence is always felt.

Once the boundary layer thickness δ is known, we can readily determine the force exerted by the fluid on the wall. In fact, referring to Fig. 3.1, when we move from $y = 0$ to $y = \delta$, the fluid velocity changes from $v = 0$ to $v = V$, and therefore the velocity gradient is approximately equal to V/δ. This implies that the shear stresses on the wall (i.e., a force per unit area directed along the direction of the fluid flow) will be $\tau \approx \mu V/\delta$. In the case of the flow past a submerged object of Fig. 3.3, these stresses will be larger in the front of the body than in the back; considering that the total force is the integral of the shear stresses τ on the body surface,[7] we see that the fluid will tend to push the body forward. When the submerged body is a sphere, from elementary symmetry considerations we see that the net transversal force must be zero and the total force exerted by the fluid on the object is a drag force, directed along the unperturbed flow velocity. In general, though, the total force will also have a transversal component, called *lift*, which is responsible of the road holding of a car, as well as of the fact that a heavy object can fly (provided it has an appropriate shape).

This analysis summarizes the basic concepts of the boundary layer theory. It shows why a resisting, dissipative force exists even at high Reynolds number, when one would expect to have instead a potential flow. Actually, as the Reynolds number increases, the boundary layer becomes thinner [see Eq. (3.3.2)], so that the velocity gradients and the shear stresses at the wall increase, thus explaining why the friction forces increase at larger velocities of the incoming fluid.

In the particular case of a fluid flowing past a flat plate of length L_1 and width L_2, the evaluation of the friction force is straightforward, yielding:

$$F = \int_0^{L_1} \int_0^{L_2} \mu \frac{dv}{dy} dx dz \approx L_2 \mu V \int_0^{L_1} \frac{1}{\delta} dx \approx \sqrt{L_2^2 \mu \rho V^3} \int_0^{L_1} \frac{dx}{\sqrt{x}} \approx \sqrt{L_1 L_2^2 \mu \rho V^3}.$$

$$(3.3.3)$$

---

[7]In addition to the viscous drag, we should also take into account the pressure drag, which can also be traced to viscous effects.

This problem was solved exactly by H. Blasius[8] in 1908, who found that Eq. (3.3.3) is correct, with a proportionality coefficient in front of the square root equal to 0.664 (see Sect. 7.2). So we can obtain basically the full result (apart from the proportionality coefficient which, most of the times, is not so important) by a simple dimensional analysis, with hardly any calculation.

In the case of a fluid flowing past a submerged object, calculations are more complex, but the basic result remains the same, that is,

$$F \approx \sqrt{L^3 \mu \rho V^3}, \tag{3.3.4}$$

where L is a characteristic linear dimension of the body. Equations (3.3.3) and (3.3.4) seem to disagree with Eq. (3.2.3), where the resistance force is proportional to the square of the mean velocity. In fact, both results are correct, but they refer to two different flow regimes: Eqs. (3.3.3) and (3.3.4) are valid for laminar flows, when there is no transversal mixing far from the particle, while Eq. (3.2.3) is valid for turbulent flows. This distinction will be clarified in Sect. 3.5.

Finally, note that for small Reynolds numbers the boundary layer analysis is not valid any more.[9] In this case, the velocity gradient at the wall is approximately equal to V/L, and therefore we find Eq. (3.2.4) for the resistance force.

3.4 Boundary Conditions

In the previous sections, we have tacitly assumed that a fluid in contact with a solid wall moves with the same speed as the wall itself. We can imagine that the molecules of the fluid penetrate the pores of the wall and therefore end up moving together with the wall. So, this, so-called, *no-slip* boundary condition, indicates that at the wall there cannot be any discontinuity in the fluid velocity field, so that we have:

$$v_w = v_0, \tag{3.4.1}$$

where v_w is the fluid velocity at the wall, while v_0 is the velocity of the wall itself. Obviously, this condition is essential to calculate the velocity profile, above all in regions near the wall.

The case of a fluid–fluid interface is more complex, because we do not know the speed of the interface. Therefore, together with the no-slip condition indicating that the velocity field must be continuous (i.e., there cannot be any slip at the interface),

---

[8]Paul Richard Heinrich Blasius (1883–1970) was a German fluid dynamicist. He was one of the first students of L. Prandtl.

[9]Actually, this analysis can be repeated also when $Re \ll 1$. In this case, however, convective and diffusive effects balance each other very far from the wall, i.e. at a distance $\Delta \gg L$, such that $Re_\Delta = V\Delta/\nu = O(1)$. The technique of separating the outer region $r > \Delta$, where convection prevails, from the inner region $r < \Delta$, where diffusion is dominant, is called *matched asymptotic expansion*.

Fig. 3.4 Boundary
conditions at an interface

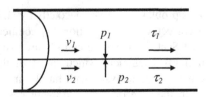

we need another boundary condition at the interface. This second condition imposes
that, in agreement with Newton's third law of dynamics, the force that phase 1
exerts on the phase 2 is opposite to that exerted by phase 2 on phase 1. Then, as
shown in Fig. 3.4, if we decompose these forces in terms of their components
parallel and normal to the interface, we see that pressure and shear stresses must
also be continuous at the interface. So, considering that the shear stresses depend on
the velocity gradients through Newton's constitutive relation (0.4.4), we conclude
that at the interface the flowing boundary conditions are to be applied:

$$v_1 = v_2; \quad p_1 = p_2; \quad \mu_1 dv_1/dy = \mu_2 dv_2/dy. \tag{3.4.2}$$

In the case of a fluid–solid interface, the velocity field, as we have seen, can be
determined applying only the first of these conditions, i.e. imposing a given velocity
at the wall. The remaining two conditions, though, are necessary to determine the
force applied by the fluid on the wall: as the wall is assumed to have a given
velocity, it will necessarily respond by applying the opposite force back to the fluid.
Naturally, this is something that we need to know to calculate the drag force of a
submerged object or the pressure drop of a fluid flowing in a pipe.

A particularly important case is that of a liquid–gas interface, which is generally
referred to a *free surface*. Think of the water flowing in a river: referring to Fig. 3.4,
phase 2 corresponds to water, while phase 1 is air. Therefore, applying the third
boundary condition and considering that the viscosity of water is much larger than
that of air, we find that at the interface $dv_{water}/dy = 0$. At this point, the velocity
field can be determined without having to use the other 2 boundary conditions;
these latter, by imposing the continuity of the velocity and the pressure at the
interface, would allow one to determine the velocity field of the air, which in
general is of no relevance.

3.5 Turbulence

Since the early days of fluid mechanics, it has always been known that there are two
distinct ways for a fluid to flow in a pipe, because at low flow rates the pressure
drop is proportional to the mean velocity, while at high flow rates it depends
approximately on its square. The distinction between the two flow regimes was
pinpointed, once again, by Reynolds, who, in 1883, injected some dye in a fluid
flowing in a pipe made of transparent glass. He observed that at low flow rates the

dye forms a steady filament, thus proving that the fluid flows along a longitudinal direction, without any transversal component. This flow regime is called *laminar*. On the other hand, progressively increasing the fluid flow rate, a critical condition is reached, where the filament starts to wave, and then disappears altogether, with the dye spreading all over the cross section of the tube. In this, so called, *turbulent* regime, the liquid does not flow exclusively along the longitudinal direction, but it moves instead randomly, continuously forming vortices that transport the dye in the transversal direction.

Reynolds established that the critical condition separating these two flow regimes depends only on the Reynolds number. Later on, observations have shown a more complex situation. For example, in the case of a fluid flowing in a conduit, the flow is always laminar when *Re* is less than 2100, and always turbulent when *Re* is larger than 4000. The interval between *Re* = 2100 and *Re* = 4000 is called transition region, where the flow can be laminar or turbulent, depending on other factors, such as pipe roughness and distance from the pipe inlet.[10] In general, however, the value of the Reynolds number is sufficient to give us an idea about the flow regime that we expect to have.

In the following, we will see some fundamental aspects of turbulence; a more detailed treatment can be found in Sect. 17. First of all, as Reynolds understood, turbulence can be interpreted as some sort of critical phenomenon, like phase transition, with a net separation between laminar and turbulent flows. In other words, there are no flows that are half laminar and half turbulent: a soon as the critical conditions are reached, laminar flow becomes turbulent. In the same way, as soon as the temperature of a fluid reaches its saturation value, bubbles start to appear and the fluid starts to boil.

The second observation is based in the fact that turbulent flow dissipates more than laminar flow, with consequently larger pressure drops. So, the first question that a person would ask is why a fluid that could flow in laminar regime with minimum dissipation should instead opt to become turbulent, enormously increasing its pressure drops. At least in part, this question has been answered by stability theory, showing that many turbulent flows are the consequence of linear instabilities in shear flows. In fact, such perturbations arise as disturbances in momentum, that can originate from the walls, as in the *wall turbulence*, or from non-homogeneities of the flow fields, as in the *free turbulence*. In turbulent flow, these disturbance, instead of being damped and disappear, as it happens in laminar flow, keep growing until the flow becomes unstable, with the formation of *eddies*. It looks as if the system finds itself with an energy surplus that in laminar conditions it cannot dissipate and so it starts to pivot, pumping the excess of kinetic energy into these eddies. The latter are initially macroscopic; for example, the wind flowing around a building starts to form eddies of the same size as the building. Obviously, these macroscopic eddies cannot slow down and disappear because the corresponding Reynolds number is quite large and viscous friction is negligible

---

[10]The smoother the pipe and the closer to the entrance, the more laminar the flow will be.

compared to convective momentum transport. Instead, eddies will split, forming smaller vortices through an inviscid flow process at constant kinetic energy. In fact, as the velocity of an eddy having size L is $V \approx \sin(kr)$, where $k = 2\pi/L$, the momentum convective flux will be $\rho V^2 \approx 1 - \cos(2kr)$ and therefore it generates new eddies with size $L/2$, which in turn will form other eddies of size $L/4$, and so on. At the end of this, so called, *energy cascade*, the eddies will have a size δ, called Kolmogorov microscale,[11] that is sufficiently small (typically, from 10 to 100 μm) that the corresponding Reynolds number is about unity, $Re_\delta \approx 1$. This implies that viscous dissipation within eddies of size δ is as important as convective transport, so that the fluid will start to decelerate and, at the end, the eddies disappear. Therefore the energy, that is initially present as kinetic energy of the large eddies, flows down a cascade wherein the big eddies split into innumerable rivulets, corresponding to ever smaller eddies, until it is converted into heat with eddies of size δ being such that $Re_\delta \approx 1$. Only now we can understand the analysis at the end of Sect. 3.2, showing that at large Reynolds numbers the drag force is proportional to the momentum convective flux, ρV^2 [see Eq. (3.2.3)]. In fact, on one hand ρV^2 represents the kinetic energy that is pumped into the macroscopic eddies, which do not dissipate, but on the other hand, it also coincides with the energy that is transported into ever smaller eddies, until it reaches the Kolmogorov scale, where dissipation occurs.

3.6 Problems

3.1 Show that from a characteristic size L, together with the viscosity μ, density ρ and mean velocity V of a fluid, it is possible to define only one independent non-dimensional group, i.e., the Reynolds number.

3.2 As indicated in Fig. 3.2, in the inlet region of a conduit a fluid in laminar flow at high Reynolds number forms a boundary layer that grows with the distance from the edge of the tube, until at distance L it occupies the entire cross section. From this point on, the fluid velocity profile remains unchanged, as it has reached its steady state, called *fully developed flow*. Supposing that the conduit has width R, evaluate the distance L. Solve this problem also considering that the *fully developed flow* is reached when the information that the velocity of the wall is zero has reached the center of the conduit.

---

[11]Named after Andrey Nikolaevich Kolmogorov (1903–1987), a Russian mathematician who worked on stochastic processes and turbulence.

Chapter 4
Macroscopic Balances

Abstract In fluid mechanics, conservation of mass, momentum and energy lead to the so-called continuity and Navier-Stokes equations. These equations can be written in either integral or differential form. In this chapter, we consider their integral formulation, which is applicable to a finite mass of fluid in motion. The more rigorous differential treatment will be the subject of Chap. 6. Here, setting up macroscopic balances and, at first, neglecting all diffusive effects, we derive the continuity equation in Sect. 4.1, the Bernoulli equation in Sect. 4.2, and the Euler equation in Sect. 4.3. Then, in Sects. 4.4, 4.5 and 4.6, the corrections to the Bernoulli equation are analyzed, discussing the pressure losses due to friction forces, in particular for the cases of flows within pipes and around submerged objects.

4.1 Mass Balance and Continuity Equation

In a stationary flow field, the mass balance is particularly simple: the mass entering per unit time in a given volume is equal to the mass leaving it, since mass cannot be created nor destroyed.

Consider a fluid flowing through a conduit of variable section, as represented in Fig. 4.1. In this section we assume that the lateral boundaries of the volume element consist of solid walls, but the analysis can be easily applied also to the case where they spin a streamtube, i.e., the surface formed by all *streamlines* that pass through a given closed curve in the fluid (recall that a *streamline* is a line in the fluid whose tangent is everywhere parallel to the fluid velocity instantaneously). In both cases, there is no mass flux leaving the volume element through its lateral boundaries. In addition, the inlet and outlet sections are chosen to be perpendicular to the momentum transport.

Now, let us write a mass balance between sections 1 and 2. First of all, define the mass *flow rate* $\dot{m}$ as the fluid mass crossing a given area S per unit time, with $[\dot{m}] = \mathrm{kg/s}$ in the SI system. Since, as we saw in Eq. (1.4.1), the mass flux, J_M,

© Springer International Publishing Switzerland 2015
R. Mauri, *Transport Phenomena in Multiphase Flows*,
Fluid Mechanics and Its Applications 112, DOI 10.1007/978-3-319-15793-1_4

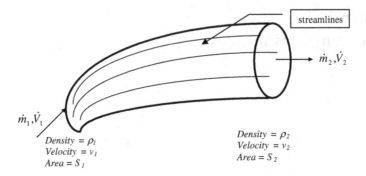

Fig. 4.1 Mass balance

defined as the ratio between $\dot{m}$ and S, is the product between the fluid density (i.e., the mass of the fluid per unit volume) and the fluid velocity, v, then we have:

$$\dot{m} = \rho v S. \tag{4.1.1}$$

Note that, starting from the mass flow rate, a quantity that is very easy to measure (only a scales and a clock are needed to perform such a measurement), we have defined first the fluid momentum, i.e., one of the fundamental physical quantities, and then the fluid velocity. Although Eq. (4.1.1) can be considered as a definition of the fluid velocity v, it is easy seen that v coincides with the usual kinematic velocity (see Problem 4.1).

Therefore, we may conclude that at steady state the mass conservation equation is

$$\dot{m} = \rho v S = \text{constant.} \tag{4.1.2}$$

This is the so-called continuity equation, which is valid at steady state. In incompressible flows, as ρ is constant, the continuity equation simplifies to

$$\dot{V} = v S = \text{constant,} \tag{4.1.3}$$

where $\dot{V}$ the volumetric flow rate, with $[\dot{V}] = \text{m}^3/\text{s}$ in the SI system.

In general, both density and fluid velocity may vary from point to point in the cross section of a stream tube. In that case, the mass balance equation takes the form,

$$\dot{m} = \int_S \rho v dS = \bar{\rho}\bar{v}S = \text{const.,} \tag{4.1.4}$$

where $\bar{\rho}$ and $\bar{v}$ are the mean density and the mean velocity, i.e.,

$$\bar{\rho} = \frac{1}{S} \int_S \rho dS \text{ and } \bar{v} = \frac{1}{\bar{\rho}S} \int_S \rho v dS. \tag{4.1.5}$$

Note that a volumetric mean velocity can also be defined as the volumetric flow rate that crosses the unit cross section, i.e.,

$$\dot{V} = \int_S v dS = v^* S; \quad v^* = \frac{1}{S} \int_S v dS. \tag{4.1.6}$$

Here, v^* represents the integral mean value over a cross section of the conduit. Equations (4.1.5) define the mass-averaged mean velocity, while Eq. (4.1.6) defines a volume-averaged velocity. Clearly, they coincide for incompressible flows.

Note that in the particular case of circular cross sections of radius R, with $S = \pi R^2$, the continuity equation for incompressible flow becomes:

$$\frac{\bar{v}_1}{\bar{v}_2} = \left(\frac{R_2}{R_1}\right)^2. \tag{4.1.7}$$

This analysis can be easily generalized to unsteady flow fields, obtaining:

$$\frac{dm}{dt} = \dot{m}_{in} - \dot{m}_{out}, \text{ i.e., } \frac{dm}{dt} = (\rho v S)_1 - (\rho v S)_2, \tag{4.1.8}$$

where $m = \int \rho dV$ is the mass of the fluid contained in the volume V between the cross sections 1 and 2. This equation follows from the general principle of mass conservation, stating that the mass accumulated in the volume is equal to the net mass influx through its boundaries.

4.2 Mechanical Energy Balance and Bernoulli Equation

Let us consider first the case of a steady flow of an inviscid fluid, writing the balance of mechanical energy for the elemental streamtube represented in Fig. 4.2. Since there is no dissipation, the mechanical energy is conserved, so that the flux of the potential and kinetic energy, $(\rho g z) dV/dt$ and $(1/2 \rho v^2) dV/dt$, must equal the total rate of working, $\dot{W} = dW/dt$ on the elemental streamtube, i.e.,

$$\dot{m} d \left(g z + \frac{1}{2} v^2 \right) = d\dot{W}.$$

Fig. 4.2 Elemental
streamtube

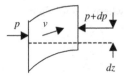

In fact, the pressure p acting at the inlet pushes the fluid in and therefore does a
positive rate of working, $p\dot{V}$, while the pressure $p + dp$ acting at the outlet does a
negative work, $-(p + dp)\dot{V}$. Therefore, we may conclude that $d\dot{W} = -\dot{V}dp$,
obtaining at the end,

$$\frac{dp}{\rho} + gdz + d\left(\frac{v^2}{2}\right) = 0.$$

In incompressible flow, this result can be recast in the form,

$$\frac{p}{\rho} + gz + \frac{v^2}{2} = \text{constant}, \tag{4.2.1}$$

showing that, in the absence of friction, the sum of the terms due to pressure, height
and velocity, is constant. This is the Bernoulli[1] equation, which basically follows
from an energy conservation statement applied to fluids in motion.

4.2.1 Example: The Pitot Tube

The Pitot[2] tube is a pressure sensitive device which enables the measurement of the
dynamic pressure of a fluid in motion. Hence, it can be used to determine the speed
of ships or airplanes relative to that of the surrounding fluid (i.e., water or air).
Suppose, for example, that the J-tube shown in Fig. 4.3 is located on a ship, moving
with a relative velocity v with respect to water. Applying the Bernoulli Eq. (4.2.2)
we obtain:

$$\frac{1}{2}v^2 - gz + \frac{p_1}{\rho} = gh + \frac{p_2}{\rho},$$

---

[1]Named after Daniel Bernoulli (1700–1782) a Swiss mathematician and physicist, who was born
into a family of very distinguished mathematicians from Holland. In fact, his father, Johann
Bernoulli, convinced Paul Euler (Leonhard's father, a pastor of the Reformed Church) that his son
was destined to become a great mathematician.
[2]Named after the French hydraulic engineer Henri Pitot (1695–1771).

Fig. 4.3 J-tube

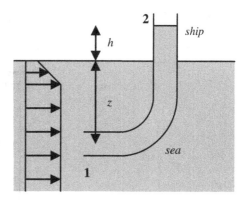

where p_2 is the atmospheric pressure, ρ is the water density and the heights h and z are measured with respect to the free surface of the sea. Then, considering that p_1 is the hydrostatic pressure,

$$p_1 = p_2 + \rho g z,$$

we obtain the so-called *Torricelli* relation,[3]

$$v = \sqrt{2gh}. \tag{4.2.2}$$

The case of a fluid in pipe flow is slightly more complicated because, unlike the previous case, the pressure inside the pipe is unknown. Then, opening a blowhole as shown in Fig. 4.4 and assuming that, as is usually the case, the pressure in the pipe, p, is larger than the atmospheric pressure, the fluid will rise to a height d that we can measure as $d = \tilde{p}/\rho g$, where $\tilde{p} = p - p_a$ is the relative pressure in the pipe. Hence, proceeding as in the previous case, we see that, positioning the Pitot tube at different z locations within the pipe and measuring the corresponding $h(z)$, we can determine the velocity profile,

$$v_1(z) = \sqrt{2gh(z)}. \tag{4.2.2a}$$

4.2.2 Generalization of the Bernoulli Equation

Let us return to the above energy conservation argument leading to the Bernoulli equation. In fact, the total energy per unit volume, ρe, is the sum of the internal, kinetic, and potential energies:

---

[3]Named after Evangelista Torricelli (1608–1647), an Italian physicist and mathematician.

Fig. 4.4 Pitot tube

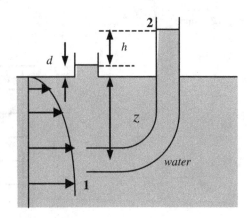

$$\rho e = \frac{\text{energy}}{\text{fluid volume}} = \rho u + \frac{1}{2}\rho v^2 + \rho g z, \tag{4.2.3}$$

where u is the internal energy per unit mass. So, the principle of energy conservation, i.e. the first law of thermodynamics, states that the energy accumulated is equal to the energy inlet minus the energy outlet, with an energy source term, i.e.,

$$\frac{dE}{dt} = \dot{E}_{in} - \dot{E}_{out} + \dot{E}_s \Rightarrow \frac{d}{dt}\int_V \rho e dv = (\rho v e S)_1 - (\rho v e S)_2 + \dot{W} + \dot{Q}. \tag{4.2.4}$$

Here, $\rho v e$ is the total energy flux, while $\dot{W}$ and $\dot{Q}$ are the work and the heat entering the volume per unit time, respectively. In turn, the total rate of working is the sum of the work done by pressure forces, PvS (the same term that we have encountered at the beginning of this section) and a so-called *shaft work* that is exchanged at the boundary of the system, for example through a pump, denoted in the following as $\dot{W}_p = \dot{m}w_p$. In addition, the heat $\dot{Q}$, that the system exchanges per unit time with a hot reservoir (in fact, we have used the convention that both heat and work are positive when entering the system) is indicated as $\dot{Q} = \dot{m}q$. Finally, we obtain:

$$\frac{d}{dt}\int_V \rho e dv = -\Delta\left(\frac{1}{2}\rho v^3 S\right) - g\Delta(\dot{m}z) - \Delta(pvS) - \Delta(\dot{m}u) + \dot{m}w_p + \dot{m}q, \tag{4.2.5}$$

where $\Delta f = f_2 - f_1$. Note that, in general, the velocity is not uniform within the section S and therefore the first term on the RHS of (4.2.5) should read as follows:

$$-\Delta\left(\frac{1}{2}\rho\int_S v^3 dS\right) = -\Delta\left(\frac{1}{2}\alpha\rho\bar{v}^3 S\right) \text{ where } \alpha = \int v^3 dS / v^{-3}S. \tag{4.2.6}$$

Naturally, α depends on the shape of the velocity profile. In pipe flows, the velocity profile is rather flat in the turbulent regime, so that $\alpha \cong 1$ in that case, while for the laminar regime $\alpha \neq 1$.

Now, let us consider the *stationary* case, when the mass flow rate $\dot{m}$ is constant. Then, reminding that $pvS = \dot{m}p/\rho$ and dividing Eq. (4.2.5) through by $\dot{m}$, we obtain:

$$\Delta H = q + w_p; \quad \text{with } H = h + \frac{1}{2}v^2 + gz, \tag{4.2.7}$$

where $h = u + p/\rho$ is the fluid enthalpy per unit mass. So, in the absence of any shaft work and when $q = 0$, we see that the total energy per unit mass, H, is constant along a streamline. The assumption $q = 0$ is satisfied for frictionless and non-conducting fluids, when no heat is exchanged with external reservoirs.

Now, assuming that the two cross sections are infinitely close to each other, consider the first and the second laws of thermodynamics, stated as,

$$dh = Tds + dp/\rho \tag{4.2.8}$$

and

$$Tds - dq = dh_f \geq 0, \tag{4.2.9}$$

where s is the entropy per unit mass, while h_f is the energy dissipated per unit mass, with the equal sign in (4.2.9) referring to reversible processes. Finally, substituting Eqs. (4.2.8) and (4.2.9) into (4.2.7) we obtain:

$$\frac{1}{2}dv^2 + gdz + dp/\rho = dw_p - dh_f, \tag{4.2.10}$$

which can be easily integrated in a finite control volume:

$$\frac{1}{2}\Delta v^2 + g\Delta z + \int_a^b \frac{dp}{\rho} = w_p - h_f. \tag{4.2.11}$$

In incompressible flow, this equation leads to the following generalized Bernoulli equation,

$$\frac{1}{2}\Delta v^2 + g\Delta z + \frac{\Delta p}{\rho} = w_p - h_f. \tag{4.2.12}$$

This equation will be discussed in Sect. 4.4.

Finally note that, in unsteady conditions, we cannot divide Eq. (4.2.5) through by the mass flow rate $\dot{m}$, so that the equation of energy conservation is:

$$\frac{d(\rho e)}{dt} dv = -\frac{1}{2} d(\dot{m}v^2) - gd(\dot{m}z) - \frac{1}{\rho} d(\dot{m}p) + d(\dot{m}w_p) - d(\dot{m}h_f). \quad (4.2.13)$$

4.3 Momentum Balance

The momentum balance states that at steady state the difference between the momentum flux through the cross sections 2 and 1 (see Fig. 4.1), $\dot{Q}_2$ and $\dot{Q}_1$, respectively, is equal to the sum of all the forces $\mathbf{F}$ exerted on the streamtube by its surroundings, i.e.,

$$\sum \mathbf{F} = \dot{\mathbf{Q}}_2 - \dot{\mathbf{Q}}_1, \quad (4.3.1)$$

As we have already seen in (1.4.2), when the diffusive momentum flux is neglected,[4] $\dot{\mathbf{Q}}$ equals the convective momentum flux through the cross sectional area S, i.e., $\dot{Q} = SJ_{Qc} = \dot{m}v$, where $J_{Qc} = \rho v^2$ in the direction of the fluid flow. The forces in Eq. (4.3.1) are of three types: (a) the pressure p exerted by the fluid element on the surrounding fluid on the cross sections 1 and 2; (b) the forces $\mathbf{F}'_w$ exerted by the surrounding fluid on the bounding surface of the volume element; (c) all volume forces, $\mathbf{F}_g$, like gravity. Therefore, we obtain:

$$\sum \mathbf{F} = p_1 \mathbf{S}_1 - p_2 \mathbf{S}_2 + \mathbf{F}'_w + \mathbf{F}_g, \quad (4.3.2)$$

where $\mathbf{S}$ is a vector having a magnitude equal to the area S and direction of the fluid flow, i.e., $\mathbf{S} = S\mathbf{v}/v$. In turn, $\mathbf{F}'_w$ is the sum between the force $\mathbf{F}_e$, exerted against the walls by the outside fluid, and the force exerted on the fluid element by the walls, $\mathbf{F}_w$. Assuming that the outside fluid is quiescent and at atmospheric pressure, since at equilibrium the sum of all the pressure forces acting on a closed surface must be zero, the product of the atmospheric pressure, p_{atm}, by the surfaces (taken with their signs and directions) of the reference volume is zero. Therefore we obtain: $\mathbf{F}_e = -p_{atm}(\mathbf{S}_1 - \mathbf{S}_2)$ and finally:

$$\dot{m}_2 \mathbf{v}_2 - \dot{m}_1 \mathbf{v}_1 = \dot{m}(\mathbf{v}_2 - \mathbf{v}_1) = \sum \mathbf{F} = \tilde{p}_1 \mathbf{S}_1 - \tilde{p}_2 \mathbf{S}_2 + \mathbf{F}_w + \mathbf{F}_g, \quad (4.3.3)$$

where $\tilde{p} = p - p_{atm}$ is the relative (or excess) pressure, that is the difference between the fluid pressure and the atmospheric pressure. Equation (4.3.3) is easily extended to non-stationary flows, obtaining:

---

[4]Here we are considering the momentum transport through a streamtube. Note that at large Reynolds numbers the diffusive contribution is indeed very small.

$$\frac{d}{dt} \int_V \rho \mathbf{v} dV = \left(\rho v^2 + \tilde{p}\right)_1 \mathbf{S}_1 - \left(\rho v^2 + \tilde{p}\right)_2 \mathbf{S}_2 - \mathbf{F}_w + \mathbf{F}_g, \qquad (4.3.4)$$

where we have considered that $\rho v^2 \mathbf{S} = \dot{m}\mathbf{v}$.

In fact, assuming that the cross sections 1 and 2 are infinitely close to each other, we see that Eq. (4.3.3) is implied by the *Euler equation* at steady state,

$$\mathbf{v} \cdot \nabla \mathbf{v} + \frac{1}{\rho} \nabla p = \mathbf{g}. \qquad (4.3.5)$$

Assuming that $\mathbf{g} = -\nabla \psi$ (clearly, $\psi = -gz$ when $\mathbf{g}$ is the gravity acceleration), and using the identity $\mathbf{v} \cdot \nabla \mathbf{v} = \nabla\left(\frac{1}{2}v^2\right) - \mathbf{v} \times (\nabla \times \mathbf{v})$, together with the first law of thermodynamics, stated as $\nabla h = T\nabla s + \nabla p/\rho$, then the Euler equation can be recast in the form of the *Crocco equation*,[5] i.e.,

$$\nabla H = T\nabla s + \mathbf{v} \times (\nabla \times \mathbf{v}), \qquad (4.3.6)$$

where, $H = h + \frac{1}{2}v^2 + \psi$ coincides with the energy per unit mass (4.2.7). Hence, it becomes clear that in irrotational and homo-entropic[6] flows, the Bernoulli equation can be generalized to:

$$\nabla H = \mathbf{0}. \qquad (4.3.7)$$

In fact, while in Eq. (4.2.7) the total energy per unit mass, H, is only constant along a streamline, Eq. (4.3.7) implies that it is constant everywhere in the fluid, i.e. it is the same for all streamlines.

4.4 Recapitulation of the Bernoulli Equation

In Sect. 4.2 we have seen that a mechanical energy balance between cross sections 1 and 2 of a streamtube leads to the Bernoulli relation, Eq. (4.2.1), i.e.,

$$\frac{p_1}{\rho} + gz_1 + \frac{v_1^2}{2} = \frac{p_2}{\rho} + gz_2 + \frac{v_2^2}{2}. \qquad (4.4.1)$$

---

[5]Luigi Crocco (1909–1986), an Italian aerospace engineer.

[6]For an iso-entropic flow, entropy is constant along any streamline, while homo-entropic flows have the entropy homogeneous everywhere. Accordingly, the flow of any inviscid and non-conducting fluid is always iso-entropic (see Eq. 6.8.4), but not necessarily homo-entropic, as, for example, there might be a temperature gradient.

In most applications, fluids flow in the vicinity of walls and interfaces, where the velocity field is non-uniform and energy dissipation becomes important. In addition, fluid flows are often forced, such as when, e.g., a pump is used to increase the pressure of a liquid in order to move it through a piping system. In such cases, Eq. (4.4.1) can be generalized by adding the following three corrections: (a) a modified kinetic energy term, accounting for the non-uniformity of the velocity field; (b) an additional term, due to viscous losses, describing the friction at walls and interfaces; (c) a shaft work term, increasing (or decreasing) the mechanical energy due to the presence of pumps (or turbines). Such corrections have already been included in the generalized Bernoulli relation, Eq. (4.2.13), i.e.,

$$\frac{p_1}{\rho} + gz_1 + \alpha_1 \frac{\overline{v_1}^2}{2} + \eta w_p = \frac{p_2}{\rho} + gz_2 + \alpha_2 \frac{\overline{v_2}^2}{2} + h_f. \qquad (4.4.2)$$

Additional discussion of each of the above contributions follows.

4.4.1 Effect of the Non-Uniformity of the Velocity Field

In the Bernoulli equation, the kinetic energy per unit mass is represented by a $v^2/2$ term, where the fluid velocity, v, is assumed to be uniform within the cross section. In order to find how this term must be modified when the velocity field is non-uniform, let us remind how the kinetic term is derived.

First, consider the mass flow rate $\rho v dS$ crossing an infinitesimal cross section dS. Since each unit of mass carries a kinetic energy $v^2/2$, the flux of kinetic energy crossing the area dS will be $\rho v^3 dS/2$. Therefore, the total kinetic energy flowing per unit time across the cross section S is:

$$\dot{E}_c = \frac{1}{2}\rho \int_S v^3 dS = \frac{1}{2}\rho S\langle v^3 \rangle,$$

where $\langle v^3 \rangle$ is the average cubic velocity, thus generalizing the first term in the RHS of Eq. (4.2.5). Now, the kinetic energy term in the Bernoulli equation is obtained by dividing $\dot{E}_c$ by the mass flow rate, which is expressed in terms of the mean velocity. Therefore, as we have seen in Eq. (4.2.6), the kinetic energy term can be written in a more convenient form, as follows, by defining a correction factor α as:

$$\frac{\dot{E}_c}{\dot{m}} = \frac{1}{2}\frac{\langle v^3 \rangle}{\langle v \rangle} = \frac{1}{2}\alpha \langle v \rangle^2, \text{ where } \alpha = \langle v^3 \rangle \Big/ \langle v \rangle^3. \qquad (4.4.3)$$

In the following, the mean velocity will be denoted as $\bar{v}$ or $\langle v \rangle$, interchangeably.

Hence, we see that the correction factor α appearing in Eq. (4.4.2) depends on the shape of the velocity profile, not on its magnitude (in fact, if we double v everywhere, α remains unchanged). In pipe flows, the velocity profile is rather flat in the turbulent regime, so that in that case we can approximate $\alpha \cong 1$. In the laminar regime, instead, velocity profiles are nearly parabolic, with α between 1.5 and 2; for example, in plane channel flows $\alpha = 1.54$, while in the case of a circular tube we find $\alpha = 2$ (see Problem 4.4). In any case, it should be kept in mind that in the laminar regimes friction losses are often dominant and the contribution of the kinetic energy term is thus negligible.

4.4.2 Effect of the Friction Forces

The correction due to friction is the most complex and important. Friction manifests itself by converting mechanical energy into heat, and, ultimately, into internal energy, so that the mechanical energy is not conserved, but instead it decreases as we move along a streamline.

As previously noted, in the Bernoulli Eqs. (4.2.12) and (4.2.13) friction is expressed through the h_f term, having the units of energy per unit mass (or a velocity squared). Unlike the other terms of the Bernoulli equation, h_f is not a local property, that is valid *at* the cross sections 1 or 2, but instead it represents a loss of mechanical energy *between* the cross sections 1 and 2. In addition, friction represents an irreversible process; therefore, h_f is always positive, and is equal to zero only for inviscid flows.

In many textbooks, friction losses are expressed in terms of a *head loss*, Δz_f, that is the height of a quiescent fluid column whose pressure drop is equal to that due to friction, i.e.,

$$h_f = g\Delta z_f. \qquad (4.4.4)$$

4.4.3 Effect of Pumps and Turbines

Pumps are used to increase the mechanical energy of the fluid. Therefore, if a pump is inserted between the cross sections 1 and 2, we should add a term w_p to the energy balance, representing the work per unit mass flow rate done by the pump on the fluid. Naturally, in a pump there are always some losses, so that only a fraction η of the power required by the pump is actually transferred to the fluid. If instead of a pump we have a turbine, absorbing energy from the fluid, then w_p will be negative.

4.5 Pressure Drops in Pipe Flow

The friction forces will be studied in detail in the next chapter. Here we intend to establish a relation between the h_f factor and the shear stresses at the wall. Consider a steady, incompressible flow of a Newtonian fluid in a horizontal circular pipe of radius R. If $p_1 = p$ and $p_2 = p - \Delta p$, where Δp is the (positive) pressure drop across a distance Δx, the Bernoulli equation reduces to:

$$\frac{\Delta p}{\rho} = h_f, \tag{4.5.1}$$

showing that h_f equals the pressure drop divided by the constant fluid density.

At steady state and within a pipe of uniform cross section, the fluid does not accelerate, so that the sum of all forces must be zero. Therefore, a simple balance of the longitudinal forces exerted on the fluid element of Fig. 4.5 leads to the following result:

$$\sum F = \pi R^2 p - \pi R^2 (p - \Delta p) - (2\pi RL)\tau_w = 0,$$

where τ_w is the shear stress at the wall. Simplifying this expression, we obtain:

$$\frac{\Delta p}{L} = \frac{2\tau_w}{R} = \frac{4\tau_w}{D}, \tag{4.5.2}$$

where $D = 2R$ is the tube diameter. Then, the shear stress at the wall can be determined once the pressure drops are known, and vice versa. Finally, from Eqs. (4.5.1) and (4.5.2) we obtain:

$$h_f = \frac{4\tau_w}{\rho} \frac{L}{D}. \tag{4.5.3}$$

Pressure drops are generally accounted for by using the so-called *friction factor f* (also called the *Fanning*[7] friction factor), defined as the ratio between the shear stress at the wall and the fluid kinetic energy density, i.e.,

$$f \equiv \frac{\tau_w}{\rho \bar{v}^2 / 2}. \tag{4.5.4}$$

Substituting Eqs. (4.5.3) into (4.5.4), we obtain the following relation between h_f, Δp and f,

$$h_f = 2f \frac{L}{D} \bar{v}^2; \quad \Delta p = 2f \frac{L}{D} \rho \bar{v}^2. \tag{4.5.5}$$

---

[7]Named after John Thomas Fanning (1837–1911), an American hydraulic engineer.

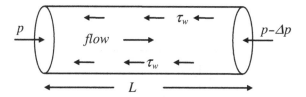

Fig. 4.5 Pressure drops in pipe flow

4.5.1 Fanning vs. Darcy Friction Factor

While the Fanning friction factor is the most commonly used in chemical engineering, civil and mechanical engineers prefer to use the *Darcy friction factor*,[8] f_D, defined as:

$$\Delta p = f_D \frac{L}{D} \frac{1}{2} \rho \bar{v}^2. \tag{4.5.5a}$$

Clearly, comparing this definition with (4.5.5), we see that the Darcy friction factor, f_D, is 4 times larger than the Fanning friction factor, f. So, for example, for pipe flow in the laminar regime we have: $f_D = 4f = 64/Re$, where $Re = \rho \bar{v} D / \mu$ is the Reynolds number. In any case, attention must be paid to which friction factor is being used in any chart or equation that we encounter.

As we have seen in the previous chapter, when fluid motion is dominated by diffusion, as it happens at low Reynolds number, the shear stresses at the wall are proportional to the mean velocity. For pipe flows, though, we also saw that the boundary layer tends to occupy the entire cross section area, so that we expect that viscous diffusion continues to be the dominant transport mechanism even for relatively large Reynolds numbers. In fact, we find that diffusion effects dominate, so that $\tau_w \propto \bar{v}$ and $f \propto 1/Re$, when $Re < 2100$. More precisely, as we will see in the next chapter, in the case of pipes with circular cross sections we have:

$$f = \frac{16}{Re} \text{ when } Re < 2100. \tag{4.5.6}$$

At larger Reynolds numbers the flow regime becomes turbulent and we see that, as predicted, the dependence of f on the Reynolds number is less pronounced (see Fig. 4.6 for smooth pipes). In the previous chapter we have seen that in the case of a fluid flowing past a submerged object in the turbulent regime the drag force is proportional to the square of the velocity, so that we would expect to find a constant value of the friction factor, i.e., f should be independent of the Reynolds number.

---

[8]Also referred to as the *Darcy–Weisbach*, the *Blasius* or the *Moody* friction factor.

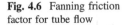

Fig. 4.6 Fanning friction factor for tube flow

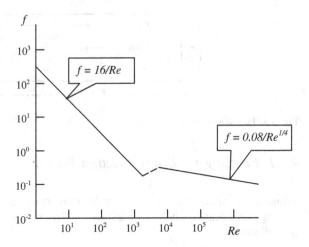

In the case of pipe flows, this does not happen[9] and the analysis is extremely complicated. So, many correlations have been proposed to correlate the innumerable experimental results. Here, we consider the following "classical" correlation, due to *Blasius* and valid for smooth pipes,[10]

$$f = \frac{0.0791}{Re^{1/4}} \text{ when } Re > 3500. \tag{4.5.7}$$

More precise correlations are based on Moody's experimental diagram, reported in Fig. 4.7, where the Darcy friction factor (4.5.5a) is plotted as a function of Re for different values of the surface roughness ε of the wall, that is the mean height of asperities in the plane of the wall. Moody's chart, published in 1944, was mostly based on previous experimental results, many of them obtained by J. Nikuradse,[11] as it was the following *Colebrook's relation*:

---

[9]A qualitative explanation of this fact is that in pipe flow we are never too distant from the boundaries, so that wall effects can never be neglected. That means that the instability vortices will move away from the walls, where they form, towards the center of the tube, until the fluid motion will be turbulent everywhere. On the other hand, in the case of the flow past a submerged object, we can always move far enough from the body, where the fluid flow is unperturbed and convective momentum fluxes are the dominant form of transport, thus explaining the quadratic dependence of the drag force on the velocity that we have seen in Eq. (2.2.3).

[10]As seen in Sect. 17.5, Blasius obtained this expression assuming that the fluid velocity depends on the distance from the center of the tube to the 1/7 power. A similar approach was followed by von Karman, who considered the self-similar logarithmic velocity profile at the wall. Other researchers simply tried to correlate the experimental data, such as Moody's diagram.

[11]Johann Nikuradse (1894–1979), a Georgian-German engineer and physicist.

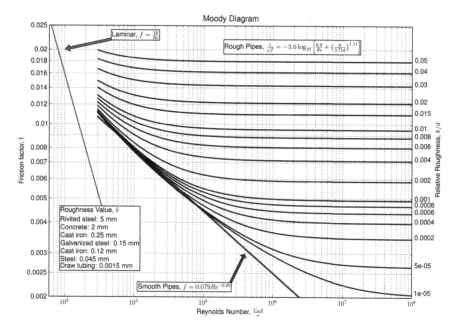

Fig. 4.7 Moody's experimental diagram, where the Darcy friction factor (4.5.5a) is plotted as a function of Re for different values of the the roughness ε of the wall

$$\frac{1}{\sqrt{f}} = 2.28 - 4.0\log\left[\frac{\varepsilon}{D} + \frac{4.65}{Re\sqrt{f}}\right] \quad \text{when } Re > 4000. \tag{4.5.8a}$$

Note that, as Re increases, at a certain point f reaches an asymptotic value, independent of Re, with,

$$\frac{1}{\sqrt{f}} = 2.28 + 4\log(D/\varepsilon). \tag{4.5.8b}$$

Also note that, when $\varepsilon/D = 10^{-4}$, a pipe can be considered to be smooth until $Re \cong 4 \times 10^4$. The dependence of the viscous resistance on the wall roughness can be explained considering that, as Re increases, the size δ of the Kolmogorov eddies decreases, until, when $\delta \approx \varepsilon$, turbulence can directly interact with the wall roughness, greatly increasing both friction and pressure drops. On the other hand, in the laminar flow regime the only characteristic dimension of the flow field is the tube radius (and, in fact, we saw that when $Re \gg 1$ there is a boundary layer occupying the whole cross section of the tube), and so the wall roughness has hardly any relevance.

Finally, it should be stressed that the transition between laminar and turbulent regimes is an extraordinarily complicated process which, until recently, has not been fully understood.[12] On one hand, repeating the experiment with the same pipe and entrance conditions, the laminar-turbulent transition takes place at the same, well defined Re. However, while $Re = 2000$ appears to be about the lowest value obtained at a rough entrance, the transition can even happen with Re as high as 40,000 for very smooth pipes and entry conditions (in the 1870s, Reynolds managed to have laminar flow at $Re \cong 13,000$). Practically, for $2100 < Re < 3500$ there is a laminar-turbulent transition region, indicated with a broken line in Fig. 4.6, that allows to connect the laminar to the turbulent regime. That line, though, should be taken *cum grano salis*.

Laminar-turbulent transition can also occur within the boundary layer. This process can occur through a number of paths, depending on the initial conditions, such as initial disturbance amplitude and surface roughness.[13]

4.6 Localized Pressure Drops

The general concept of pressure drop due to friction remains valid even when the variation of pressure occurs in a very small region, where the velocity field is not known. Think, for example, of the case where the fluid passes through a valve, or more simply, when the cross section of a conduit changes abruptly. In these cases, we can still describe the pressure drop (called *localized* or *concentrated* pressure drop) using the following *localized friction factor, k,*

$$h_f = k \frac{\bar{v}^2}{2},   (4.6.1)$$

where $\bar{v}$ is the mean velocity in the downstream cross section. Clearly, this expression is valid only at large Re, when the convective momentum transport is much larger than its diffusive counterpart, so that we expect that k does not depend on the Reynolds number. We can think of k as a sort of coefficient of ignorance, similar to Newton's heat transfer coefficient, defined as the heat flux at the wall divided by the temperature drop. Just like the heat transfer coefficient is not a function of the temperature, the localized pressure drop does not depend on the velocity. In both cases, these transfer coefficients have to be measured

---

[12]In 1932, in an address to the British Association for the Advancement of Science, Sir Horace Lamb, a famous British applied mathematician and physicist, reportedly said, "I am an old man now, and when I die and go to heaven [Sir Horace was a hopeless optimist] there are two matters on which I hope for enlightenment. One is quantum electrodynamics, and the other is the turbulent motion of fluids. And about the former I am rather optimistic."

[13]See S.B. Pope, *Turbulent Flows*, Cambridge Univ. Press (2000), Par. 7.3.

Table 4.1 Values of the localized friction factor, k, with β = (smaller cross section)/(larger cross section)

	k
Smooth entrance in a tube	0.05
Restriction	$0.45(1 - \beta)$
Enlargement	$[(1/\beta) - 1]^2$
Orifice	$2.7(1 - \beta)(1 - \beta^2)/\beta^2$
Smooth elbow at 90°	0.4–0.9
Sharp elbow at 90°	1.3–1.9
Elbow at 45°	0.3–0.4

Fig. 4.8 Flow through a sudden enlargement

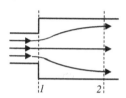

experimentally. In Table 4.1 some values of the localized friction factor are listed. More data can be found in Perry's handbook.

4.6.1 Example: Flow Through a Sudden Enlargement

An example of a localized pressure drop occurs for incompressible flow in a pipe with a sudden enlargement (see Fig. 4.8). Consider the volume between cross section 1, located just upstream of the enlargement, and cross section 2, located downstream, where the flow is again fully developed, and therefore uni-directional. First of all, mass conservation requires that

$$v_1 = \frac{S_2}{S_1} v_2 = \frac{1}{\beta} v_2, \tag{4.6.2}$$

where, in agreement with the notation of Table 4.1, we have defined,

$$\beta = \frac{\text{area of smaller cross section}}{\text{area of larger cross section}} = \frac{S_1}{S_2}. \tag{4.6.3}$$

Now, apply the Bernoulli equation,

$$\frac{1}{2} \left(v_1^2 - v_2^2 \right) + \frac{1}{\rho} (p_1 - p_2) = h_f, \tag{4.6.4}$$

showing that the pressure difference $(p_1 - p_2)$ has to be known in order to determine h_f, and then k, defined through Eq. (4.6.1). To that end, consider the momentum balance equation along the longitudinal direction,

$$\dot{m}v_1 - \dot{m}v_2 + \tilde{p}_1 S_1 - \tilde{p}_2 S_2 = F_w, \tag{4.6.5}$$

where F_w is the force exerted by the fluid against the walls. This force equals the sum between the force acting on the annulus of area $(S_2 - S_1)$ of the enlargement cross section and the friction force on the lateral wall, that here we assume to be negligible as the distance between the two cross sections is quite small. On that annulus, there is a positive external force due to the atmospheric pressure and a negative force exerted by the fluid inside, at pressure p_1, so that at the end we find a force $F_w = -\tilde{p}_1 (S_2 - S_1)$ in a direction opposite to the fluid flow. Therefore, the momentum balance gives:

$$(\tilde{p}_2 - \tilde{p}_1) S_2 = (p_2 - p_1) S_2 = \dot{m}(v_1 - v_2) = \rho v_2 S_2 (v_1 - v_2). \tag{4.6.6}$$

Finally, substituting Eq. (4.6.5) into the Bernoulli Eq. (4.6.4) we conclude:

$$h_f = \frac{1}{2}v_1^2 - \frac{1}{2}v_2^2 - v_1 v_2 + v_2^2 = \frac{1}{2}(v_1 - v_2)^2. \tag{4.6.7}$$

Defining the friction factor k through Eq. (4.6.1) in terms of the downstream velocity v_2, i.e., $h_f = kv_2^2/2$, we obtain:

$$k = \left(\frac{1}{\beta} - 1\right)^2, \tag{4.6.8}$$

in agreement with the value reported in Table 4.1.

4.7 Flow Around a Submerged Object

Consider the incompressible flow around a submerged object. In general, the flow exerts on the object a resistance force F_f, that is convenient to write as:

$$F_f = S\left(\frac{1}{2}\rho\bar{v}^2\right)f. \tag{4.7.1}$$

Here, S is a characteristic cross sectional area of the body and f is the friction factor,[14] in general depending on the Reynolds number and on the shape of the object. In particular, at high Reynolds number, as the convective momentum flux is larger than its diffusive counterpart, we expect that the resistance force must be proportional to the square of the mean velocity, implying a friction coefficient independent of the Reynolds number. Equation (4.7.1) can be considered as a

---

[14]In many textbooks, the friction factor for submerged objects is denoted by C_D.

generic definition of the drag coefficient, f, which reduces to the friction factor in the case of flow in a conduit, where the characteristic area S is that of the wetted surface, and $F_f/S = \tau_w$ is the shear stress at the wall, and therefore Eq. (4.7.1) reduces to (4.5.4). In the case of flow past a submerged obstacle, instead, we prefer to take S as a cross sectional area of the body perpendicular to the unperturbed flow direction. In addition, $\bar{v}$ is the unperturbed flow velocity, far from the object. For example, consider a solid sphere of diameter D and density ρ_s sedimenting in a quiescent fluid of density ρ. Two opposing forces are exerted on the sphere: drag and buoyancy; their difference determines the acceleration of the body based on its inertia. At steady state, the sphere reaches a constant, so-called, *terminal velocity*, v_∞, when the buoyancy equals the drag force, obtaining:

$$F_f - F_g = \frac{1}{4}\pi D^2 \left(\frac{1}{2}\rho v_\infty^2\right)f - \frac{1}{6}\pi D^3 g \Delta\rho = 0, \tag{4.7.2}$$

where $\Delta\rho = \rho_s - \rho$. Thus, the friction factor can be determined from terminal velocity measurements, i.e.,

$$f = \frac{4gD}{3v_\infty^2}\frac{\Delta\rho}{\rho}. \tag{4.7.3}$$

The total resistance force of the body can be split into a drag component, $\mathbf{F}_D$, having the same direction as the unperturbed flow velocity, and a lift force, $\mathbf{F}_L$, which is the component of the resistance force perpendicular to the flow direction. Obviously, a sphere has no lift, while the drag force is a strong function of the Reynolds number, defined as $Re = v_\infty D/v$, where v is the kinematic viscosity of the fluid.

For small Re, the flow field of a fluid past a sphere of radius R admits an analytical solution, due to Stokes,[15] who obtained the following seminal result see Eq. 22.2.14):

$$F_D = 6\pi\mu R v_\infty, \tag{4.7.4}$$

where μ is the dynamic fluid viscosity. Then, using the definition (4.7.1) with $S = \pi R^2$, we obtain:

$$f = \frac{24}{Re}; \text{ when } Re < 0.1. \tag{4.7.5}$$

For larger Re, the problem is more complicated. From the experimental values reported in Fig. 4.9, the following correlations are obtained:

---

[15]George Gabriel Stokes (1819–1903) was a mathematician, physicist, politician and theologian. Born in Ireland, Stokes spent all of his career at the University of Cambridge, where he served as Lucasian Professor of Mathematics.

Fig. 4.9 Friction factor for a flow around a sphere

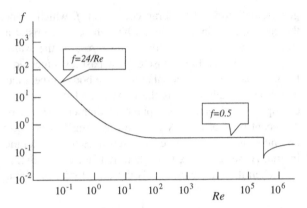

$$f = \frac{18.5}{Re^{3/5}}; \text{ when } 2 < Re < 5 \times 10^2;$$ (4.7.6)

$$f = 0.5; \text{ when } 5 \times 10^2 < Re < 2 \times 10^5.$$ (4.7.7)

Note that, for intermediate values of Re, we expect that a boundary layer will form which, if it covered the whole surface of the sphere, would cause F_D to be proportional to $v_\infty^{3/2}$ (see 3.3.4), so that $f \propto Re^{-1/2}$. In reality, as shown in Sect. 7.3, the boundary layer separates from the surface in the back of the sphere, thus decreasing the drag, with a friction factor $f \propto Re^{-3/5}$, as shown in Eq. (4.7.6). As for Eq. (4.7.7), we already mentioned that for turbulent flows, the resistance forces are dominated by the transport of convective momentum fluxes, estimated to be of $O(\rho v_\infty^2)$, and therefore the drag coefficient turns out to be roughly constant. For even larger Reynolds numbers, first the drag coefficient decreases sharply because of a *drag crisis*, i.e., a sudden transition to turbulence in the boundary layer. Such a transition occurs earlier as Re further increases, which finally leads to increasing values of f.

4.8 Problems

4.1 Derive the expression for the convective flux of kinetic energy.

4.2 In deriving the Bernoulli equation, we considered that the sum of all pressure forces acting on the pipe walls equals $p\Delta S$, where p is the (mean) pressure of the fluid, while ΔS is the difference between the cross section area at the exit and at the entrance. Prove this result in the static case of Fig. 4.10, considering that the total force here is zero.

4.3 This problem is the continuation of the previous one. In the dynamical case, it is sufficient to replace $p + \Delta p$ to p at the cross section $S + \Delta S$ (see Fig. 4.10), while the mean inner pressure will be equal a $p + \Delta p/2$. Show that

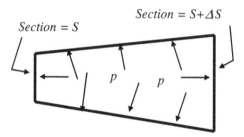

Fig. 4.10 Pressure within a pipe

Fig. 4.11 Channel flow

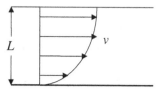

the correction introduced in this case, compared to the static case considered
in the previous problem, is negligible.

4.4 Consider a fluid flowing in a channel of height L, with velocity v
$(y) = V_{max}(1 - y^2/L^2)$, where $0 < y < L$ (see Fig. 4.11). Calculate the correction
factor α defined in Eq. (4.4.3). Calculate α for a circular pipe of radius R,
where $v(r) = V_{max}(1 - r^2/R^2)$.

4.5 Air at 300 K flows at steady state in a straight tube having a 10 cm internal
diameter. The inlet pressure is 7×10^5 Pa, while at the outlet the mean
velocity is 30 m/s and the pressure is atmospheric. Calculate the friction
force exerted by the air on the inner surface of the tube.

4.6 Water flows in a horizontal U-tube with a 10 cm inner diameter, with a 20 m/
s mean velocity. The relative inlet and outlet pressures are 2 bar and 1.6 bar,
respectively. Calculate the friction force exerted by the water on the tube.

4.7 Calculate the outlet velocity of an ideal (i.e., inviscid) incompressible fluid
leaving a reservoir as a function of the hydrostatic head. Discuss the conditions where this result can be applied to a real case.

4.8 Calculate the time it takes to empty a reservoir containing, initially, a known
quantity of an incompressible and inviscid fluid.

4.9 Evaluate the friction losses per unit length of an oil having 1.0 g/cm^3 density
and 10 cP viscosity, flowing at a 0.01 m^3/s volumetric flow rate in a smooth
pipe with 5 cm diameter.

4.10 A viscous oil, having a 1.0 g/cm^3 density and a 10 cP viscosity, flows in a
smooth pipe with a 5 cm diameter. Knowing that 0.066 bar/m is the pressure
drop per unit length, calculate the volumetric flow rate.

4.11 We want to transport 0.01 m^3/s of an oil having a 1.0 g/cm^3 density and a 10 cP
viscosity, through a horizontal circular pipe 5 km long. Assuming a maximum
pumping power of 10 kW, what is the minimum diameter of the pipe?

Fig. 4.12 Water jet leaving a faucet

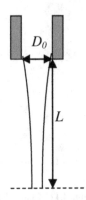

4.12 Consider a water jet leaving a faucet, as shown in Fig. 4.12. Assuming a turbulent regime, determine how the diameter of the jet varies as a function of the distance, L.

4.13 A 150 kW pump with 70 % efficiency induces water to flow between two reservoirs located at a 39 m height difference, as indicated in Fig. 4.13, where L denotes the total length of the pipe and D its diameter. Calculate the volumetric flow rate, stressing the contributions due to potential energy, kinetic energy and friction.

4.14 At the end of a pipe, with an 8 in. inner diameter, water flows from a pressurized tank into a smaller, 4 in. pipe, as shown in Fig. 4.14. Evaluate the force that the fluid exerts on the flange.

4.15 (a) Calculate the force that water with a 3.5 m³/s volumetric flow rate exerts on a barrier, having 1 m width (see Fig. 4.15). (b) Calculate the torque with

Fig. 4.13 Pipe-line flow between two reservoirs

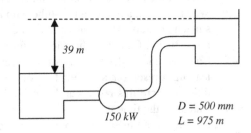

Fig. 4.14 Gasket between two unequal pipes

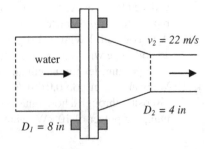

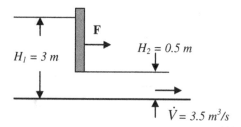

Fig. 4.15 Example of a pressure force

respect to the center of mass of the barrier. (c) Determine the height H_2 of the barrier when the force changes sign.

4.16   Consider a horizontal laminar water jet flowing out of a capillary having diameter D_0. Explain why the jet forms a filament of diameter $D_1 = 0.87D_0$.

4.17   A very viscous fluid is pumped in the system represented in Fig. 4.16, where all the segments 1–2, 1–3, 3–4, 3–5 and 5–6 represent circular tubes of length L; all tubes have a diameter D, with the exception of the tubes 1–2 and 3–4, having diameters D_1 and D_2, respectively. Knowing that pressures p_2, p_4 and p_6 are atmospheric, determine the ratios D_1/D and D_2/D, so that the volumetric flow rates in the pipes 1–2, 3–4 and 5–6 are equal to each other.

4.18   A fluid is pumped into the piping system of Fig. 4.17, where all pipes (1–2, 2–3, 2–4 and 3–5) are identical, with length L and circular cross section of diameter D. Note that pipes 1–2 and 2–3 are horizontal, while 2–4 and 3–5 are vertical. Pressures p_4 and p_5 are atmospheric. Assuming that the fluid has a $\mu = 10$ Poise viscosity and a $\rho = 1$ g/cm^3 density, with $L = 1$ m and $D = 1$ cm, and that the volumetric flow rate at the inlet (that is in 1–2) is equal to 10^{-4} m^3/s, determine: (a) the ratio between the volumetric flow rates in 3–5 and 2–4; (b) The inlet pressure p_1; (c), solve the problem assuming that the fluid is water.

4.19   We want to circulate 0.01 m^3/s of water at 20 °C between two open reservoirs, using a pump as indicated in Fig. 4.18, where the two horizontal pipes are identical, with length $L = 10$ m and diameter $D = 4$ cm. At steady state, find:

Fig. 4.16 Flow in a manifold

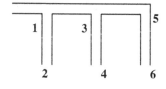

Fig. 4.17 Flow in a manifold

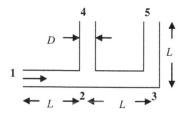

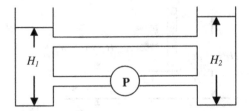

Fig. 4.18 Fluid flow circulating between two reservoirs

    (a) The difference $\Delta H = H_2 - H_1$ between the water levels in the two reservoirs.

    (b) The power of the pump.

4.20 A fluid leaks out of a reservoir of cross section S_1 through a tube of cross section $S_2 = \beta S_1$, with a hydrostatic head H. Determine the outlet velocity v_2 assuming that the localized pressure drops at the outlet are equal to $kv_2^2/2$, with $k = 0.45(1 - \beta)$. Compare the result with Torricelli's law.

4.21 Water is conveyed from a reservoir through a house and out, as in Fig. 4.19. All tubes have the same inner diameter, $D = 12$ cm, and total length $L = 91.5$ m. Knowing that the volumetric flow rate is 0.11 m³/s, check if there is a pump inside the house.

4.22 Water is conveyed from a reservoir (height $H_R = 10$ cm and diameter $D_R = 50$ cm) through an elbow pipe to a capillary, as shown in Fig. 4.20. The pipes have diameters $D_1 = D_2 = 1$ cm, $D_3 = 3$ mm and lengths $L_1 = L_2 = 0.2$ m, $L_3 = 6$ cm. Calculate the pressure Δp that must be applied to the water of the reservoir, so that the fluid velocity in the capillary is $v_3 = 30$ m/s. Is it an absolute or a relative pressure? Data: the localized pressure drops are given by $k\rho v^2/2$, where ρ is the fluid density, v is the downstream velocity, while (a) in the elbow $k_g = 0.16$; (b) in the restrictions $k_c = 0.45 (1 - \beta)$, where β = (smaller area)/(larger area).

4.23 A very viscous fluid flows, due to gravity, in the two systems shown in Fig. 4.21, where the length of all the pipes (1–2, 2–3, 2–4, and 4–5) and also the height of the free surface are equal to L. All tubes have diameters D, with the exception of 2–3, having diameter D_1. Pressures p_3 and p_5 are

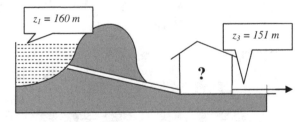

$z_1 = 160$ m

$z_3 = 151$ m

?

Fig. 4.19 Example of pipe flow

Fig. 4.20 Flow out of a
cylindrical tank

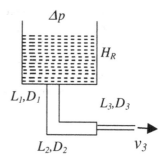

Fig. 4.21 Example of
gravity-driven viscous flows

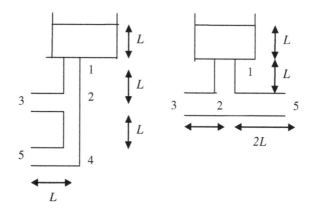

atmospheric. Neglecting all localized pressure drops, determine the ratio D_1/D, so that the volumetric flow rates in the pipes 2–3 and 4–5 are equal to each other.

4.24 Consider the piping system shown in Fig. 4.22, constituted by a cylindrical reservoir with cross section S, initially filled with water up to a height h_0. The

Fig. 4.22 Pipe flow out of a
reservoir

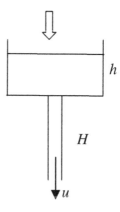

water flows out of the reservoir through a vertical pipe having diameter d and height H, while the reservoir is partially refilled with a water flow rate that is 80 % of that of the outlet pipe. Determine (a) the equation (without solving it) that allows to determine the velocity of the water at the outlet; (b) solve the equation when $h \ll H$ and assuming that the pressure drops are due mainly to friction; (c) determine the time needed to empty the reservoir.

Chapter 5
Laminar Flow Fields

Abstract In this chapter we apply the macroscopic balance equations that have been developed in Chap. 3 to study a few important problems. First, the pipe flow of a Newtonian fluid is considered in Sect. 5.1; then, in Sects. 5.2 and 5.3, this case is generalized to non-Newtonian fluids, stressing how the velocity profiles and the consequent pressure drops are functions of the fluid constitutive equations. In Sect. 5.4. we analyze the flow of a fluid across porous media, stressing when and how a fixed bed becomes fluidized. Then, in Sect. 5.5 we start considering non-stationary flows, introducing the Quasi Steady State (QSS) assumption that will be used extensively in the following. Finally, in Sect. 5.6, we study capillary flows, i.e. the fluid flows that are driven by surface tension effects.

5.1 Fully Developed Flow of a Newtonian Fluid in a Pipe

Consider the elemental control volume in the circular conduit shown in Fig. 5.1. Since the pressure decreases as we move along the axis in the positive z direction, here, as in a previous Section (cf. Figure 5.1 and Fig. 4.5.1), we choose dp to be positive. In general, the sum of all forces applied to the volume element is equal to the time rate of change of its momentum which, at steady state, is zero. Therefore, we have: $(\pi r^2)dP = (2\pi r\, dz)\tau$, where τ is the shear stress, that is:

$$\frac{dp}{dz} = \frac{2\tau}{r}. \qquad (5.1.1)$$

In this equation, the term on the LHS is a function of the z variable only, while the term on the RHS depends only on r; therefore, the two terms must be equal to a constant. In fact, from a physical point of view, as the fluid flow is stationary and the cross section is uniform along z, the pressure drop per unit length must be a constant, so that $dp/dz = \Delta p/L$. Therefore, we see from Eq. (5.1.1) that $\tau/r = \tau_w/R = \Delta p/2L$, that is:

© Springer International Publishing Switzerland 2015
R. Mauri, *Transport Phenomena in Multiphase Flows*,
Fluid Mechanics and Its Applications 112, DOI 10.1007/978-3-319-15793-1_5

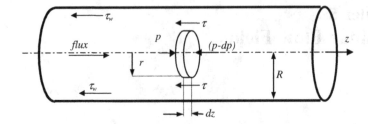

Fig. 5.1 Force distribution on a viscous fluid flowing in a circular conduit

$$\tau = \frac{\tau_w}{R} r = \frac{1}{2} \frac{\Delta p}{L} r, \tag{5.1.2}$$

showing that the shear stress varies linearly with distance from the centerline: in particular, $\tau(r = 0) = 0$ at the centerline, while $\tau(r = R) = \tau_w$ at the wall (see Fig. 5.1).

Now, let us determine the velocity field assuming that (a) the flow regime is laminar, with a uni-directional velocity v along the z-direction; (b) the fluid is Newtonian, so that we can apply the constitutive relation (1.4.4), i.e.,

$$\tau = -\mu \frac{dv}{dr} \tag{5.1.3}$$

Therefore, the velocity field is obtained by integrating the expression of the shear stress: when τ is constant, then v is linear, while when τ is linear, as it happens here, v is quadratic. In fact, substituting Eq. (5.1.3) into (5.1.2) and integrating between $r = R$, where $v = 0$, and a generic radius r, we obtain:

$$\int_0^v dv = -\frac{\tau_w}{\mu R} \int_R^r r dr.$$

Integrating this expression we obtain the, so called, *Poiseuille* velocity profile,[1]

$$v = \frac{\tau_w}{2\mu R} (R^2 - r^2) = \frac{\Delta p}{L} \frac{R^2}{4\mu} \left[1 - \left(\frac{r}{R}\right)^2 \right], \tag{5.1.4}$$

where we have substituted Eq. (5.1.2). This equation shows that the velocity profile of a fluid in laminar flow within a circular conduit is parabolic, with the following maximum velocity, v_{max}, corresponding to the velocity at the centerline,

---

[1]Named after Jean Léonard Marie Poiseuille (1797–1869), a French physicist and physiologist.

$$v_{max} = \frac{\tau_w R}{2\mu} = \frac{\Delta p}{L}\frac{R^2}{4\mu}. \tag{5.1.5}$$

Therefore, Eq. (5.1.4) becomes,

$$\frac{v}{v_{max}} = 1 - \left(\frac{r}{R}\right)^2. \tag{5.1.6}$$

Like all parabolic profiles in cylindrical geometries, here also its mean value equals half its maximum value, i.e.,

$$\bar{v} = \frac{1}{2}v_{max} = \frac{\tau_w R}{4\mu} = \frac{\Delta p}{L}\frac{R^2}{8\mu}. \tag{5.1.7}$$

Now, substituting Eq. (5.1.7) into (4.5.4), we obtain,

$$f = \frac{16\mu}{D\bar{v}\rho} = \frac{16}{Re}. \tag{5.1.8}$$

This expression coincides with Eq. (3.5.6), showing that the friction factor for laminar flows can be derived analytically, unlike Eqs. (3.5.7)–(3.5.8) for turbulent flows, which are found experimentally. Thus, the pressure drops in laminar flow between two cross sections at distance L are:

$$\Delta p = \frac{32L}{D^2}\mu\bar{v}. \tag{5.1.9}$$

Note that the pressure drop per unit length is proportional to the mean velocity (and consequently to the Reynolds number and the mass flow rate as well) and the viscosity while it is inversely proportional to the square of the diameter, i.e. to the cross sectional area. These are fundamental characteristics of the laminar flow of a Newtonian fluid in a conduit and do not depend on the shape of the cross section. Therefore, if the pipe is not circular, we prefer to keep using Eq. (5.1.9), where D is the so-called hydraulic diameter, $D_h = 4S/P_w$, where S and P_w denote the area of the cross section and its perimeter (also called wetted perimeter).

Finally, note that sometimes it is more convenient to rewrite Eq. (5.1.9) in terms of the volumetric flow rate, $\dot{V} = \bar{v}S$, where $S = \pi R^2$ is the area of the cross section, i.e.,

$$\dot{V} = \frac{\pi R^4}{8\mu}\frac{\Delta p}{L} = \frac{\pi D^4}{128\mu}\frac{\Delta p}{L}. \tag{5.1.10}$$

5.1.1 Thermodynamic and Modified Pressure

Let us assume that in the analysis presented above the pipe is arranged vertically. Then, the force balance will include gravity as well, so that the term ρg should be added to the RHS of Eq. (5.1.1). In incompressible flow, that is equivalent to replacing the thermodynamic pressure p with the so-called *modified pressure* $P = p + \rho g z$, where z is measured along a vertical axis directed upward. Therefore, P is constant in a static fluid, so that any spatial changes in P are associated with fluid motion; in fact, P is also referred to as the *dynamic pressure*.[2] Clearly, for compressible fluids, the interpretation of P is less simple.

5.1.2 Couette Flow

The laminar flow of a viscous fluid confined in the narrow space between two parallel plates, one of which is moving relative to the other, is termed *Couette flow*.[3] Practically, a Couette device consists of two long, coaxial cylinders, whose annular gap is much smaller than the radii of the cylinders. A typical example is the Couette viscometer, where the inner cylinder is fixed and the outer one is rotated.

The fluid velocity field can be easily determined considering that, as the fluid streamlines form closed trajectories, there cannot be any pressure drop, so that the shear stress is constant throughout the flow domain. Consequently, neglecting curvature effects, the velocity profile is linear, with $v = 0$ on one wall and $v = V$ on the other (see Problem 5.5 and Sect. 7.1 for the complete solution of this problem).

5.2 Fluid Rheology

Fluids are called *non Newtonian* when presenting rheological behaviors that are more complicated than the simple linear relation between shear stresses and shear rates, characterizing Newtonian fluids. In Fig. 5.2 the constitutive properties of three of the most common non-Newtonian fluids are represented, namely the shear-stress relations of *Bingham* fluids, *shear-thinning* fluids and *shear-thickening* fluids.

Bingham[4] fluids are visco-plastic materials that behave as rigid bodies at low stresses but flow as viscous fluids at high stresses. A common example is tooth-paste, which will not be extruded until a certain pressure is applied to the tube, then it flows out as a solid plug. Other examples are slurries, such as mud, and many

---

[2]This notation can be misleading, as in aerodynamic applications the *dynamic pressure* is the kinetic energy per unit volume of a uniform stream.

[3]Named after Maurice Marie Alfred Couette (1858–1943), a French physicist.

[4]Named after Eugene Cook Bingham (1878–1945), an American chemist.

Fig. 5.2 Rheological behavior of Newtonian and non-Newtonian fluids

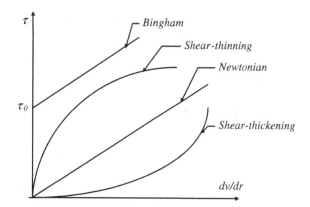

suspensions of small particles. As shown in Fig. 5.2, Bingham fluids do not exhibit any shear rate (and consequently no velocity and no flow), until a certain yield stress τ_0 is reached. Beyond this point, the shear stresses increase linearly with increasing shear rates, and the slope of the line, μ_0, is called *plastic viscosity*. So, the constitutive relation for Bingham fluids reads:

$$\begin{aligned} \tau = \tau_0 + \mu_0 dv/dr, \quad &\text{when} \quad \tau > \tau_0; \\ dv/dr = 0, \quad &\text{when} \quad \tau < \tau_0. \end{aligned} \tag{5.2.1}$$

Shear-thinning fluids (also referred to as *pseudoplastic fluids*) present an *effective viscosity* (that is the slope of the shear-stress curve of Fig. 5.2) that decreases with increasing shear rate (and shear stress). A modern paint is a typical example of a shear-thinning fluid: when they are applied on a wall, the shear created by the brush or roller will allow them to thin and wet out the surface evenly; then, once applied, the paint regains its higher viscosity which avoids drips and runs. Another example is ketchup: when we squeeze it out of a bottle, it goes from being thick like honey to flow like a low viscosity fluid. The simplest model of shear-thinning behavior is that of a suspension composed of micron-size, mutually attracting particles immersed in a Newtonian fluid. When the suspension is quiescent, or at low shear rates, the suspended particles will tend to aggregate, forming clusters that oppose the shearing and therefore the fluid is viscous. On the other hand, as the shear rate increases, clusters will gradually dissolve, so that the effect of the attracting forces will decrease, thus decreasing the viscosity.

Shear-thickening fluids (also referred to as *dilatant fluids*) behave in the opposite manner, compared to shear-thinning fluids, i.e., their effective viscosity increases at increasing shear rate (and shear stress). A simple example of a shear-thickening fluid consists of a mixture of water and cornstarch (or potato flour), which acts in counterintuitive ways: when we apply a force the complex fluid thickens and tends to behave as a solid, while otherwise it acts as a low viscosity liquid. So, we can easily insert a finger in the cornstarch, provided that we do it slowly, while if we push the finger rapidly in (or we decide to walk rapidly on it), the cornstarch acts as a solid. Another example of a shear-thickening material is sand, when it is completely soaked with water: this is the reason why when one walks on wet sand, a dry area appears underneath her foot, and also why it is easy to jog on the shoreline, while it is much harder to jog on dry sand. In general, shear thickening behavior occurs when a colloidal suspension transitions from a stable state to a state of flocculation.

Shear-thinning and shear-thickening fluids can be described by the following relation:

$$\tau = K(-dv/dr)^n, \tag{5.2.2}$$

where K and n are characteristic constant of the fluid. In particular, $n < 1$ for shear-thinning fluids, $n > 1$ for shear-thickening fluids, and $n = 1$ for Newtonian fluids.[5]

5.2.1 Time-Dependent Rheology

In general, materials with time-dependent rheological properties are called visco-elastic, as they exhibit both viscous and elastic characteristics when undergoing deformation. Many materials, in response to a step change in shear rate, take a finite time to reach their equilibrium state, while at a very short times they exhibit a solid-like behavior. Such materials, where the viscosity decreases when the shear rate remains constant, are called *thixotropic*. Examples are clays, muds, honey and many kinds of paints.

For some other fluids, called *rheopectic*, applying a constant shear stress in time causes an increase in the viscosity, or even solidification. Fluids exhibiting this property are much less common; examples include gypsum pastes and printer inks.

The simplest thixotropic materials are the *viscoelastic Maxwell liquids*; their rheological behavior is described by the Maxwell constitutive relation,

$$\dot{\gamma} = \frac{1}{\mu}\tau + \frac{1}{G}\frac{d\tau}{dt}, \tag{5.2.3}$$

where $\dot{\gamma} = dv/dr$ is the shear rate, μ is the kinematic viscosity, characterizing a perfectly viscous fluid, while G is the elastic modulus, that is typical of a perfectly elastic material. Now, if the material is put under a constant shear rate, with $\dot{\gamma}(t<0) = 0$ and $\dot{\gamma}(t > 0) = \dot{\gamma}_0$, solving Eq. (5.2.3) we obtain:

$$\tau = \tau_0(1 - e^{-t/\vartheta}), \quad \text{with} \quad \vartheta = \mu/G \tag{5.2.4}$$

---

[5]The exponent in Eq. (4.2.2) is generally indicated by n, despite the fact that it is not an integer. Also, note that K is not a viscosity, unless $n = 1$.

Here, $\tau_0 = \mu \dot{\gamma}_0$ is the asymptotic viscous shear stress, while ϑ is the characteristic relaxation time of the material. Therefore, for long times, when $t \gg \vartheta$, the material behaves like a Newtonian fluid, with viscosity μ, while for short times it exhibits a solid-like behavior. In fact, when $t \ll \vartheta$, we find: $\tau = \tau_0 t/\vartheta = G\dot{\gamma}_0 t$, i.e., $\tau = G\gamma$, indicating that the shear stresses are proportional to the deformation $\gamma = \dot{\gamma}_0 t$ through the elastic modulus G. In reality, most thixotropic materials at steady state behave like shear-thinning liquids, with their viscosity decreasing with the shear rate.

Rheopectic materials, on the contrary, are viscoelastic solids. In the simplest case, their rheological behavior is described by the *Kelvin-Voigt* constitutive relation,

$$\tau = G\gamma + \mu \dot{\gamma}. \tag{5.2.5}$$

Imposing a step change in shear stress τ_0, the solution of this equation has the same form as Eq. (5.2.4), i.e.,

$$\gamma = \gamma_0 \left(1 - e^{-t/\vartheta}\right); \quad \text{with} \quad \gamma_0 = \tau_0/G \quad \text{and} \quad \vartheta = \mu/G \tag{5.2.6}$$

indicating that for long times, when $t \gg \vartheta$, the material behaves like an ideal elastic solid, with modulus G, while at short times it exhibits a liquid-like, dissipative behavior, with viscosity μ.

5.3 Flow of Non-Newtonian Fluids in Circular Pipes

The analysis of velocity fields in non-Newtonian fluids is very similar to the one for Newtonian fluids, that we have described in Sect. 5.1. In fact, a force balance leads the same Eq. (5.1.2), i.e., $\tau = (r/2)(\Delta P/L)$, showing that the shear stress grows linearly with the distance from the centerline, r. Then, for Newtonian fluids these stresses are proportional to the velocity gradient via Eq. (5.1.3), so that the velocity field turns out to be a quadratic function of r, obtaining the parabolic Poiseuille velocity profile. Naturally, using a different constitutive relation, the velocity field will be different.

Consider, for example, a shear-thinning, or shear-thickening fluid that follows the constitutive relation (5.2.2). Substituting Eq. (5.2.2) into the force balance result, Eq. (5.1.2), we obtain the following velocity profile,

$$v = \left(\frac{\tau_w}{RK}\right)^{1/n} \frac{n}{n+1} \left(R^{(n+1)/n} - r^{(n+1)/n}\right), \tag{5.3.1}$$

and pressure drop,

$$\frac{\Delta p}{L} = 2K \left(\frac{3n+1}{n} \right)^n \frac{\bar{v}^n}{R^{n+1}}. \qquad (5.3.2)$$

Obviously, when $n = 1$ and $K = \mu$, we obtain again the results (5.1.4) and (5.1.9) for Newtonian fluids.

Different velocity profiles, corresponding to the same volumetric flow rate, are represented in Fig. 5.3, showing that shear-thickening fluids have more pointed velocity profiles, with smaller velocity gradients at the wall than Newtonian fluids, while shear-thinning fluids have a more step-like velocity profile, with larger velocity gradients at the wall. Therefore, if we fix the volumetric flow rate, and consequently the mean velocity, Eq. (5.3.2) shows that shear-thinning fluids have a smaller pressure drop than the shear-thickening fluids. Vice versa, if we fix the pressure drop, the flow rate of shear-thinning fluids will be larger than that of the shear-thickening fluids. In fact, the pressure drop is proportional to the shear stress at the wall, which in turn depends on the velocity gradient through the constitutive relation (5.2.2). Accordingly, a shear-thinning fluid can "afford" to have a larger velocity gradient at the wall, as that will have a smaller effect on pressure drops, as compared with the shear-thickening case. So, we can say that the optimal velocity profile is step-like, corresponding to a plug flow, as it maximizes the flow rate for a given pressure drop.

For Bingham fluids, the constitutive relation (5.2.1) is characterized by a yield stress τ_0: when $\tau < \tau_0$, the fluid moves in plug flow, while when $\tau > \tau_0$, the fluid behaves as a Newtonian fluid, with constant viscosity μ_0, also known as the plastic viscosity. Again, the shear stress is a linear function of r, as shown in Fig. 5.4. So, near the centerline, $\tau < \tau_0$ when $r < r_0$, where $r_0 = R\tau_0/\tau_w$, while in the annulus, when $r > r_0$, the fluid behaves as a Newtonian fluid, with parabolic velocity profile. Comparing this result with the velocity profile of Fig. 5.3, we see that, as expected, the behavior of Bingham fluids is similar to that of shear-thinning fluids. Finally, imposing the shear stress at the wall, the following analytical solution is found:

Fig. 5.3 Velocity profiles in laminar flow for Newtonian and non-Newtonian fluids

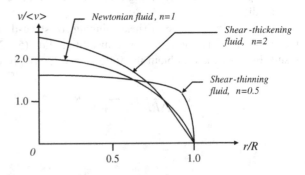

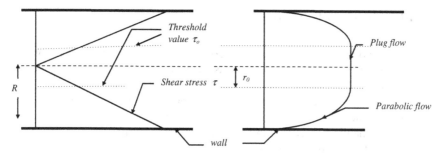

Fig. 5.4 Velocity field and shear stress of a Brigham fluid in laminar flow within a circular tube

$$v = \frac{R-r}{\mu_0}\left[\frac{\tau_w}{2}\left(1+\frac{r}{R}\right) - \tau_0\right] \quad \text{when} \quad r \geq r_0 = \frac{\tau_0}{\tau_w}R; \tag{5.3.3a}$$

$$v = v_0 = \frac{\tau_0}{2\mu_0 r_0}(R - r_0)^2 \quad \text{when} \quad r \leq r_0, \tag{5.3.3b}$$

where τ_w is the shear stress at the wall, which is proportional to the pressure drop through Eq. (5.1.1), i.e.,

$$\tau_w = \frac{R\,\Delta p}{2\,L}. \tag{5.3.5}$$

Integrating Eq. (5.3.3a, 5.3.3b) over the cross sectional area, we obtain the following volumetric flux:

$$\dot{V} = \frac{\pi R^4}{8\mu_0}\frac{\Delta p}{L}\left[1 - \frac{4}{3}\left(\frac{\tau_0}{\tau_w}\right) + \frac{1}{3}\left(\frac{\tau_0}{\tau_w}\right)^4\right].$$

which is valid only when $\tau_w \geq \tau_0$. This result can be inverted, expressing the pressure drop in terms of the friction factor as:

$$f = \frac{16}{Re}\left[1 + \frac{1}{6}\left(\frac{He}{Re}\right) - \frac{1}{3}\frac{1}{(f\,Re)^3}\left(\frac{He}{Re}\right)^4\right], \tag{5.3.4}$$

where $He = \rho D^2 \tau_0/\mu_0^2$ denotes the *Hedstrom* number. Clearly, when $\tau_0 = 0$, we find the previous results for Newtonian fluids.

5.4 Flow in Porous Media

A porous material is composed of a continuous solid phase, containing pores, which are typically filled with a fluid (liquid or gas). Typical examples of porous materials are many natural substances, such as rocks, soil, zeolites and biological tissues, e.g. bones, wood, cork. Other examples consists of many manmade materials, such as cements, ceramics, paper and, in general, fixed beds and membranes acting as catalyzers of chemical reactions. In this paragraph we consider the case where the morphology of the medium is bi-continuous, that is both solid and fluid phases are continuous, so that the fluid can cross the porous material. Obviously, when the fluid phase is not continuous, the porous material is not permeable.

Now, consider the steady flow of a Newtonian fluid flowing through the interstices of the medium. First of all, we assume that the porous material is uniform, so that no channeling[6] is observed. In this case, the typical dimension of the medium is the porous size, a, corresponding, in most applications, to a very small Reynolds number. Then, consider that: (1) due to translational invariance, the pressure drop Δp within a uniform material at steady state is proportional to its thickness, L; (2) at low Re, diffusion is the leading mechanism of momentum transport, so that we expect that the shear stresses at the wall, and the pressure drops as well, is proportional to the mean fluid velocity, $\bar{v}$, and viscosity, μ. Therefore, we expect to find a relation similar to Eq. (5.1.9) for the Poiseuille flow, where the typical flow dimension is the pore size, a. So, at the end, we expect the following relation:

$$\bar{v} = \frac{\dot{V}}{S} = \frac{\kappa}{\mu} \frac{\Delta p}{L}, \tag{5.4.1}$$

which is named Darcy's equation,[7] where κ is a property of the porous medium named *permeability*,[8] with the units of an area, and indicates how easily a fluid penetrates the medium. This is a phenomenological equation and therefore should not include any microscopic quantity (which H. Darcy could not measure anyway), apart from the permeability, where all the morphological properties of the material are condensed. As such, the mean velocity $\bar{v}$ must be derivable from a macroscopic quantity, and this is why in Eq. (5.4.1) it is defined as the ratio between the volumetric flow rate and the cross section area.[9] As we saw, the permeability κ has the units of an area and thus $\kappa \approx a^2$, since the characteristic dimension of the

---

[6]It refers to the fluid flowing through preferential channels or rivulets within the medium, where it encounters a smaller resistance than elsewhere.

[7]Named after Henry Philibert Gaspard Darcy (1803–1858), a French engineer.

[8]The coefficient κ is sometimes denoted as *Darcy permeability*, to distinguished it to the mass transfer coefficient, having the units of a velocity, that in the biomedical literature is, unfortunately, also called permeability.

[9]The mean velocity in porous media is often called *superficial velocity* and denoted by v_s.

medium is its pore size, a. Accordingly, since in most cases $a = O(\mu m)$, we expect that $\kappa = O(\mu m^2)$, and therefore the permeability is often measured in darcy, with:

$$1 \text{ darcy} = \frac{(\text{cm/s})\text{cP}}{\text{bar/cm}} = 10^{-8} \text{cm}^2 = 1 \ \mu m^2. \tag{5.4.2}$$

Note that, in general, Darcy's equation should be written in vector form, as $\bar{\mathbf{v}} = \mu^{-1}\boldsymbol{\kappa} \cdot \nabla p$, where the permeability of the medium is a tensor, as it represents the proportionality term between two vectors, i.e., the mean velocity and the pressure gradient (see Problem 5.10). Only when the porous medium is isotropic, it does not present any preferential direction and therefore it reduces to a scalar.

Now, let consider the microscopic structure of the porous material. The most easily measured physical quantity that characterizes the morphology of a porous material is its *porosity*, or *void fraction*, ε, defined as the ratio of the volumes of the pores or interstices to the total volume, i.e.,

$$\varepsilon = \frac{V_w}{V} = 1 - \frac{V_s}{V}, \tag{5.4.3}$$

where V is the total volume, V_s is the volume occupied by the solid phase and $V_w = (V - V_s)$ is the volume occupied by the fluid. Naturally, there are many way how a given porosity can be achieved. The simplest model of a homogeneous porous material is that of a solid crisscrossed by a network of small tubes, whose *hydraulic radius* is a $R_h = 2V_w/S_w$, where $V_w = (V - V_s)$ is the volume occupied by the fluid, while S_w is the wetted surface.[10] Now, suppose that the medium is composed of N_p particles having volume V_p and surface S_p. Then $S_w = N_p S_p$ while $V_s = N_p V_p = (1 - \varepsilon)V$, and therefore we obtain,

$$R_h = 2\frac{V - V_s}{S_w} = \frac{2V\varepsilon}{N_p S_p} = \frac{2V_p\varepsilon}{(1 - \varepsilon)S_p} = \frac{D_p\varepsilon}{3(1 - \varepsilon)}, \tag{5.4.4}$$

where $D_p = 6V_p/S_p$ is the equivalent diameter of the particles composing the porous medium. Clearly, when the particles are spherical, the equivalent diameter coincides with the sphere diameter; in that case, when the spheres are distributed randomly, it is known that $\varepsilon \approx 0.6$, so that, from Eq. (5.4.4), we see that the hydraulic radius is approximately half the sphere diameter, as it should be.

Now, for laminar flows, the mean velocity of the fluid, of viscosity μ_f, flowing inside the small tubes is called *interstitial velocity*, v_i, given by Eq. (5.1.7),

$$v_i = \frac{(\Delta p)R_h^2}{8\mu_f L}. \tag{5.4.5}$$

---

[10]With this definition, the hydraulic radius coincides with the radius of the cylindrical tubes. Note that in many textbooks the hydraulic radius is defined as V_w/S_w.

In addition, for random, isotropic porous media, the ratio between fluid volume and total volume equals the ratio between wetted and total cross section surface area, so that:

$$\varepsilon = \frac{V_w}{V} = \frac{S_w}{S},\qquad(5.4.6)$$

Therefore, considering that by continuity $\bar{v}S = v_i S_w$, we find a simple relation between this interstitial velocity, v_i, and the mean velocity, $\bar{v}$, i.e.,

$$\bar{v} = \varepsilon v_i.\qquad(5.4.7)$$

Finally, substituting Eqs. (5.4.4) and (5.4.5) into (5.4.7), we obtain the di *Blake-Kozeny* equation:

$$\bar{v} = \frac{\Delta p}{L}\frac{D_p^2}{150\mu_f}\frac{\varepsilon^3}{(1-\varepsilon)^2},\qquad(5.4.8)$$

where the numerical constant has been modified (150 instead of 72), to account for the experimental results, due to the fact that the fluid trajectories are very tortuous, and so the length of the small tubes is larger than the thickness L of the porous medium.[11] As usual, we can define a friction coefficient,

$$\Delta p = 2f_{BK}\frac{L}{D_p}\rho_f\bar{v}^2,\qquad(5.4.9)$$

where,

$$f_{BK} = \frac{\alpha}{2Re_p},\quad \text{with} \quad \alpha = 150\frac{(1-\varepsilon)^2}{\varepsilon^3},\qquad(5.4.9a)$$

Here, ρ_f is the fluid density, $Re_p = \rho_f\bar{v}D_p/\mu_f$ is the Reynolds number evaluated in terms of the mean particle diameter. This result is valid in the laminar region, when $Re_p < 10(1-\varepsilon)$ and for porosity $\varepsilon < 0.5$.

More properly, we can express the Blake-Kozeny result in terms of Darcy's Eq. (5.4.1),

$$\bar{v} = \frac{1}{\mu_f}\kappa\frac{\Delta p}{L},\quad \text{with} \quad \kappa = \frac{1}{\alpha}D_p^2,\qquad(5.4.10)$$

---

[11]Sometimes it is preferable to use the *Carman-Kozeny* equation, where the coefficient 150 is replaced by 180.

showing that, as expected, the square root of the permeability gives a measure of the pore size.

In turbulent regimes, when $Re_p > 1000(1 - \varepsilon)$, the Burke-Plummer equation is used, assuming that the small tubes of a packed bed correspond to very rough pipes, so that the friction factor is constant [see Eq. (3.5.8b)], finding: $\Delta p/L \propto \rho_f v_i^2/R_h$. Then, proceeding as for laminar flows, we obtain:

$$\Delta p = 2 f_{BP} \frac{L}{D_p} \rho_f \bar{v}^2, \tag{5.4.11}$$

with,

$$f_{BP} = 0.875 \frac{1 - \varepsilon}{\varepsilon^3} = \frac{\beta}{2}, \tag{5.4.11a}$$

where the numerical constant fits the experimental data.

For intermediate Re_p, the two correlations by Blake-Kozeny and Burke-Plummer are simply added together, obtaining the so-called Ergun equation,

$$\frac{\Delta p}{L} = \frac{150 \mu_f \bar{v} (1 - \varepsilon)^2}{D_p^2 \varepsilon^3} + \frac{1.75 \rho_f \bar{v}^2}{D_p} \frac{1 - \varepsilon}{\varepsilon^3} = \alpha \frac{\mu_f \bar{v}}{D_p^2} + \beta \frac{\rho_f \bar{v}^2}{D_p}, \tag{5.4.12}$$

where α and β are defined in (5.4.10) and (5.4.11), respectively. This equation puts into evidence the two contributions to momentum transport, the viscous one, due to diffusion, and the kinetic one, due to convection. The former is linear with the velocity and depends on viscosity, the latter is quadratic and depends on density. The Ergun equation can also be written in non-dimensional form as follows:

$$\left(\frac{\Delta p}{\rho_f \bar{v}^2} \right) \left(\frac{D_p}{L} \right) \left(\frac{\varepsilon^3}{1 - \varepsilon} \right) = \frac{150(1 - \varepsilon)}{Re_p} + 1.75. \tag{5.4.12a}$$

It should be noticed that the Ergun equation is but one of the many that have been proposed to model the pressure drop across packed beds; its agreement with experimental results is nevertheless remarkable.

5.4.1 Packed Beds and Fluidized Beds

Consider a vessel or a pipe that is filled with packing material; the packing can be randomly filled or else it can be a specifically designed structured packing. The solid substrate (the catalytic material upon which chemical species react) material in the fluidized bed reactor is typically supported by a porous plate, known as a distributor. A fluid is then forced through the distributor up through the solid material. At lower fluid velocities, the solids remain in place as the fluid passes through the voids in the material. This is known as a *packed bed* reactor (PBR). The purpose of a packed bed is typically to improve contact between two phases, due to the large interfacial surface, by letting a fluid flow through the particles. Packed beds are widely used in chemical reactors, for example in filters that absorb one or more solutes, or in distillation processes, by using a packing of catalyst particles.

As the mass flow rate of the inlet fluid increases, en ever increasing portion of the bed weight is equilibrated by the viscous resistance, $S\Delta p$, where S is the bed cross section, while Δp is the pressure drop of the fluid flowing across the bed, which can be evaluated through the Ergun correlation (5.4.12) (the remaining part of the bed weight is obviously supported by the distributor). At the end, when drag equilibrates gravity, the bed particles are suspended by the fluid stream and swirl around, much like in an agitated tank or in a boiling pot of water. This is known as a *fluidized bed* reactor (FBR), indicating that the fluid/solid mixture behaves as a real fluid. In addition to the advantages offered by fixed beds, fluidized beds present uniform particle mixing and uniform temperature, thus allowing for a uniform product that can often be hard to achieve in other reactor designs. In fact, fluidized beds have been used extensively in industry; its most important application is in the catalytic cracking process, where catalyst are used to reduce petroleum to simpler compounds through a process known as cracking.

The minimum fluid velocity that equilibrates gravity is indicated as v_{mf} and is known as the *minimum fluidization velocity*. It can be evaluated imposing $\Delta p = Mg/S$ in (5.4.12), where M is the buoyed mass, $M = (\rho_s - \rho_f)(1 - \varepsilon)SH$, obtaining:

$$(\rho_s - \rho_f)(1 - \varepsilon_{mf})gH_{mf} = \Delta p(v_{mf}), \qquad (5.4.13)$$

where H_{mf} is the bed height at minimum fluidization, ρ_f is the fluid density (negligible, when it is a gas), ρ_s the solid density, and ε_{mf} is the bed porosity at minimum fluidization. Since at minimum fluidization the bed has not expanded, yet, we may assume that H_{mf} and ε_{mf} coincide with their unperturbed value, H and ε. Therefore, solving Eq. (5.4.13) in the two limit cases corresponding to using the Blake-Kozeny and the Burke-Plummer correlations, we find:

$$Re_{mf} = \frac{1}{150} \frac{\varepsilon^3}{(1 - \varepsilon)} Ar \quad \text{when} \quad Re_{mf} < 10(1 - \varepsilon), \qquad (5.4.14a)$$

and

$$Re_{mf} = \frac{1}{\sqrt{1.75}} \varepsilon^{3/2} Ar^{1/2} \quad \text{when} \quad Re_{mf} > 1000(1 - \varepsilon), \qquad (5.4.14b)$$

where $Re_{mf} = \rho_f v_{mf} D_p / \mu_f$ is the Reynolds number at minimum fluidization, while $Ar = g D_p^3 \rho_f (\rho_s - \rho_f) / \mu_f^2$ is the Archimedes number. For intermediate Reynolds numbers, we may solve Eq. (5.4.13) using the full Ergun correlation, obtaining:

$$Re_{mf} = 42.85(1 - \varepsilon)\left\{-1 + \left[1 + 3.1 \times 10^{-4}Ar\frac{\varepsilon^3}{(1-\varepsilon)^2}\right]^{1/2}\right\}. \qquad (5.4.14c)$$

When $v > v_{mf}$, the pressure drop remains constant, equal to the bed weight, while the behavior of the bed strongly depends on the fluid employed. When it is a liquid, the bed remains homogeneous and gradually expands, increasing ε, while when the fluid is a gas, soon the bed becomes unstable, meaning that the porosity ceases to be uniform; in this case, bubbles are observed to form at the exit of the distributor plate, and then increase their size as they rise, until they reach the upper free surface, where they explode, sputtering around the particles located in their wake, just like it happens in boiling. A fundamental explanation of this instability is still missing.

Finally, at a larger entrainment velocity, v_t, the bed particles are entrained, carried away by the fluid flow, and then re-circulated via an external loop back into the reactor bed. This is called *circulating fluidized bed* reactor (CFBR).

5.4.2 Filters

Filters are porous materials that are used to "capture" particles that are suspended in a fluid. Trivially, the pores of a filter must be smaller than the particles to be captured. Often a cake, composed of the filtered particles, forms on the filter (see Fig. 5.5); this on one hand acts as an additional filter, thus improving the filter performance, but on the other hand it has the undesirable effects of increasing the pressure drops. Examples of natural filters are soil, sand and wood, while in industry filters are used to remove undesirable chemicals, biological contaminants or suspended solids.

Applying the Darcy equation across the filter shown in Fig. 5.5, we obtain:

$$\bar{v} = \frac{1}{S}\dot{V} = \frac{1}{S}\frac{dV}{dt} = \frac{\kappa_c}{\mu}\frac{p_1 - p_2}{H_c} = \frac{\kappa_f}{\mu}\frac{p_2 - p_3}{H_f}. \qquad (5.4.15)$$

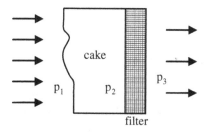

Fig. 5.5 Flow through a filter cake

Here, V is the volume of the filtered fluid, S the cross section area, κ_c, κ_f and H_c, H_f are the permeability and the thickness of cake and filter, respectively, while p_1, p_2 and p_3 are the pressures indicated in the figure. Note that $p_3 = p_{atm}$ and H_f are constant, while p_1, p_2 and H_c increase with time. Thus, we obtain (see Problem 5.7):

$$\bar{v}(t) = \frac{1}{S}\frac{dV}{dt} = \frac{\kappa_{eff}}{\mu}\frac{(p_1(t) - p_3)}{H_f}; \quad \kappa_{eff} = \left[\frac{H_c/H_f}{\kappa_c} + \frac{1}{\kappa_f}\right]^{-1}, \tag{5.4.16}$$

where κ_{eff} is an effective permeability of the filter, indicating its ability to filter. Now, consider that the cake volume is $V_c = SH_c = V\gamma/(1 - \varepsilon)$, where γ is the volume of the filtered material divided by the volume of the filtered fluid, while ε is the cake porosity. Finally, we obtain:

$$\kappa_{eff} = \left[\frac{V\gamma}{(1 - \varepsilon)V_f\kappa_c} + \frac{1}{\kappa_f}\right]^{-1}, \tag{5.4.17}$$

where $V_f = SH_f$ to be solved with initial condition $V(0) = 0$. The two most important cases correspond to having one of the two resistances (i.e., that of the cake and that of the fluid) negligible with respect to the other.

(a) In all processes to purify air or weakly polluted liquids, cake resistance is generally negligible. Therefore, $H_c/\kappa_c \ll H_f/\kappa_f$, so that $\kappa_{eff} = \kappa_f$, and we obtain:

$$\frac{dV}{dt} = \frac{S\kappa_f}{\mu H_f}(p_1 - p_3) \quad \Rightarrow \quad V(t) = \frac{S\kappa_f}{\mu H_f}(p_1 - p_3)t, \tag{5.4.18}$$

corresponding to the removal of a volume $V\gamma$ of particles (with mass $\rho_s V\gamma$).

(b) When filtering liquids at high concentrations of suspended particles, the filter resistance is negligible. Then $H_c/\kappa_c \gg H_f/\kappa_f$, so that $\kappa_{eff} = \kappa_c H_f/H_c$, and we obtain:

$$V\frac{dV}{dt} = \frac{S^2(1 - \varepsilon)\kappa_c}{\mu\gamma}(p_1 - p_3) \quad \Rightarrow \quad V^2(t) = \frac{2S^2(1 - \varepsilon)\kappa_c}{\mu\gamma}(p_1 - p_3)t. \tag{5.4.19}$$

5.5   Quasi Steady Fluid Flows

In this Section we want to calculate the characteristic time that is needed to empty the reservoir represented in Fig. 5.6, through a capillary of radius R and length L. The outlet flux is time-dependent, as it depends on the height H of the free surface, which in time, decreases with time.

Fig. 5.6 Flow out of a
reservoir through a capillary

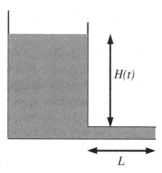

From an elementary mass balance, we obtain: $A_T(dH/dt) = -\dot{V}$, where A_T is the
cross section area of the reservoir, and $\dot{V}$ the volumetric flux in the capillary.
Assuming that the flow is laminar, $\dot{V}$, in turn, depends on the pressure drop,
$\Delta p = \rho g H$, through Eq. (5.1.10) and so the following equation is obtained:

$$\frac{dH}{dt} = -\frac{\pi R^4}{8\mu}\frac{\rho g H}{A_T L} \quad \text{with} \quad H(0) = H_0, \tag{5.5.1}$$

i.e.,

$$H = H_0 e^{-t/\tau} \quad \text{where} \quad \tau = \frac{8\mu L A_T}{\pi \rho g R^4}. \tag{5.5.2}$$

In this analysis we have used Eq. (5.1.10), i.e., a steady state expression for the
volumetric flow rate, substituting it into a time-dependent mass balance. This, so
called, Quasi Steady State (QSS) approximation is valid when the characteristic
emptying time, τ, is much larger than the time required to reach steady state, τ_{ss}.
This latter is the characteristic time that the system will take to adjust to the new
condition, when H decreases abruptly to $H - \Delta H$. Whenever the QSS approximation is valid, τ_{ss} is so short, compared to the time τ characterizing the variation of
H, that it can be considered as instantaneous. Since here $\tau_{ss} = R^2/\nu$, the QSS
approximation and Eq. (5.5.2) are valid when,

$$\tau_{ss} \ll \tau \quad \Rightarrow \quad \frac{R^6 g}{A_T L \nu^2} \gg 1. \tag{5.5.3}$$

Note: the characteristic emptying time does not depend on the initial height H_0
of the free surface, i.e. it is independent of the initial quantity of fluid contained in
the reservoir. So, doubling the initial height, the emptying time remains invariant.
This is due to the fact that in laminar flow the volumetric flow rate is proportional to
the pressure drop and therefore to the free surface height. Consequently, doubling
the initial fluid volume, the volumetric flow rate also doubles, so that the emptying
time does not change.

5.6 Capillary Flow

Capillary flow consists of the motion of a liquid along a solid surface, driven by forces associated with surface tension, that is the attraction of liquid to solid molecules. The easiest example is the capillary action in paper towels and trees, where a liquid rises in narrow spaces, such as a capillary tube, in opposition to gravity. As we saw in Sect. 1.4, applying the Young-Laplace equation we find the height, h_0, that the meniscus will eventually reach in a cylindrical capillary, i.e., (Fig. 5.7)

$$h_0 = (2\sigma \cos \theta)/(R_c g \Delta \rho) = R_c(2 \cos \theta / Bo) \qquad (5.6.1)$$

where θ is the contact angle, while $Bo = gR^2\Delta\rho/\sigma$ is the Bond number defined in terms of the capillary radius, R_c. Now, the velocity, dh/dt, of the rising meniscus can be evaluated under the quasi stationary hypothesis, so that, applying Eq. (5.1.7), we obtain:

$$\frac{dh}{dt} = \bar{v} = \frac{R^2}{8\mu} \frac{\Delta p}{h}, \qquad (5.6.2)$$

with $h < h_0$. The pressure drop Δp across the liquid column is equal to the difference between the pressure drop at the meniscus, evaluated applying the Young-Laplace equation, and the hydrostatic pressure, i.e.,

$$\Delta p = \frac{2\sigma \cos \theta}{R} - \rho g h. \qquad (5.6.3)$$

Fig. 5.7 Capillary flow

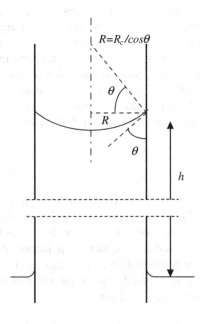

Substituting (5.6.3) into (5.6.2), we obtain:

$$\frac{dh}{dt} = \frac{\sigma Bo}{8\mu}\left[\frac{2}{Bo}\left(\frac{R}{h}\right)\cos\theta - 1\right] = V\left(\frac{h_0}{h} - 1\right), \qquad (5.6.4)$$

where $V = (\sigma Bo)/(8\mu)$ is a characteristic velocity. Now, rewrite this equation as,

$$dt = \tau\left(\frac{h/h_0}{1 - h/h_0}\right)d(h/h_0); \quad \tau \equiv \frac{h_0}{V} = \frac{8\mu h_0}{\rho g R^2}, \qquad (5.6.5)$$

where τ is the characteristic time that it takes to reach the final height. The final solution, with initial condition $h(t = 0) = 0$, is:

$$t = -\tau\left[\ln\left(1 - \frac{h}{h_0}\right) + \frac{h}{h_0}\right]. \qquad (5.6.6)$$

When we approach equilibrium, i.e., as h/h_0 tends to 1, we find:

$$h = h_0\left(1 - e^{-t/\tau}\right). \qquad (5.6.7)$$

Finally, let check when the QSS approximation is valid. Again, $\tau_{ss} = R^2/\nu$, so that the condition $\tau_{ss} \ll \tau$ yields:

$$\tau_{ss} \ll \tau \quad \Rightarrow \quad \frac{Bo R^3 g}{\nu^2} \ll 1. \qquad (5.6.8)$$

Note that in general the characteristic time τ is rather short. In fact, when $R = 0.1$ mm, $\sigma = 73$ dyn/cm and $\mu = 1$ cP (so that $Bo \approx 10^{-3}$ and $h_0 \approx R/Bo \approx 10$ cm), we obtain: $\tau \cong 12$ s (with $\theta = 0$) and thus the typical velocity is $V = h_0/\tau \approx 1$ cm/s. In this case, the QSS condition is satisfied.

5.7 Problems

5.1 Consider an incompressible Newtonian fluid that is pumped in a canal with a constant mean velocity, $\bar{v}$ (i.e., a given volumetric flow rate), represented in Fig. 5.8. Repeating all the steps of Sect. 5.1, with boundary conditions $v = 0$ at the wall, $y = -H$, and $dv/dy = 0$ at the free surface, $y = 0$, show that the velocity profile and the pressure drops per unit length have the same form as for circular pipes, i.e.,

$$v = \frac{3}{2}\bar{v}\left[1 - \left(\frac{y}{H}\right)^2\right], \frac{\Delta p}{L} = \frac{3\mu\bar{v}}{H^2}$$

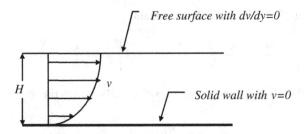

Free surface with dv/dy=0

H v

Solid wall with v=0

Fig. 5.8 Channel flow

5.2 This problem must be solved after Problem 5.1. Consider a Newtonian fluid, flowing with as constant mass flow rate per unit width Γ (expressed in $kg\ m^{-1}\ s^{-1}$) down a plate, inclined by an angle β with respect to the vertical. Show that the film thickness is,

$$H = \left(\frac{3\mu\Gamma}{\rho^2 g \cos\beta}\right)^{1/3},$$

where ρ and μ are, respectively, the density and the viscosity of the fluid, while g is the gravity acceleration. Note: $\Gamma = \rho H \bar{v}$.

5.3 A very viscous Newtonian polymeric solution is pumped into the device represented in Fig. 5.9 (left), with given pressure drop Δp. Determine the volumetric flow rate.

5.4 A Newtonian fluid is pumped into the device of Fig. 5.9 (right), with a given volumetric flow rate, supporting the weight Mg of a slab. Determine the thickness H for (a) laminar flow, assuming that $\Delta P/L = C_1\mu v/H^2$; (b) turbulent flow, assuming that $\Delta P/L = C_2\rho v^2/H$, where v is the mean velocity, while C_1 and C_2 are constant.

5.5 Consider the flow of a Newtonian fluid in the Couette device represented in Fig. 5.10, consisting of two concentric cylinders of height L. The outer cylinder, having radius R_o, is held still, while the inner cylinder, having radius R_i, rotates with angular velocity ω. Calculate the torque that must be applied.

5.6 Calculate the velocity profile and the pressure drops per unit length of a shear-thickening fluid with constitutive Eq. (5.2.2) with $n = 2$, flowing in a circular pipe. Verify that the results agree with Eqs. (5.3.1)–(5.3.2).

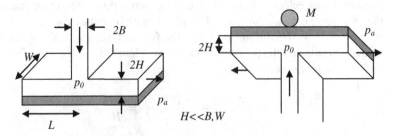

Fig. 5.9 Viscous flows through splitting channel devices

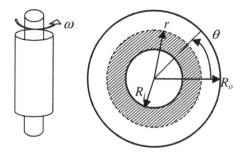

Fig. 5.10 Couette device

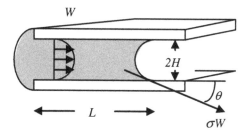

Fig. 5.11 Flow of a drop in a channel

5.7 In the clean rooms of semiconductor industries, air must be extremely pure, with no suspended particulate. Submicron particles can be filtered using filters composed of very fine fibers with radius $R \approx 5$ μm. Assuming laminar flow, determine the non-dimensional pressure drops $\Phi = (\Delta P/L)(R^2/4a\mu v)$, with v denoting the mean velocity, as a function of the solid fraction, $\alpha = 1 - \varepsilon$.

5.8 Consider a piece of land, with width $W = 500$ m and inclined by an angle $\phi = 5°$, that is irrigated with a water flow rate $q_0 = 0.01\,\mathrm{m^3/s}$ per meter width. As the water flows down, due to gravity, it evaporates at a rate $v_e = 10^{-10}\,\mathrm{m^3/s}$ per square meter of surface (therefore, v_e is expressed in m/s). Knowing the soil permeability, $\kappa = 10$ darcy, determine (a) the maximum irrigation distance; (b) the water penetration in the soil.

5.9 Determine the effective permeability of a composite material of cylindrical shape, having radius R and length L, consisting of m little tubes of radius r and length L, cemented together.

5.10 Determine the effective permeability of a composite porous material consisting of two layered species with volume fraction $\varepsilon_1 = \varepsilon$ and $\varepsilon_2 = 1 - \varepsilon$ and permeability κ_1 and κ_2, assuming that the two species are arranged in layers (a) along the flow direction; (b) along a direction perpendicular to the flow.

5.11 A drop of volume V_g is placed at the entrance of a channel, as shown Fig. 5.11 and thereby it is sucked by the capillary forces. Knowing the contact angle θ (with, naturally, $\theta < \pi/2$), evaluate the time needed to enter the channel.

Chapter 6
The Governing Equations of a Simple Fluid

Abstract In this chapter we derive formally the fundamental equations describing the transport of mass, momentum and energy of a one component, single phase fluid. Here, the balance equations are derived based on both an Eulerian and a Lagrangian approach. In the former, the conservation principles are enforced using a fixed volume element, as opposed to the Lagrangian approach, focused on a material volume element as it moves in time, with the fluid velocity along the fluid path lines. An easier, albeit less rigorous, derivation of the governing equations can be found in Appendix E, where an Eulerian approach is adopted, by performing the balance of mass, momentum and energy.

6.1 General Microscopic Balance Equation

Consider a fixed control volume V, bounded by a closed surface S, expressing any extensive quantity $F(t)$ as

$$F(t) = \int_V \rho(\mathbf{x}, t) f(\mathbf{x}, t) dV, \qquad (6.1.1)$$

where $f(\mathbf{x}, t)$ represents the mass specific density of F and $\rho(\mathbf{x}, t)$ the mass per unit volume at position $\mathbf{x}$ and time t. Defining, at any point of the (closed) surface S a unit normal $\mathbf{e}_n$ directed outward from V, and normal to the surface, the most general balance equation describing the variation of $F(t)$ is:

$$\frac{dF(t)}{dt} = \int_V \frac{\partial}{\partial t}(\rho f) dV = -\oint_S J_{Fn} dS + \int_V \sigma_F dV, \qquad (6.1.2)$$

where σ_F is the source density, i.e. the amount of F generated per unit volume per unit time, while J_{Fn} is the flux of F leaving the volume, that is the amount of F crossing the surface S per unit time and having the direction of the unit vector $\mathbf{e}_n$,

© Springer International Publishing Switzerland 2015
R. Mauri, *Transport Phenomena in Multiphase Flows*,
Fluid Mechanics and Its Applications 112, DOI 10.1007/978-3-319-15793-1_6

perpendicular to S and directed outward from V. Note that the time rate of change
within the volume integral is a partial derivative, since the volume V remains fixed.
In addition, note the minus sign in the RHS of Eq. (6.1.2), as J_{Fn} indicates an
outward mass flux. At this point, we want to prove that J_{Fn} is not just any scalar
quantity, but it can be written as the inner product between two vectors: the unit
vector e_n and a mass flux vector, J_F, i.e.,

$$J_{Fn} = \mathbf{e_n} \cdot \mathbf{J}_F \tag{6.1.3}$$

This is the so-called *Cauchy lemma*, and its proof is basically the same as that in
Sect. 1.3. Let us consider the mass balance on a volume element consisting of the
tetrahedron of Fig. 6.1, built along the Cartesian axes x_1, x_2, x_3:

$$\frac{\partial}{\partial t}(\rho f dV) = J_{F1} dS_1 + J_{F2} dS_2 + J_{F3} dS_3 - J_{Fn} dS_n,$$

where J_{Fn}, as we have seen, denotes the mass flux leaving the area dS_n, while J_{F1},
J_{F2} and J_{F3} are the fluxes of F along the positive directions of the three Cartesian
axes. In addition, dS_1, dS_2, dS_3 and dS_n are, respectively, the areas of the triangles
OAB, OAC, OBC and ABC which, based on elementary geometric considerations,
are related to each other as:

$$dS_1 = dS_n(\mathbf{e_n} \cdot \mathbf{e_1}), \ dS_2 = dS_n(\mathbf{e_n} \cdot \mathbf{e_2}), \ dS_3 = dS_n(\mathbf{e_n} \cdot \mathbf{e_3}), \tag{6.1.4}$$

where e_1, e_2, e_3 are the unit vectors along x_1, x_2, x_3. Now, let Δx_1, Δx_2 and Δx_3
squeeze to zero in Eq. (6.1.3), neglecting the RHS because volumes tend to zero

Fig. 6.1 Fluxes on a
material volume element

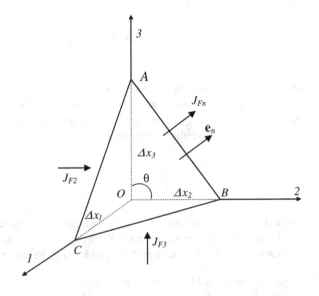

more rapidly than surfaces, we obtain Eq. (6.1.3), where we have defined the following mass flux vector,

$$\mathbf{J}_F = J_{F1}\mathbf{e}_1 + J_{F2}\mathbf{e}_2 + J_{F3}\mathbf{e}_3 \tag{6.1.5}$$

Now, the surface integral on the RHS of Eq. (6.1.2) can be rewritten as a volume integral using the divergence theorem (also called Gauss-Green theorem; see Appendix F). Hence, Eq. (6.1.2) can be rearranged to the form,

$$\int_V \left[\frac{\partial(\rho f)}{\partial t} + \nabla \cdot \mathbf{J}_F - \sigma_F \right] dV = 0. \tag{6.1.6}$$

Now, since this integral must vanish independently of the control volume that we have chosen,[1] we obtain the following conservation for F:

$$\frac{\partial}{\partial t}(\rho f) + \nabla \cdot \mathbf{J}_F = \sigma_F \tag{6.1.7}$$

Clearly, this equation must be coupled with an expression for the source density, σ_F, and a constitutive relation relating the flux, $\mathbf{J}_F$, to f and its gradients. In the following, as we discuss the conservation equations for a particular thermodynamic system consisting of a pure substance, we will specify F as a set of extensive variables comprising mass, momentum, angular momentum and energy.

6.2 Mass Balance: The Continuity Equation

For mass transport, we set $F = M$ in Eq. (6.1.1), where M is the mass, so that $f = 1$. As mass is conserved, $\sigma_M = 0$; in addition, we have seen in Eq. (1.4.1) that the fluid velocity $\mathbf{v}$ can be defined in terms of the mass flux, $\mathbf{J}_M$, as,

$$\mathbf{J}_M = \rho\mathbf{v}. \tag{6.2.1}$$

Finally, Eq. (6.1.7) reduces to the so-called *continuity equation*:

$$\frac{\partial \rho}{\partial t} + \nabla \cdot (\rho\mathbf{v}) = 0. \tag{6.2.2}$$

Now, let us consider another approach to the mass balance problem: instead of choosing a control volume fixed in space, let us consider a material element of fluid

---

[1]It is not a trivial statement: an integral can vanish even if its integrand is not zero. Think, for example, of the integral of x between -1 and $+1$. When, however, an integral is zero for any choice of the integration interval, then its integrand is identically null.

during its motion as it translates, rotates and deforms. The volume occupied by this material element changes in time and is called *material control volume* $V_m(t)$, where the subscript m indicates that the volume contains a fixed quantity of material.

Note that, for incompressible fluids, the fluid velocity will be parallel to the surface S_m of the material volume, otherwise the fluid would penetrate V_m, thus changing its mass and contradicting the hypothesis of conservation of the material element mass. Mathematically, that means:

$$\mathbf{e}_n \cdot \mathbf{v} = 0 \quad \text{on} \ S_m. \tag{6.2.3}$$

Therefore, the continuity equation referred to the material control volume is very simple, indicating that the mass M_m contained within the material volume V_m is constant:

$$\frac{DM_m}{Dt} = 0, \quad \text{where} \ M_m = \int\limits_{V_m(t)} \rho dV. \tag{6.2.4}$$

Here, the symbol D/Dt indicates the *material derivative* with respect to time. In general, for any local property $f(\mathbf{x})$, the material derivative Df/Dt represents the time rate of change of f moving with the fluid velocity along the fluid velocity path. Suppose, for example, that we are interested in observing the variation in time of the water temperature, T, in a river. If we remain fixed at a place, we measure $\partial T/\partial t$: that is the intrinsic time rate of change of the water temperature due, for example, to the fact that the water heats up as the day progresses. On the other hand, if we move together with the flow velocity, following the current, we measure DT/Dt: in addition to the intrinsic heating of the river, this time rate of change of temperature might be due to the fact that some portion of the river are sunny and some others are in a shadow.

The relation between the partial derivative, $\partial T/\partial t$, and the material derivative, Df/Dt, of a quantity $f(\mathbf{x}, t)$ depends on the time dependence of the position $\mathbf{x}$, where the time derivative is taken. In fact, following a fluid particle path located at $\mathbf{x} = \mathbf{x}(t)$, then, expanding the total time derivative f through a multivariate chain rule we obtain:

$$\frac{df}{dt} = \frac{\partial f}{\partial t} + \frac{\partial f}{\partial \mathbf{x}} \cdot \frac{d\mathbf{x}}{dt} = \frac{\partial f}{\partial t} + \frac{d\mathbf{x}}{dt} \cdot \nabla f,$$

where $\partial f/\partial \mathbf{x} = \nabla f = (\partial f/\partial x_1, \partial f/\partial x_2, \partial f/\partial x_3)$ is the gradient of f. Considering that the substantial timer derivative is denoted as D/Dt, we have:

$$\frac{Df}{Dt} = \frac{\partial f}{\partial t} + \mathbf{v} \cdot \nabla f. \tag{6.2.5}$$

It is useful to apply these results to the transport of any extensive quantity F, expressed in terms of the intensive density f defined in (6.1.1). Then, consider the following equality:

$$\frac{\partial}{\partial t}(\rho f) = \rho \frac{\partial f}{\partial t} - f \nabla \cdot (\rho \mathbf{v}) = \rho \frac{\partial f}{\partial t} - \nabla \cdot (\rho f \mathbf{v}) + \rho \mathbf{v} \cdot \nabla f.$$

that is,

$$\rho \frac{Df}{Dt} = \frac{\partial(\rho f)}{\partial t} + \nabla \cdot (\rho f \mathbf{v}), \tag{6.2.6}$$

where we have applied the continuity Eq. (6.2.2). This is the differential form of the integral *Reynolds transport theorem*, shown in Appendix F,

$$\frac{D}{Dt} \int_{V_m(t)} (\rho f) dV = \int_{V_m(t)} \left[\frac{\partial(\rho f)}{\partial t} + \nabla \cdot (\rho f \mathbf{v}) \right] dV = \int_{V_m(t)} \rho \frac{Df}{Dt} dV. \tag{6.2.7}$$

Obviously, when $f = 1$, this equation reduces to Eq. (6.2.4).

Substituting Eq. (6.1.7) into (6.2.6), we can rewrite the general balance equation for f as follows:

$$\rho \frac{Df}{Dt} = \frac{\partial(\rho f)}{\partial t} + \nabla \cdot (\rho f \mathbf{v}) = -\nabla \cdot \mathbf{J}_F^{(d)} + \sigma_F, \tag{6.2.8}$$

where

$$\mathbf{J}_F^{(d)} = \mathbf{J}_F - \rho f \mathbf{v}. \tag{6.2.9}$$

Since, as we saw in Sect. 1.4, $\rho f \mathbf{v}$ is the *convective* contribution to the flux, $\mathbf{J}_F^{(d)}$ is the flux of F that takes place *in the absence of convection*, and is therefore referred to as the *diffusive* flux. The above viewpoints, i.e., that which follows the fluid velocity along the fluid particle paths, and the other one, wherein control volumes are fixed in space, are generally called the *Lagrangian* and *Eulerian* approaches, respectively.

Note that, using Eq. (6.2.8) in addition to (6.2.7), and applying the divergence theorem, we see that the balance equation for any quantity f can be formulated in its integral form using a Lagrangian approach, obtaining:

$$\frac{D}{Dt} \int_{V_m(t)} (\rho f) dV = - \oint_{S_m(t)} J_{Fn}^{(d)} dS + \int_{V_m(t)} \sigma_F dV. \tag{6.2.10}$$

As we saw, this equation is totally equivalent to the Eulerian balance Eq. (6.1.7).

The Lagrangian and Eulerian approaches are totally equivalent to each other. For example, in the case of mass transport, mass conservation is ensured in the Eulerian approach by imposing that the time rate of change for the mass contained in a fixed control volume be equal to the net mass flux into the control volume, while, in the Lagrangian approach, by imposing that the mass flux of a material element remain constant.

Before concluding this Section, note that the continuity equation can be written as,

$$\frac{D\rho}{Dt} + \rho \nabla \cdot \mathbf{v} = 0, \tag{6.2.11}$$

that is, since $\rho = 1/\tilde{V}$, where $\tilde{V}$ is the specific volume,

$$\nabla \cdot \mathbf{v} = \frac{1}{\tilde{V}} \frac{D\tilde{V}}{Dt}. \tag{6.2.12}$$

This equation indicates that the divergence of the velocity, $\nabla \cdot \mathbf{v}$, is the inverse of a characteristic dilatation time of the fluid. Obviously, for incompressible fluids, the continuity equation simplifies greatly and becomes:

$$\nabla \cdot \mathbf{v} = 0, \tag{6.2.13}$$

indicating that the velocity field is solenoidal, i.e., divergence-free.

Indicating by $\beta = \tilde{V}^{-1}(\partial \tilde{V}/\partial T)_P$ and $\kappa_T = -\tilde{V}^{-1}(\partial \tilde{V}/\partial P)_T$ the coefficient of thermal expansion and the coefficient of isothermal compressibility, respectively, the variations of the specific volume are related to changes in temperature and pressure through the relation:

$$\frac{d\tilde{V}}{\tilde{V}} = \beta \, dT - \kappa_T \, dP. \tag{6.2.14}$$

Therefore, Eq. (6.2.12) becomes:

$$\beta \frac{DT}{Dt} - \kappa_T \frac{dP}{Dt} = \nabla \cdot \mathbf{v}. \tag{6.2.15}$$

6.3 Momentum Balance: Cauchy's Equation

As in the case of mass conservation, the principle of momentum conservation can be formulated using either a fixed control volume (Eulerian approach) or a material volume, moving with the fluid velocity along a fluid path lines (Lagrangian approach). In the following, we will follow this latter approach, because we believe that in this way the physical implications of the momentum balance are better explained.

Momentum conservation requires that, in agreement with Newton's second law of dynamics, the time rate of change of the momentum contained in a material volume equal the sum of all the forces that are applied to it.[2] In turn, these forces consist of body forces (i.e., forces per unit volume) such as, e.g., $\rho \mathbf{g}$ (here we consider only gravity; extension to other body forces is straightforward) and superficial forces, $\mathbf{f}_n$, exerted on the bounded surface by the surrounding fluid. Therefore, we obtain:

$$\frac{D}{Dt} \int_{V_m(t)} \rho \mathbf{v} dV = \int_{S_m(t)} \mathbf{f}_n dS + \int_{V_m(t)} \rho \mathbf{g} dV, \qquad (6.3.1)$$

where $\mathbf{f}_n$ is the force per unit area exerted on the surface of the material element that has the direction of the unit vector $\mathbf{e}_n$, perpendicular to S and directed outward from the volume element. This equation coincides with (6.2.10), in the special case where the conserved physical quantity F is the momentum $\mathbf{Q}$. Therefore, $f = \mathbf{v}$ is the momentum mass density,[3] $\sigma^{(F)} = \rho \mathbf{g}$ is the momentum source per unit time and unit volume, while $\mathbf{J}_{Qn}^{(d)} = -\mathbf{f}_n$ (note the minus sign) is the diffusive momentum flux, that is the momentum crossing, per unit area per unit time, the surface of the material element, in the absence of any net convection. Now, $\mathbf{f}_n$ is a function of both the position, $\mathbf{r}$, and the orientation, $\mathbf{e}_n$, of the point on the surface of the volume element where the force is applied. In agreement with the Cauchy lemma (6.1.3), we can define a tensor $\mathbf{T}$ as the proportionality coefficient between $\mathbf{f}_n$ and $\mathbf{e}_n$, called the *stress tensor* (see Fig. 6.2),

$$\mathbf{f}_n = \mathbf{e}_n \cdot \mathbf{T}, \quad \text{where } \mathbf{T} = \mathbf{e}_1 \mathbf{f}_1 + \mathbf{e}_2 \mathbf{f}_2 + \mathbf{e}_3 \mathbf{f}_3, \qquad (6.3.2)$$

where $\mathbf{T}$ depends only on $\mathbf{x}$ and is independent of the orientation of the surface. In fact, proceeding as in the previous Section, a force balance on the tetrahedron represented in Fig. 6.2 yields:

$$\mathbf{f}_n S(ABC) = \mathbf{f}_1 S(AOB) + \mathbf{f}_2 S(AOC) + \mathbf{f}_3 S(BOC), \qquad (6.3.3)$$

where $dS_n = S(ABC)$ is the area of the triangle ABC, $dS_1 = S(OAB)$, $dS_2 = S(OAC)$ and $dS_3 = S(OBC)$, and where the volumetric forces have been neglected, since they vanish more rapidly than surface forces when the size of the tetrahedron tends to

---

[2]Note that the second law can be applied because, although the material volume translates and deforms continuously as it follows the fluid motion, it corresponds to a well-defined portion of matter, with constant mass.

[3]Since momentum is a vector, Eq. (6.2.1) corresponds to 3 equations, with f indicating, v_1, v_2, and v_3.

Fig. 6.2 Surface forces on a
material volume element

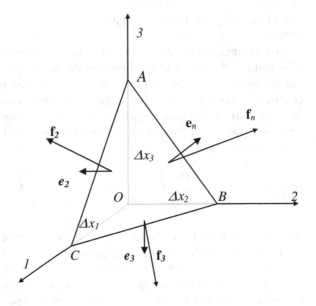

zero. Therefore, considering Eq. (6.1.4), we finally obtain the Cauchy theorem
(6.3.2), where the stress tensor,

$$\mathbf{T} = -\mathbf{J}_Q^{(d)} \tag{6.3.4}$$

is the negative of the diffusive momentum flux tensor: here, the minus sign is due to
the definition of surface force in Eq. (6.3.1). Therefore, we see that the component
T_{ji} of the stress tensor is the force along the direction i that is exerted on a surface
having direction j, as indicated in Fig. 6.3. Note that these forces are positive when
they are referred to a "positive" face, i.e., a face whose outward normal is directed
along the positive coordinate direction. Thus, for example, the force (per unit area)
T_{12} is positive (i.e., it is directed along the positive x_2 axis) when it refers to the face
on the right in Fig. 6.3, because its outward normal is directed along the positive x_1
direction, while it is negative when referred to the face on the left.

Fig. 6.3 Shear stresses on a
material volume element

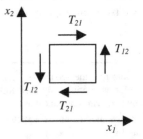

Substituting Eq. (6.3.2) into (6.3.1) and applying the divergence theorem, we finally obtain Eq. (6.2.8), i.e., the so-called Cauchy equation,

$$\rho \frac{D\mathbf{v}}{Dt} = \rho \left(\frac{\partial \mathbf{v}}{\partial t} + \mathbf{v} \cdot \nabla \mathbf{v} \right) = \nabla \cdot \mathbf{T} + \rho \mathbf{g}. \tag{6.3.5}$$

This is basically Newton's equation of motion, stating that the product of mass (per unit volume) and acceleration equals the sum of surface and body forces.

Had we considered a fixed volume element, adopting an Eulerian approach, we would have obtained a momentum conservation equation of the same form as Eq. (6.1.7), with a momentum flux tensor $\mathbf{J}_Q = \mathbf{J}_Q^{(d)} + \rho \mathbf{vv}$, in agreement with Eq. (6.2.9), i.e., based on Eq. (6.3.4),

$$\frac{\partial (\rho \mathbf{v})}{\partial t} + \nabla \cdot (\rho \mathbf{vv}) = \nabla \cdot \mathbf{T} + \rho \mathbf{g}. \tag{6.3.6}$$

6.4 Angular Momentum Balance

The momentum balance equation must be coupled to a corresponding balance equation for the angular momentum. To that end, consider the material volume element of Fig. 6.3: imposing that the total torque applied is zero, we easily obtain that $T_{12} = T_{21}$, and in general $T_{ij} = T_{ji}$, indicating that the stress tensor is symmetric.

Let us prove this result more rigorously. Adopting a Lagrangian approach, the conservation of angular momentum states that the time rate of change of the angular momentum of a material volume element equals the applied total torque, i.e.,

$$\frac{D}{Dt} \int_{V_m(t)} (\mathbf{x} \times \rho \mathbf{v}) dV = \int_{S_m(t)} [\mathbf{x} \times (\mathbf{e}_n \cdot \mathbf{T})] dS + \int_{V_m(t)} (\mathbf{x} \times \rho \mathbf{g}) dV, \tag{6.4.1}$$

where $\mathbf{a} \times \mathbf{b}$ indicates the vector (or cross) product. First of all, applying the Reynolds transport theorem (6.2.7), we obtain:

$$\frac{D}{Dt} \int_{V_m(t)} (\mathbf{x} \times \rho \mathbf{v}) dV = \int_{V_m(t)} \mathbf{x} \times \left(\rho \frac{D\mathbf{v}}{Dt} \right) dV. \tag{6.4.2}$$

Note that,[4]

---

[4]Here and in the following we adopt the Einstein convention, meaning that when an index appears twice in a single term (sometimes called a dummy index) that implies summation over all values of the index. See Appendix F.

$$
\left(\int_{S_m(t)} [\mathbf{x} \times (\mathbf{n} \cdot \mathbf{T})] dS \right)_i = \int_{S_m(t)} \varepsilon_{ijk} x_j n_m T_{mk} dS = \int_{V_m(t)} \varepsilon_{ijk} \nabla_m \left(x_j T_{mk} \right) dV
$$

$$
= \int_{V_m(t)} \varepsilon_{ijk} T_{jk} + \varepsilon_{ijk} x_j (\nabla_m T_{mk}) dV \tag{6.4.3}
$$

$$
= \left(\int_{V_m(t)} [\varepsilon : \mathbf{T} + \mathbf{x} \times (\nabla \cdot \mathbf{T})] dS \right)_i,
$$

where $\mathbf{A:B} = A_{ij} B_{ij}$. It should be noted that the vector product can be expressed in terms of the anti-symmetric third-order Levi-Civita tensor[5] $(\mathbf{a} \times \mathbf{b})_i = \varepsilon_{ijk} a_j b_k$. Substituting Eqs. (6.4.2) and (6.4.3) into (6.4.1), we obtain:

$$
\int_{V_m(t)} \left[\mathbf{x} \times \left(\rho \frac{D\mathbf{v}}{Dt} - \nabla \cdot \mathbf{T} - \rho \mathbf{g} \right) + \varepsilon : \mathbf{T} \right] dV = 0. \tag{6.4.4}
$$

Hence, substituting for the fluid acceleration from Eq. (6.3.5) and considering that the material volume element is arbitrary, we obtain at any point in the fluid:

$$
\varepsilon : \mathbf{T} = 0 \quad \Rightarrow \quad \mathbf{T} = \mathbf{T}^T, \tag{6.4.5}
$$

where the superscript "T" indicates the transpose, i.e., $T_{ij}^T = T_{ji}$. Therefore, the conservation of angular momentum amounts to a requirement of symmetry for the stress tensor.

6.5   The Constitutive Equation for Newtonian Fluids

For the PDE system comprising the continuity and the Cauchy equations to be solvable, we must first express the stress tensor in terms of the velocity field. Now, as previously noted I Chap. 1, for a fluid at rest the surface force $\mathbf{f}_n$ acting on a surface oriented along the unit outward normal vector $\mathbf{e}_n$ is $\mathbf{f}_n = -p e_n$, where p is the local pressure, i.e., a force perpendicular to the surface which tends to compress the volume element. Therefore, since $\mathbf{f}_n = \mathbf{e}_n \cdot \mathbf{T}$ [see (6.3.2)], we have:

$$
\mathbf{T} = -p\mathbf{I}, \tag{6.5.1}
$$

---

[5]Named after Tullio Levi-Civita (1873–1941), an Italian mathematician. ε_{ijk} is actually a pseudo-tensor, as it is invariant with respect to a rigid rotation of the Cartesian axes, but it changes sign upon a mirror reflection.

and the Cauchy equation reduces to

$$\rho \mathbf{g} = -\nabla \cdot \mathbf{T} = \nabla \cdot (p\mathbf{I}) = \nabla p, \tag{6.5.2}$$

which coincides with the hydrostatic balance, Eq. (2.1.1).

In general, the tensor $\mathbf{T} + p\mathbf{I}$ at any point $\mathbf{x}$ is a function of the velocity field $\mathbf{v}(\mathbf{x}')$ at a nearby point $\mathbf{x}'$. Now, expanding

$$\mathbf{v}(\mathbf{x}') = \mathbf{v}(\mathbf{x}) + (\mathbf{x}' - \mathbf{x}) \cdot \nabla \mathbf{v}(\mathbf{x}) + \cdots, \tag{6.5.3}$$

and considering that a uniform velocity field cannot generate any stress within the fluid body, we may conclude that,

$$\mathbf{T} + p\mathbf{I} = f(\nabla \mathbf{v}, \nabla\nabla \mathbf{v}, \nabla\nabla\nabla \mathbf{v}, \ldots). \tag{6.5.4}$$

Now, we have to determine the relation between the stress tensor (which, as we saw, is symmetric) and the velocity gradients. First of all, let us assume that the fluid is *Newtonian*, meaning that the stress tensor depends linearly on the velocity gradient, while the effects of higher-order gradients are negligible. Now, let us analyze the type of deformation caused by a generic velocity gradient, $\nabla \mathbf{v}$. From Eq. (6.5.3) we see that the difference $d\mathbf{v}$ in velocity between two points at a distance $d\mathbf{x}$ is $d\mathbf{v} = d\mathbf{x} \cdot \nabla \mathbf{v}$. Now, $\nabla \mathbf{v}$, like all tensors, can be written as the sum of symmetric and anti-symmetric components, $\mathbf{S}$ and $\mathbf{A}$, respectively, with

$$\nabla_i v_j = S_{ij} + A_{ij}; \quad S_{ij} = \frac{1}{2}\left(\nabla_i v_j + \nabla_j v_i\right); \quad A_{ij} = \frac{1}{2}\left(\nabla_i v_j - \nabla_j v_i\right). \tag{6.5.5}$$

Note that $\mathbf{S} = \mathbf{S}^T$, $\mathbf{A} = -\mathbf{A}^T$ and $A_{11} = A_{22} = A_{33} = 0$. Thus, we may write:

$$d\mathbf{v} = (d\mathbf{v})_s + (d\mathbf{v})_a; \quad (d\mathbf{v})_s = d\mathbf{x} \cdot \mathbf{S}; \quad (d\mathbf{v})_a = d\mathbf{x} \cdot \mathbf{A}. \tag{6.5.6}$$

Hence, since,

$$(d\mathbf{v})_a \cdot d\mathbf{x} = d\mathbf{x} \cdot \mathbf{A} \cdot d\mathbf{x} \equiv 0, \tag{6.5.7}$$

we may conclude that $(d\mathbf{v})_a$, being proportional to $|d\mathbf{x}|$ and perpendicular to $d\mathbf{x}$, describes a rigid body rotation, and thus it may not cause any stress within the fluid. So, we conclude that $(\mathbf{T} + p\mathbf{I})$ must be independent of $\mathbf{A}$ and proportional to $\mathbf{S}$, as one would expect, considering that $\mathbf{T}$ is a symmetric tensor.

As for the type of linear dependence occurring between $\mathbf{T}$ and $\mathbf{S}$, we must account for the isotropy of the fluid, meaning that $(\mathbf{T} + p\mathbf{I})$ is an isotropic tensor function of $\mathbf{S}$. Considering that:

(a) the most general linear relation between the two symmetric tensors is
$T_{ij} + p\delta_{ij} = \eta_{ijkl}S_{kl};$

(b) the most general isotropic fourth-order tensor is $\eta_{ijkl} = \lambda \delta_{ij}\delta_{kl} + \lambda''\delta_{ik}\delta_{jl} + \lambda''\delta_{il}\delta_{jk}$. Then, since $T_{ij} = T_{ji}$ and $S_{kl} = S_{lk}$ we have: $\eta_{ijkl} = \eta_{jikl} = \eta_{ijlk}$, and so we see that $\lambda' = \lambda'' = \mu$ and so we obtain the general constitutive relation:

$$T = [-p + \lambda(\nabla \cdot \mathbf{v})]\mathbf{I} + 2\mu\mathbf{S}, \qquad (6.5.8)$$

where we have considered that $\nabla \cdot \mathbf{v} = \mathbf{I}:\mathbf{S}$. Here, μ represents the dynamic viscosity that we have already seen in Sect. 1.4.2, while λ is a constant. In general, Eq. (6.6.8) is often written as:

$$T = [-p + \zeta(\nabla \cdot \mathbf{v})]\mathbf{I} + 2\mu\tilde{\mathbf{S}}. \qquad (6.5.9)$$

Here, $\tilde{\mathbf{S}} = \mathbf{S} - \frac{1}{3}\mathbf{I}(\nabla \cdot \mathbf{v})$ denotes the traceless velocity gradient, with $\mathbf{I}:\tilde{\mathbf{S}} = 0$, while ζ denotes the *dilatational*, or *bulk viscosity*, $\zeta = \lambda + \frac{2}{3}\mu$. The physical meaning of ζ becomes clear if we consider the mean normal stresses, $\bar{p}$, also referred to as *mechanical pressure*, defined as the negative mean value of the diagonal elements of the stress tensor, obtaining,

$$\bar{p} = -\frac{1}{3}\mathbf{I}:T = p - \zeta(\nabla \cdot \mathbf{v}). \qquad (6.5.10)$$

Hence we see that the bulk viscosity induces an increase in the normal stresses (i.e., a sort of additional pressure) when the flow field tends to compress the fluid, with a negative velocity divergence. The bulk viscosity is very difficult to measure; however, it is generally very small and therefore neglected. In addition, for incompressible flows, bulk viscosity is irrelevant and the constitutive relation for Newtonian fluids becomes:

$$T = -p\mathbf{I} + 2\mu\mathbf{S}. \qquad (6.5.11)$$

6.6 Energy Balance

The energy contained in a material volume element is the sum of the macroscopic mechanical (kinetic plus potential) energy and the internal energy. The latter accounts for, amongst others, the potential energy associated with short-range molecular interactions, and the kinetic energy of the molecular velocity fluctuations. Let us consider a material point, that is a volume element that is sufficiently large so that fluctuations are small compared to the mean, but also sufficiently small that the changes due to macroscopic gradients are smaller than those due to the thermal fluctuations (see Sect. 1.2). As the kinetic energy, E_K, of such material element depends on the square of the particle velocities, considering that the velocity of the *i*th particle, $\mathbf{v}_i$, is the sum of an average value, $\bar{\mathbf{v}}$, and a fluctuating component, $\tilde{\mathbf{v}}_i$, i.e., $\mathbf{v}_i = \bar{\mathbf{v}} + \tilde{\mathbf{v}}_i$, we find:

$$E_K = \sum_{i=1}^{N} \frac{1}{2} m v_i^2 = \sum_{i=1}^{N} \frac{1}{2} m (\bar{\mathbf{v}} + \tilde{\mathbf{v}}_i)^2 = \frac{1}{2} M \bar{\mathbf{v}}^2 + \frac{1}{2} M \langle \tilde{\mathbf{v}}^2 \rangle, \qquad (6.6.1)$$

where the brackets denoting ensemble averaging, with $\langle \tilde{\mathbf{v}} \rangle = \frac{1}{N} \sum_{i=1}^{N} \tilde{\mathbf{v}}_i = 0$ by definition, $\langle \tilde{\mathbf{v}}^2 \rangle = \frac{1}{N} \sum_{i=1}^{N} \tilde{\mathbf{v}}_i^2$, and we have considered that $\langle \tilde{\mathbf{v}}_i \cdot \bar{\mathbf{v}} \rangle = \langle \tilde{\mathbf{v}}_i \rangle \cdot \bar{\mathbf{v}} = 0$. Obviously, the velocity $\bar{\mathbf{v}}$ of the material point $\mathbf{x}$ is what has been previously indicated as the macroscopic velocity $\mathbf{v}(\mathbf{x})$. Similar considerations apply to the potential energy as well; in particular, in solid crystals, where atoms can be modeled as harmonic oscillators, an equation almost identical to Eq. (6.6.1) can be obtained. Therefore, we see that the total specific (per unit mass) energy, e, is the sum of a coherent mechanical energy and an internal energy, u, that accounts for internal molecular fluctuations, so that

$$e = e_{me} + u, \quad \text{where } e_{me} = \frac{1}{2} v^2 + \psi, \qquad (6.6.2)$$

where we have considered that e_{me} is the sum of a macroscopic kinetic energy, $\frac{1}{2} v^2$, and its potential counterpart, ψ.

Now, the mechanical energy balance can be derived from the continuity and the Cauchy equation. In fact, taking the dot product of Eq. (6.3.5) by $\mathbf{v}$ and rearranging, we obtain:

$$\rho \mathbf{v} \cdot \frac{D\mathbf{v}}{Dt} = \rho \frac{D(\frac{1}{2} v^2)}{Dt} = \mathbf{v} \cdot (\nabla \cdot \mathbf{T}) + \rho \mathbf{g} \cdot \mathbf{v} = \nabla \cdot (\mathbf{v} \cdot \mathbf{T}) - \mathbf{T} : \mathbf{S} + \rho \mathbf{g} \cdot \mathbf{v}.$$
$$(6.6.3)$$

In addition, considering that $\mathbf{g} = -\nabla \psi$, where ψ is the potential energy, and assuming that ψ does not depend explicitly on time, we have:

$$\rho \frac{D(\psi)}{Dt} = \rho \mathbf{v} \cdot \nabla \psi = -\rho \mathbf{g} \cdot \mathbf{v}. \qquad (6.6.4)$$

Note that the source term in the potential energy balance represents the conversion of potential energy into kinetic energy, as an equal but opposite term appears in the kinetic energy equation as well. Now, summing Eqs. (6.6.3) and (6.6.4), we obtain a conservation equation for the mechanical energy that has the form (6.2.8), i.e.,

$$\rho \frac{De_{me}}{Dt} + \nabla \cdot \mathbf{J}_{ME}^{(d)} = \sigma_{ME}, \tag{6.6.5}$$

with

$$\mathbf{J}_{ME}^{(d)} = -\mathbf{T} \cdot \mathbf{v}; \quad \sigma_{ME} = -\mathbf{T}{:}\mathbf{S}, \tag{6.6.6}$$

where we have considered that $T_{ij}\nabla_i v_j = T_{ij}S_{ij}$, due to the symmetry[6] of $\mathbf{T}$.

In the same way as for the mechanical energy, the internal energy, u and the total energy, e, will each satisfies a balance equation of the same form as Eq. (6.2.8), i.e.,

$$\rho \frac{Du}{Dt} + \nabla \cdot \mathbf{J}_q = \dot{q}, \tag{6.6.7}$$

and

$$\rho \frac{De}{Dt} + \nabla \cdot \mathbf{J}_E^{(d)} = 0, \tag{6.6.8}$$

where $\mathbf{J}_q = \mathbf{J}_U^{(d)}$ is the diffusive internal energy flux, commonly referred to as the *heat flux*, $\dot{q} = \sigma_U$ is the internal energy, or heat, source, and we have considered that $\sigma_E = 0$, since energy is a conserved quantity. Since the sum of Eqs. (6.6.5) and (6.6.7) must be equal to Eq. (6.6.8), we may conclude:

$$\mathbf{J}_E^{(d)} = \mathbf{J}_q - \mathbf{T} \cdot \mathbf{v}, \tag{6.6.9}$$

and

$$\dot{q} = -\sigma_{ME} = \mathbf{T}{:}\mathbf{S}. \tag{6.6.10}$$

From Eqs. (6.6.7) and (6.6.10), we may conclude that changes in the internal energy of a material volume are due to the source term, $\mathbf{T}{:}\mathbf{S}$, or to the net heat flux, $\mathbf{J}_q$ into the material element. The source term represents the conversion of mechanical energy into heat, a process that is, improperly, called *energy dissipation*. In particular, applying the constitutive Eq. (6.6.9) for the stress tensor, this term assumes the following form:

$$\dot{q} = \mathbf{T} : \mathbf{S} = 2\mu \breve{\mathbf{S}} : \breve{\mathbf{S}} - \bar{p}(\nabla \cdot \mathbf{v}), \tag{6.6.10a}$$

where $\bar{p}$ represents the negative mean normal stress (also called the mechanical pressure), and $\breve{\mathbf{S}}$ is the traceless strain rate. Therefore, we see that the internal energy

---

[6]In fact, $T_{ij}\nabla_i v_j = T_{ij}S_{ij} + T_{ij}A_{ij} = T_{ij}S_{ij}$. In general, the double inner product between a symmetric and an anti-symmetric tensor is identically zero.

source term (6.6.10) is a deterministic quantity, as it equals the dissipation of mechanical energy; on the other hand, the diffusive internal energy flux necessitates a new constitutive equation.

We could reach the same conclusions by considering an integral balance over a material volume element, i.e.,

$$\frac{D}{Dt} \int_{V_m(t)} \rho e \, dV = \left(\begin{array}{c} \text{net energy flux into the material} \\ \text{volume through its boundaries} \end{array} \right). \tag{6.6.11}$$

Now, the total energy flux at the boundaries can be considered as the sum of two contributions: one coherent and the other incoherent. The former, $\mathbf{f}_n \cdot \mathbf{v} = \mathbf{e}_n \cdot \mathbf{T} \cdot \mathbf{v}$, is the work per unit time and unit surface made by the surface forces defined in (6.3.2); the latter, $-\mathbf{e}_n \cdot \mathbf{J}_U$, is a flux of incoherent energy, or heat. Here, $\mathbf{e}_n$ is a unit vector normal to the surface and outward from the material volume. Note that we have used the convention that both the heat flux and the rate of working by the surface forces are positive when they enter the volume element. Therefore, the energy balance equation becomes:

$$\frac{D}{Dt} \int_{V_m(t)} \rho e \, dV = \oint_{S_m(t)} \mathbf{e}_n \cdot \mathbf{T} \cdot \mathbf{v} dS - \oint_{S_m(t)} \mathbf{e}_n \cdot \mathbf{J}_U dS. \tag{6.6.12}$$

Now, applying the Reynolds transport theorem (6.2.7) to the LHS of Eq. (6.6.12) and the divergence theorem to the RHS, we obtain:

$$\rho \frac{De}{Dt} = \nabla \cdot \left(\mathbf{T} \cdot \mathbf{v} - \mathbf{J}_q \right). \tag{6.6.8a}$$

This results coincides with Eqs. (6.6.8) and (6.6.9). Naturally, substituting for the mechanical energy balance from Eqs. (6.6.5) to (6.6.6), we obtain again the internal energy balance, Eq. (6.6.7).

As for the heat flux, it is generally assumed that it depends on the temperature distribution through the Fourier constitutive equation,

$$\mathbf{J}_q = -k \nabla T, \tag{6.6.13}$$

where k is the thermal conductivity of the medium. Again, we stress that here we have considered a Lagrangian point of view, i.e., one moving with the fluid velocity along a fluid velocity paths. Had we considered a fixed reference frame, the internal energy flux would be [see Eq. (6.2.9)]:

$$\mathbf{J}_U = \rho u \mathbf{v} - k \nabla T. \tag{6.6.14}$$

Before closing this Section, it is worth showing that the energy balance (6.6.8) assumes a particularly interesting form at steady state. In fact, applying the continuity equation, we find the following identity:

$$-\nabla \cdot (p\mathbf{v}) = -\rho \frac{D(p/\rho)}{Dt} + \frac{\partial p}{\partial t}.$$

Therefore, considering that $\mathbf{T} = -p\mathbf{I} + \mathbf{T}^d$, when the pressure field is steady Eq. (6.6.8) becomes:

$$\rho \frac{DH}{Dt} = \nabla \cdot \left(\mathbf{T}^d \cdot \mathbf{v} - \mathbf{J}_q\right), \quad \text{with } H = \frac{1}{2}v^2 + \psi + h, \qquad (6.6.15)$$

where $h = u + p/\rho$ is the fluid enthalpy per unit mass. In particular, for steady, frictionless flow of a non-conducting fluid, we obtain:

$$\frac{DH}{Dt} = 0 \quad \text{with } H = \frac{1}{2}v^2 + \psi + h, \qquad (6.6.15a)$$

indicating that the quantity H has the same value for all points along a path line which, at steady state, is the same as a streamline. Equation (6.6.15a) coincides with Eq. (4.2.7) and therefore is another statement of the Bernoulli equation. In fact, although pressure acts as a normal stress at the surface bounding a fluid element, its gradient, ∇p, acts as a body force per unit volume; so, it is not surprising that the pressure under certain conditions may play the role of a potential energy.

6.6.1 Temperature Dependence of the Energy Equation

For an ideal gas, we have: $du = c_v dT$, where c_v is the specific heat at constant volume, while $\mathbf{T} = -p\mathbf{I}$, as the fluid is inviscid. Therefore, the internal energy balance Eq. (6.6.7) becomes:

$$\rho c_v \frac{DT}{Dt} = \nabla \cdot (k\nabla T) - p\nabla \cdot \mathbf{v}. \qquad (6.6.16)$$

For a generic medium, instead, the internal energy can be expressed as a function of temperature, T, and specific volume, ρ^{-1}, as:

$$du = c_v dT + \left(\frac{\beta T}{\kappa_T} - p\right) d\rho^{-1}, \qquad (6.6.17)$$

where β e κ_T are the coefficients of isobaric expansion and isothermal compressibility, defined in (6.2.14). Substituting Eqs. (6.6.13) and (6.6.17) into Eq. (6.6.7), applying the continuity Eq. (6.2.11) and defining $\mathbf{T}^d = \mathbf{T} + p\mathbf{I}$, we obtain,

$$\rho c_v \frac{DT}{Dt} = \nabla \cdot (k\nabla T) + \mathbf{T}^d : \mathbf{S} - \frac{\beta T}{\kappa_T}(\nabla \cdot \mathbf{v}). \qquad (6.6.18)$$

This equation, though, is not very useful because c_v is difficult to measure. Therefore, let us rewrite Eq. (6.6.7) by substituting for $(\nabla \cdot \mathbf{v})$ from Eq. (6.2.12):

$$\rho \frac{Du}{Dt} = \mathbf{T}^d : \mathbf{S} - p(\nabla \cdot \mathbf{v}) - \nabla \cdot \mathbf{J}_U$$

$$= \mathbf{T}^d : \mathbf{S} + \rho \left[-\frac{D}{Dt}(p\rho^{-1}) + \rho^{-1}\frac{Dp}{Dt} \right] - \nabla \cdot \mathbf{J}_U.$$

Then, we obtain:

$$\rho \frac{Dh}{Dt} = \mathbf{T}^d : \mathbf{S} + \frac{Dp}{Dt} - \nabla \cdot \mathbf{J}_U. \qquad (6.6.19)$$

Expressing h as a function of temperature and pressure we have:

$$dh = c_p dT + \rho^{-1}(1 - \beta T)dp, \qquad (6.6.20)$$

where c_p is the specific heat per unit mass at constant pressure, which is much easier to measure than c_v. At the end, we obtain:

$$\rho c_p \frac{DT}{Dt} = \nabla \cdot (k\nabla T) + \mathbf{T}^d : \mathbf{S} + \beta T \frac{Dp}{Dt}. \qquad (6.6.21)$$

This equation can also be derived directly from (6.6.18), by expressing c_v in terms of c_p [see Eq. (6.7.7)]; note that it is very convenient to use, because the term Dp/Dt can be often neglected.

6.7 Governing Equations for Incompressible Flow of Newtonian Fluids

The continuity equation for incompressible flow reduces to a divergence-free constraint, i.e.,

$$\nabla \cdot \mathbf{v} = 0. \qquad (6.7.1)$$

Also, substituting for the constitutive equation from Eq. (6.6.11) into the Cauchy Eq. (6.3.5), we obtain:

$$\rho \frac{D\mathbf{v}}{Dt} = \rho \left[\frac{\partial \mathbf{v}}{\partial t} + \mathbf{v} \cdot \nabla \mathbf{v} \right] = -\nabla p + \mu \nabla^2 \mathbf{v} + \rho \mathbf{g}, \qquad (6.7.2)$$

where we have considered that $\nabla_j(\nabla_i v_j + \nabla_j v_i) = \nabla^2 v_i$, since $\nabla_j v_j = 0$. This is the *Navier-Stokes* equation,[7] valid when the fluids can be assumed to be Newtonian and, at least approximately, isothermal. In fact, in Eq. (6.7.2) the viscosity has been taken to be constant, while in general it is a function of temperature. In that case, the Navier-Stokes equation is easily generalized to:

$$\rho \left[\frac{\partial \mathbf{v}}{\partial t} + \mathbf{v} \cdot \nabla \mathbf{v} \right] = -\nabla p + \mu \nabla^2 \mathbf{v} + \nabla \mu \cdot \left[\nabla \mathbf{v} + \nabla \mathbf{v}^T \right] + \rho \mathbf{g}, \qquad (6.7.3)$$

with $\nabla \mu = (d\mu/dT)\nabla T$.

The Navier-Stokes equation is vectorial (i.e., it consists of 3 equations) while the continuity equation is scalar; the unknowns, assuming that density ρ and viscosity μ are constant, are pressure p and velocity $\mathbf{v}$. Thus, we have 4 equations in 4 unknowns, and the problem is closed.

The *kinematic viscosity* $\nu = \mu/\rho$ is a very important quantity; as we saw in Chap. 1, it represents the ability for momentum to diffuse. In fact, from Eq. (6.7.2) we see that it can also be estimated in terms of the ratio between the fluid acceleration $D\mathbf{v}/Dt$ and the spatial change in the velocity gradient, $\nabla^2 \mathbf{v}$.

Often, it is convenient to introduce the concept of *modified pressure*, defined as the driving force of the fluid motion, while it is identically zero when the fluid is at rest. Therefore, since for a stationary fluid $\nabla p = \rho \mathbf{g}$, the modified pressure P is defined as

$$\nabla P = \nabla p - \rho \mathbf{g}, \quad \text{that is,} \quad P = p + \rho g z, \qquad (6.7.4)$$

where $z = -\mathbf{g} \cdot \mathbf{x}/g$ is the coordinate measured along a direction opposite to $\mathbf{g}$ (i.e., when $\mathbf{g}$ is the gravity field, z is the vertical coordinate, directed upward). At the end, the Navier-Stokes equation becomes,

$$\frac{D\mathbf{v}}{Dt} = \frac{\partial \mathbf{v}}{\partial t} + \mathbf{v} \cdot \nabla \mathbf{v} = -\frac{1}{\rho} \nabla P + \nu \nabla^2 \mathbf{v} \qquad (6.7.5)$$

Note that the modified pressure P exists only when ρ is constant, that is for incompressible flows; otherwise, the integration of P in Eq. (6.7.4) cannot be performed. The Navier-Stokes Eq. (6.7.5) indicates clearly that the gravity force

---

[7]Named after Claude-Louis Navier (1785–1836), a French engineer and physicist, who was a professor of calculus and mechanics at the École Polytechnique in Paris, and Sir George Gabriel Stokes (1819–1903), an Irish mathematician, physicist, politician and theologian, who for 54 years held the position of Lucasian professor of mathematics at Cambridge.

(like any other potential force) does not have any effect on the flow field, i.e., we would obtain the same velocity field even if the magnitude of gravity were doubled, or disappeared altogether.

Finally, let us comment on the energy balance equation. For incompressible flows, we know that the variation of the specific (i.e., per unit mass) internal energy, du, is proportional to the temperature variation,

$$du = c_v dT, \tag{6.7.6}$$

where c_v is the specific heat (per unit mass) at constant volume. Now, considering that,

$$c_v = c_p - T \left(\frac{\partial P}{\partial T} \right)_\rho \left(\frac{\partial \rho^{-1}}{\partial T} \right)_P, \tag{6.7.7}$$

where c_p is the specific heat (per unit mass) at constant pressure, we see that $c_v = c_p = c$ for incompressible flows. Therefore, the internal energy balance (6.6.7) for incompressible flows of a Newtonian fluid, in cases where the thermal conductivity k, density ρ and specific heat c are all constant, becomes

$$\frac{DT}{Dt} = \frac{\partial T}{\partial t} + \mathbf{v} \cdot \nabla T = \alpha \nabla^2 T + \frac{\dot{q}}{\rho c}, \tag{6.7.8}$$

where $\alpha = k/\rho c$ is thermal diffusivity, while

$$\dot{q} = \mathbf{T}^d : \mathbf{S} = 2\mu \mathbf{S} : \mathbf{S} = \mu (\nabla_i v_j) \left[(\nabla_i v_j) + (\nabla_j v_i) \right] \tag{6.7.9}$$

is the source term, that is the heat "generated" per unit volume and unit time. Here, $\dot{q}$ is due exclusively to viscous dissipation (i.e. mechanical energy converted into heat); in general, however, it also includes other mechanisms of heat "generation", such the conversion into heat of chemical or nuclear energy.

Equation (6.7.8) coincides with Eq. (6.6.21), as $\beta = 0$ for incompressible flow. In general, Eq. (6.7.8) is valid when the last term of Eq. (6.6.21) can be neglected. That occurs when (a) $\beta = 0$, i.e., the density is independent of temperature (but still depends on pressure); (b) when the pressure is constant, as it happens in low Mach number flows.

Equations (6.7.5) and (6.7.8) are very similar to each other: on the LHS there is a convective term, while on the RHS we have the sum of a diffusive term and a source term. Note, however, that in the momentum balance equation the convective term is quadratic in $\mathbf{v}$, while the source term is a body force that includes the pressure gradient term.

6.8 The Entropy Equation

Assuming local equilibrium, we may assume that within a small fluid element we can apply the first law of thermodynamics as:

$$T\frac{Ds}{Dt} = \frac{Dh}{Dt} - \frac{1}{\rho}\frac{Dp}{Dt}, \tag{6.8.1}$$

where s and h are the fluid entropy and enthalpy per unit mass, respectively. Then, substituting Eq. (6.6.19) and rearranging, we obtain an entropy balance equation, which can be recast in the general form (6.2.8), with $f = s$, as

$$\rho\frac{Ds}{Dt} + \nabla \cdot \mathbf{J}_s^{(d)} = \sigma_S, \tag{6.8.2}$$

where $\mathbf{J}_s^{(d)} = \mathbf{J}_U^{(d)}/T$ is the entropy diffusive flux, while

$$\sigma_S = \mathbf{J}_U^{(d)} \cdot \nabla\left(\frac{1}{T}\right) + \frac{1}{T}\mathbf{T}^d : \nabla\mathbf{v}, \tag{6.8.3}$$

represents the entropy production term. Therefore, it appears that the entropy generation is due to the diffusive fluxes of energy and of momentum and therefore it is zero in the absence of diffusion.

This latter expression reveals that the entropy production term is the sum of a series of products between generalized fluxes and generalized forces. Then, it appears natural to relate these conjugated quantities with each other, thus obtaining the constitutive relations (6.6.8) and (6.6.13). This approach is typical of the so-called Non-Equilibrium Thermodynamics.[8] At the end, Eq. (6.8.3) becomes,

$$\sigma_S = \frac{1}{T^2}k(\nabla T)\cdot(\nabla T) + \frac{1}{T}\zeta(\nabla\cdot\mathbf{v})^2 + \frac{1}{T}\mu\breve{\mathbf{S}} : \breve{\mathbf{S}}, \tag{6.8.4}$$

where $\breve{\mathbf{S}} = \mathbf{S} - \frac{1}{3}\mathbf{I}(\nabla\cdot\mathbf{v})$ is the traceless velocity gradient, with $\mathbf{I} : \breve{\mathbf{S}} = 0$, k is the fluid thermal conductivity, while ζ and μ are the dynamic and bulk viscosities, respectively. Accordingly, as expected, inviscid and non-conducting fluids are necessarily iso-entropic, i.e., their entropy per unit mass is constant along a streamline.

---

[8]See R. Mauri, *Non-Equilibrium Thermodynamics in Multiphase Flows*, Springer (2013).

Chapter 7
Unidirectional Flows

Abstract In this Chapter we consider cases where the fluid velocity is directed along a single direction, so that the continuity equation is identically satisfied and the inertial term in the Navier-Stokes equation is identically null. Consequently, these fluid flows are much easier to study, as the four equations of mass and momentum balance reduce to a single, linear equation. In Sect. 7.1 we consider the most important example of unidirectional flow, namely the pipe flow of a Newtonian fluid in laminar regime, where the fluid velocity is directed along the axial direction of the pipe. This case is generalized in Sects. 7.2 and 7.3, where the fluid velocity is directed along the azimuthal or the radial directions. In the following Sect. 7.4 the velocity field induced by the sudden movement of a wall is studied, deriving the classical self-similar solution. Then, in Sect. 7.5, the first-order correction to the unidirectional pipe flow solution is determined, by studying the slider bearing problem with its related lubrication approximation. Other approximated solutions are presented at the end of the Chapter, considering first the quasi steady state hypothesis (Sect. 7.6) and then an integral approach (Sect. 7.7) that will be studied further in later Sections.

7.1 Flow in Pipes and Channels

Consider the flow of a Newtonian fluid in a circular pipe of radius R. Using cylindrical coordinates, we see that the fluid velocity is directed along the axis of the tube, z, and depends only on the distance r from the centerline, i.e., $\mathbf{v} = v(r)\mathbf{e}_z$, where $\mathbf{e}_z$ is a unit vector along the z-axis. Repeating the same reasoning as in Sect. 4.1, we see that the continuity equation is satisfied identically, while in the Navier-Stokes the inertial term is identically null and the dynamic pressure depends only on z. Finally, the Navier-Stokes equation along the longitudinal z-direction becomes,

© Springer International Publishing Switzerland 2015
R. Mauri, *Transport Phenomena in Multiphase Flows*,
Fluid Mechanics and Its Applications 112, DOI 10.1007/978-3-319-15793-1_7

$$\frac{dP}{dz} = \mu \frac{1}{r} \frac{d}{dr}\left(r\frac{dv}{dr}\right) = -\frac{\Delta P}{L}, \qquad (7.1.1)$$

where P is the modified pressure, to be solved with boundary condition:

$$v(R) = 0. \qquad (7.1.2)$$

Integrating twice we obtain:

$$v(r) = -\frac{(\Delta P)r^2}{4\mu L} + C_1 \ln r + C_2, \qquad (7.1.3)$$

where C_1 and C_2 are two constants to be determined. At first sight, the problem looks ill-posed, because we have two constants and only one boundary condition. However, as usual for problems in cylindrical and spherical coordinates, the problem presents another implicit boundary condition, stating that the velocity remains finite everywhere. Therefore, as the ln r term in (7.1.3) diverges at the centerline, where $r = 0$, we obtain[1]: $C_1 = 0$. Alternatively, we can also apply the implicit condition $dv/dr = 0$ at $r = 0$, which is due to elementary symmetry properties, obtaining the same conclusion. Finally, imposing that the boundary condition $v(R) = 0$ is satisfied, we find the usual Poiseuille velocity profile,

$$v(r) = -\frac{(\Delta P)R^2}{4\mu L}\left[1 - \left(\frac{r}{R}\right)^2\right]. \qquad (7.1.4)$$

Consider now a stationary flow of a Newtonian fluid in a conduit of height H, subjected to two driving forces: (a) there is pressure drop, ΔP, between the ends of the conduit; (b) the upper wall of the conduit moves with velocity V (see Fig. 7.1).

Assuming that the flow is laminar, the fluid velocity within the conduit has direction x and depends only on y, that is $\mathbf{v} = v(y)\mathbf{e}_x$, where $\mathbf{e}_x$ is a unit vector along the x-axis. Consequently, we see that the continuity equation is satisfied identically. As for the Navier-Stokes equation, its y-component reduces to $\partial P/\partial y = 0$, so that $P = P(x)$, showing that the dynamic pressure is only a function of the longitudinal coordinate, x. Here we remind that the thermodynamic pressure p depends also on the height z though the relation $P = p - \rho g z$, where g is the gravity field directed along $(-z)$-direction; However, the hydrostatic part, $\rho g z$, simply balances gravity and does not cause any fluid movement. Finally, the Navier-Stokes equation in the x-direction has its inertial part identically null, so that it reduces to the following simple relation,

$$\frac{dP}{dx} = \mu \frac{d^2v}{dy^2}. \qquad (7.1.5)$$

---

[1]We remind that $\lim_{x\to 0} (x \ln x) = 0$.

Fig. 7.1 Forced convection
of a sheared fluid

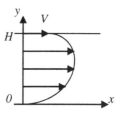

(Note that all derivatives are total) Now, the LHS of this equation is a function of
x only, while the RHS is a function of y only; therefore, both sides must be equal to
a constant, equal to the mean pressure gradient, $-\Delta P/L$, where the minus sign
indicates that ΔP is positive. Finally, solving Eq. (7.1.5) with boundary conditions

$$v(0) = 0; \quad v(H) = V, \tag{7.1.6}$$

we obtain:

$$v(y) = \frac{V}{H}y + \frac{\Delta P}{2L\mu}y(H - y) = V\left\{\frac{y}{H} + \frac{(\Delta P)H^2}{2LV\mu}\left[\left(\frac{y}{H}\right)\left(1 - \frac{y}{H}\right)\right]\right\}. \tag{7.1.7}$$

We see that the velocity field is simply the sum of a linear velocity profile resulting
from the motion of the upper wall (Couette flow) and a parabolic velocity profile,
due to the pressure drop (Poiseuille flow). This behavior is typical of all linear
systems, where the *superposition principle* can be applied, stating that the net
response at a given place and time caused by two or more inputs is the sum of the
responses which would have been caused by each input individually. In general, the
superposition principle can be applied in fluid mechanics only when, as in this case,
the non-linear, inertial term of the Navier-Stokes is identically null, so that the input
(i.e., the pressure drop) is proportional to the response (i.e., the velocity field).

7.1.1 Falling Cylinder Viscometer

The falling cylinder viscometer consists of two coaxial cylinders: a stationary outer
cylinder of radius R_o, containing the viscous fluid whose viscosity we want to
measure, and an inner cylinder, or rod, driven vertically by gravity (see Fig. 7.2).
We want to determine the terminal velocity V, assuming that the inner cylinder
remains at the center of the outer cylinder as it falls.

Assume for sake of simplicity that the clearance $H = R_o - R_i$ is small compared
to both radii, i.e., $\varepsilon = H/R_i \ll 1$. Therefore, in the clearance region, the curvature

Fig. 7.2 The falling cylinder
viscometer

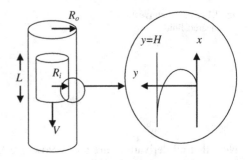

of the cylinder surfaces can be neglected, so that the velocity field is given by
Eq. (7.1.7), i.e.,

$$v(y) = V\left[-\frac{y}{H} + \Phi\left(\frac{y}{H}\right)\left(1 - \frac{y}{H}\right)\right] \quad \Phi = \frac{(\Delta P)H^2}{2LV\mu}, \tag{7.1.8}$$

where the minus sign is due to the fact that V is negative, as it is directed downward,
while Φ is assumed to be positive, i.e., it is directed upward.

Let us study the problem: we need to find two conditions to evaluate V and ΔP.
These conditions can be determined from a mass balance and a momentum balance,
which, considering that the problem is stationary and the system is closed, mean
that the total mass flow rate is null, and that the total force acting on the inner
cylinder is null.

The first condition translates into imposing that, as the inner rod moves downward, the displaced fluid must necessarily move upward through the clearance. In
other words, the volume flow rate of the descending rod must balance the volume
flow rate of the rising fluid, that is,

$$(2\pi R_i) \int_0^H v(y)dy = V\left(\pi R_i^2\right), \tag{7.1.9}$$

obtaining:

$$\Phi = 3\left(1 + \frac{1}{\varepsilon}\right) \cong \frac{3}{\varepsilon}. \tag{7.1.10}$$

The second condition states that gravity is balanced by the sum of pressure and
the upward friction force applied to the rod by the rising fluid:

$$|\tau|(2\pi R_i L) + (\Delta P)\left(\pi R_i^2\right) = (\Delta\rho)g\pi R_i^2 L, \tag{7.1.11}$$

where:

$$|\tau| = \mu \left.\frac{dv}{dy}\right|_{y=H} = \frac{\mu V}{H}(-1+\Phi) \cong \frac{3\mu V}{\varepsilon H}. \tag{7.1.12}$$

Now, considering that the first term on the LHS of Eq. (7.1.11) is negligible, we obtain:

$$\mu V = \frac{1}{6}\varepsilon^3(\Delta\rho)gR_i^2 = K. \tag{7.1.13}$$

where K is a constant, characteristic of the device. Naturally, that means that this device can be used as a viscometer, as μ is the ratio between K and the measured free fall terminal velocity V of the rod.

7.2 Parallel Plates Viscometer

Consider the viscometer of Fig. 7.3, composed of two parallel disks rotating with respect to one another; without loss of generality, here we assume that the lower plate is fixed, while the upper disk rotates with angular velocity Ω. Assuming that the rotation is sufficiently slow so that centrifugal forces are negligible, we want to determine the relation between the applied torque and the resulting angular velocity.

Due to symmetry considerations, we see that the fluid velocity is azimuthal (as we said, we neglect centrifugal effects, so that there is no radial component of the velocity), depending on the axial coordinate, z and the distance r from the axis, that is $\mathbf{v} = v_\phi(r,z)\mathbf{e}_\phi$. Again, as the velocity is unidirectional, we see that the continuity equation is satisfied identically, while the Navier-Stokes equation becomes,

$$0 = \mu\left\{\frac{\partial}{\partial r}\left[\frac{1}{r}\frac{\partial}{\partial r}(rv_\phi)\right] + \frac{\partial^2 v_\phi}{\partial z^2}\right\}, \tag{7.2.1}$$

to be solved with the following boundary conditions,

$$v_\phi(z=0) = 0 \quad\text{and}\quad v_\phi(z=H) = r\Omega. \tag{7.2.2}$$

Fig. 7.3 Parallel plate viscometer

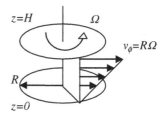

Note that when, as in this case, the streamlines are closed lines, the pressure must be constant along the flow (i.e., the azimuthal, in this case) direction. The Navier-Stokes equations in the other two directions state simply that $\partial P/\partial z = 0$ and $\partial P/\partial r = 0$; so, we may conclude that the modified pressure is constant everywhere.

This problem can be easily solved applying the method of the separation of variables, that is assuming that v_ϕ is the product of two functions, depending separately r and z. At the end, we obtain:

$$v_\phi(r, z) = rz\Omega/H. \tag{7.2.3}$$

Now, we can determine the stress exerted by the fluid on the upper plate,

$$\tau_{\phi z} = \mu \frac{\partial v_\phi}{\partial z} = \frac{\mu\Omega r}{H}, \tag{7.2.4}$$

and the force, $dF = \tau_{\phi z}dA$, exerted by the fluid in the azimuthal direction on the elementary surface area dA of the upper plate; therefore, integrating on the surface of the upper plate, we find the total torque,

$$\Gamma = \int_0^R rdF = \int_0^R r\tau_{\phi z}2\pi rdr = C\mu; \qquad C = \frac{\pi\Omega R^4}{2H}. \tag{7.2.5}$$

So, even in this case, we can determine the fluid viscosity by measuring the torque Γ that is necessary to maintain a given angular velocity Ω of the upper plate.

7.3 Radial Flux Between Two Parallel Disks

Plastic extrusion is a common industrial process, where raw material from a top mounted hopper is pumped into the barrel of the extruder through a feed throat, as shown in Fig. 7.4 (the case of Cartesian geometry has been considered in Problems 4.3 and 4.4). Here, a very viscous fluid is pumped in a tube of radius R_0 and then exits in a region bounded by two parallel disks of radius $R \gg R_0$. The fluid velocity has radial direction and, in general, depends on r and z, that is $\mathbf{v} = v(r, z)\mathbf{e}_r$. Therefore, the continuity equation in cylindrical coordinates become,

$$\frac{1}{r}\frac{\partial(rv)}{\partial r} = 0, \tag{7.3.1}$$

yielding

$$v(r, z) = \frac{1}{r}C(z), \tag{7.3.2}$$

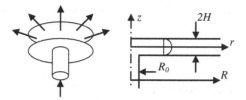

Fig. 7.4 Plastic extrusion process

where $C(z)$ is a function to be determined. This result could be obtained also considering that at steady state the volumetric flow rate at any cylindrical surface of radius r must be constant and must equal the volumetric flow rate at the inlet, that is,

$$(2\pi r)(2H)\bar{v}(r) = \dot{V} \quad \Rightarrow \quad \bar{C} = r\bar{v} = \frac{\dot{V}}{4\pi H}. \tag{7.3.3}$$

The Navier-Stokes equations along directions z and ϕ state that $\partial P/\partial z = 0$ and $\partial P/\partial\phi = 0$, showing that the modified pressure P depends only on r, that is $P = P(r)$. In the radial direction, instead, the Navier-Stokes equation becomes,

$$\rho v \frac{\partial v}{\partial r} = -\frac{dP}{dr} + \mu \left[\frac{\partial}{\partial r} \left(\frac{1}{r} \frac{\partial (rv)}{\partial r} \right) + \frac{\partial^2 v}{\partial z^2} \right]. \tag{7.3.4}$$

Now, the first viscous term, i.e. the first term within the square parenthesis, is identically zero out of the continuity (7.3.2). In addition, as the fluid is very viscous, the Reynolds number is very small, so that the inertial term on the LHS can be neglected. Note that in pipe flow we can neglect the inertial term whenever the flow regime is laminar, and that happens when the Reynolds number, Re, is less than 2100. Here, on the other hand, the inertial term can be neglected only when $Re \ll 1$. Finally, the Navier-Stokes becomes:

$$r \frac{dP(r)}{dr} = \mu \frac{d^2 C(z)}{dz^2} = A. \tag{7.3.5}$$

Here we have considered that, as the first term is a function of r, while the second is a function of z, they must be both equal to a constant. At this point, since C, like v, is zero at the walls, that is $C(\pm H) = 0$, we obtain: $C(z) = -(AH^2/2\mu) [1 - (z/H)^2]$. The value of A can be determined considering that $\bar{C} = -A/3$ and imposing that Eq. (7.3.3) be satisfied, obtaining,

$$A = -\frac{3\mu\dot{V}}{4\pi H^3}. \tag{7.3.6}$$

So, we see that the velocity profile is parabolic along z, but decays like $1/r$, i.e.,

$$v(r, z) = \frac{3}{2}\bar{v}\left[1 - \left(\frac{z}{H}\right)^2\right]; \quad \bar{v} = \frac{\dot{V}}{4\pi Hr}. \tag{7.3.7}$$

Finally, we can determine the pressure drop from Eq. (7.3.5), obtaining:

$$dP = A\frac{dr}{r} \quad \Rightarrow \quad P - P_a = \left(\frac{3\mu\dot{V}}{4\pi H^3}\right)\ln\left(\frac{R}{r}\right), \tag{7.3.8}$$

where we have imposed that $P(R) = P_a$, i.e., the outer pressure is equal to the atmospheric pressure. In particular, we find the inlet pressure $P_0 = P(R_0)$,

$$P_0 - P_a = \left(\frac{3\mu\dot{V}}{4\pi H^3}\right)\ln\left(\frac{R}{R_0}\right). \tag{7.3.9}$$

7.4 Fluid Flow Due to the Rapid Movement of a Wall

Consider a fluid occupying a semi-infinite region of space, $y \geq 0$. Assume that the fluid is initially quiescent and that, at time $t = 0$, the wall at $y = 0$ is set into motion, with constant velocity V along the x-direction. The resulting fluid velocity will be directed along x and will depend on the time t and the distance y from the wall, i.e., $\mathbf{v} = v(y, t)\mathbf{e}_x$. Naturally, we expect to find for $v(y, t)$ a solution of the type represented in Fig. 7.5, showing, for growing time t, the velocity profiles as functions of y, indicating that the fluid is set gradually in motion. However, as we will see from the formal solution of this problem, all the curves here collapse into a single curve when the fluid velocity is represented in terms of a "smart" coordinate, depending on both space and time. These types of problems are called *self-similar* and are very important because they offer a very clear physical interpretation of the phenomenon.

In our case, the governing, Navier-Stokes equation is the following:

$$\frac{\partial v}{\partial t} = \nu\frac{\partial^2 v}{\partial y^2}, \tag{7.4.1}$$

Fig. 7.5 Transient velocity field, $v(y,t)$, near a wall suddenly set in motion

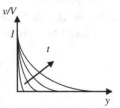

where v is the kinematic viscosity. This is the so called *diffusion equation*; it consists of a partial differential equation of first-order in time and second-order in space, and therefore it must be solved with one initial condition and two boundary conditions, that are,

$$v(y, t = 0) = 0; \quad v(y = 0, t) = V; \quad v(y \to \infty, t) = 0. \tag{7.4.2}$$

Now, suppose that a self-similar solution exists: in this case, it is possible to express the velocity v as a function of a single, self-similar coordinate, i.e., depending on both y and t, so that,

$$v(y, t) = V\tilde{v}(\eta); \quad \eta = \frac{y}{g(t)}, \tag{7.4.3}$$

where the function $g(t)$ has to be determined.[2] Considering that,

$$\frac{\partial}{\partial t} = \frac{d}{d\eta}\left(-\frac{y}{g^2}\frac{dg}{dt}\right) = \frac{d}{d\eta}\left(-\frac{\eta}{g}\frac{dg}{dt}\right); \quad \frac{\partial}{\partial y} = \frac{d}{d\eta}\left(\frac{1}{g}\right); \quad \frac{\partial^2}{\partial y^2} = \frac{d^2}{d\eta^2}\left(\frac{1}{g^2}\right),$$

the governing equation becomes,

$$\frac{d^2\tilde{v}(\eta)}{d\eta^2} + \left(\frac{1}{v}g\frac{dg}{dt}\right)\eta\frac{d\tilde{v}(\eta)}{d\eta} = 0. \tag{7.4.4}$$

Now, if a self-similar solution exists, then $\tilde{v}$ is a function of η only. That means that the term in parenthesis, which in general is a function of t, here must be a constant. As we will see in the following, the value of this constant is arbitrary and so, for sake of convenience, we choose the value 2, i.e.,

$$g\frac{dg}{dt} = 2v \quad \Rightarrow \quad \frac{1}{2}\frac{dg^2}{dt} = 2v \quad \Rightarrow \quad g^2 = 4vt + K \quad \Rightarrow \quad g(t) = \sqrt{4vt + K}. \tag{7.4.5}$$

Therefore, Eq. (7.4.4) becomes:

$$\frac{d^2\tilde{v}}{d\eta^2} + 2\eta\frac{d\tilde{v}}{d\eta} = 0. \tag{7.4.6}$$

At this point, the problem, at first sight, seems ill-posed: we have a second-order differential Eq. (7.4.6), requiring two conditions to be solved, while, on the contrary, we have the three conditions (7.4.2). However, in the particular case that we

---

[2]More generally, we should have assumed $\eta = h(y,t)$, where h is a function to be determined. However, in our case, as in the large majority of cases, we want to rescale a coordinate, y, in terms of a function of the other quantity, t, as we do in Eq. (6.4.3).

are considering, expressing these three conditions in terms of the self-similar coordinate, we obtain:

$$
\begin{aligned}
v(y, t = 0) = 0 \quad &\Rightarrow \quad \tilde{v}\left(y / \sqrt{K}\right) = 0; \\
v(y = 0, t) = V \quad &\Rightarrow \quad \tilde{v}(0) = 1; \\
v(y \rightarrow \infty, t) = 0 \quad &\Rightarrow \quad \tilde{v}(\infty) = 0.
\end{aligned}
\tag{7.4.7}
$$

Therefore, choosing $K = 0$, the first condition ends up coinciding with the third condition, and therefore we are left with two conditions, so that the problem is well-posed.

This part is very important: the diffusion Eq. (7.4.1) can always be transformed into the second-order differential Eq. (7.4.6); however, only very rarely it happens that the thee (one initial and two boundary) original conditions collapse into two. For example, if the fluid is bounded within a finite domain, with $0 < y < Y$, then the boundary condition imposing, for example, the value of $v(Y, t)$, does not collapse into one of the other conditions, as it happens in (7.4.7). That shows why self-similar solutions are so rare.

Now, denoting $f = d\tilde{v}/d\eta$, Eq. (7.4.7) reduces to $f' + 2\eta f = 0$, which can be solved, obtaining $f = b_1 \exp(-\eta^2)$, so that:

$$
\tilde{v}(\eta) = b_2 + b_1 \int_0^\eta e^{-\xi^2} d\xi.
$$

Then, applying the conditions $\tilde{v}(0) = 1$ and $\tilde{v}(\infty) = 0$, considering that $\int_0^\infty e^{-\xi^2} d\xi = \sqrt{\pi}/2$, we obtain the self-similar solution:

$$
\tilde{v}(\eta) = \frac{v(y, t)}{V} = erfc(\eta) = 1 - erf(\eta); \quad \eta = \frac{y}{\sqrt{4\nu t}},
\tag{7.4.8}
$$

where $erf(\eta)$ is the *error* function, while $erfc(\eta)$ is the *conjugated error function*,

$$
erf(\eta) = \frac{2}{\sqrt{\pi}} \int_0^\eta e^{-\xi^2} d\xi. \quad erfc(\eta) = 1 - erf(\eta) = \frac{2}{\sqrt{\pi}} \int_\eta^\infty e^{-\xi^2} d\xi
\tag{7.4.9}
$$

Note that $erf(\eta)$ is an odd function, i.e., $erf(-\eta) = erf(\eta)$, with $erf(0) = 0$ and $erf(\infty) = 1$.

From Fig. 7.6 we see that $v/V = 0.5$ when $\eta \cong 0.5$, which means that at a distance L from the wall the fluid velocity will be equal to one half the wall velocity after a time $t_L = L^2/\nu$. Therefore, once again, it appears that the square of the distance travelled by momentum is proportional to time and inversely proportional to kinematic viscosity.

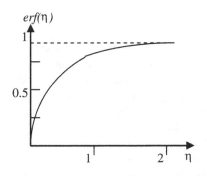

z	erf(z)	z	erf(z)
0.00	0.00	0.50	0.520
0.04	0.045	0.60	0.604
0.08	0.090	0.70	0.678
0.12	0.135	0.80	0.742
0.16	0.179	0.90	0.797
0.20	0.223	1.00	0.843
0.24	0.266	1.20	0.910
0.28	0.308	1.40	0.952
0.32	0.349	1.60	0.976
0.36	0.389	1.80	0.989
0.40	0.428	2.00	0.995

Fig. 7.6 The error function

Note that, expanding the exact solution (7.4.8) in a power series, we have:

$$v = U\left[1 + K_1\eta + K_2\eta^2 + K_3\eta^3 + \cdots\right], \quad \text{with } \eta = \frac{y}{2\sqrt{2vt}}, \tag{7.4.8a}$$

where $K_1 = -1.596$, $K_2 = 0$ and $K_3 = 0.532$.

Finally, let us evaluate the shear stress at the wall, that is the force per unit area exerted by the fluid on the wall along the axial x-direction, which coincides with the momentum flux in the transversal y-direction,

$$\tau_w(t) = -\mu\left(\frac{\partial v}{\partial y}\right)_{y=0} = -\mu V\left(\frac{d[erfc(\eta)]}{d\eta}\right)_{\eta=0}\frac{\partial \eta}{\partial y} = \frac{\mu V}{\sqrt{4vt}}\frac{2}{\sqrt{\pi}}\left(e^{-\eta^2}\right)_{\eta=0} = V\sqrt{\frac{\mu\rho}{\pi t}}. \tag{7.4.10}$$

We can find, at least qualitatively, this same result considering that the fluid velocity decays by 85 % over a distance δ, so that $\tau_w \approx \mu V/\delta$, with $\delta \approx \sqrt{(4vt)}$.

From Eq. (7.4.10) we see that the momentum flux decreases as $t^{-1/2}$, so that for $t = 0$ we find an infinite stress. However, this condition corresponds to having a discontinuous velocity profile at the wall, when the self-similar solution is not applicable. In any case, since this discontinuity decays very rapidly in time, the mean stress over a finite time interval t_m remains finite,

$$\bar{\tau}_w = \frac{1}{t_m}\int_0^{t_m}\tau_w dt = \frac{V}{t_m}\sqrt{\frac{\mu\rho}{\pi}}\int_0^{t_m}\frac{dt}{\sqrt{t}} = 2V\sqrt{\frac{\mu\rho}{\pi t_m}} = 2\tau_w(t_m). \tag{7.4.11}$$

7.5 Lubrication Approximation

In the previous Sections, we have considered cases where the fluid velocity is uni-directional, so that the inertial term within the Navier-Stokes equation is identically null and an exact solution can be obtained. In this Section, we consider instead cases where a second velocity component is present, although very small. The prototype of these problems is the slider bearing, represented in Fig. 7.7, that is a bearing designed to provide free motion in one direction.

The main feature here is that the flow results entirely from the relative motion of the surfaces (above all the longitudinal velocity U, as in most cases $V = 0$), with the gap height monotonically decreasing in the flow direction, while the pressure drop between the two ends is imposed (often, it is zero).

In general, this type of problems presents three unknowns, namely two velocity components, v_x and v_y, and the modified pressure, P, which can be determined through the two components of the Navier-Stokes equation, coupled with the continuity equation, i.e.,

$$\rho\left(v_x\frac{\partial v_x}{\partial x}+v_y\frac{\partial v_x}{\partial y}\right)=-\frac{\partial P}{\partial x}+\mu\left(\frac{\partial^2 v_x}{\partial x^2}+\frac{\partial^2 v_x}{\partial y^2}\right),\qquad(7.5.1)$$

$$\rho\left(v_x\frac{\partial v_y}{\partial x}+v_y\frac{\partial v_y}{\partial y}\right)=-\frac{\partial P}{\partial y}+\mu\left(\frac{\partial^2 v_y}{\partial x^2}+\frac{\partial^2 v_y}{\partial y^2}\right),\qquad(7.5.2)$$

$$\frac{\partial v_x}{\partial x}+\frac{\partial v_y}{\partial y}=0.\qquad(7.5.3)$$

This problem has to be solved with appropriate boundary conditions; in this case we assume that the lower wall moves horizontally with velocity U, the upper wall is fixed, i.e., $V = 0$, and the pressures at the two ends are the same, i.e.,

$$v_x = U;\ v_y = 0\ \text{at}\ y = 0;\ v_x = 0;\ v_y = 0\ \text{at}\ y = H(x);\quad P = P_0\ \text{at}\ x = 0, L.$$
$$(7.5.4)$$

Fig. 7.7 Slider bearing geometry

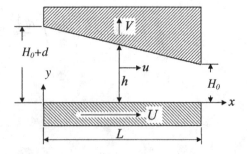

In addition, we assume to know how the height varies with x; For example, in this case:

$$H(x) = H_0\left[1 + \kappa\left(1 - \frac{x}{L}\right)\right], \quad \text{with} \quad \kappa = \frac{d}{H_0}. \tag{7.5.5}$$

Considering that L and d are the length scales for velocity variations in the x and y directions, the lubrication approximation is based on the following geometrical condition on the aspect ratio, ε:

$$\varepsilon = d/L \ll 1. \tag{7.5.6}$$

Therefore, $\partial^2 v_x/\partial y^2 \ll \partial^2 v_x/\partial x^2$. In addition, as U and V are the respective scales for v_x and v_y, it follows from continuity that $V/U \approx d/L \ll 1$, so that: $v_x \partial v_x/\partial x \approx v_y \partial v_x/\partial y \approx U^2/L$. As for the pressure contribution, comparing term by term Eq. (7.5.1) with Eq. (7.5.2), we conclude that $\partial P/\partial y \ll \partial P/\partial x$, so that P is approximately a function of x only. Finally, we see that the inertial term can be neglected when $v_x \partial v_x/\partial x \ll v \partial^2 v_x/\partial y^2$, i.e.,

$$\left(\frac{d}{L}\right)\left(\frac{Ud}{v}\right) = \varepsilon \, Re \ll 1. \tag{7.5.7}$$

This condition shows that the inertial term can be neglected even when the Reynolds number is not small, provided that the aspect ratio is small enough.

Equations (7.5.6) and (7.5.7) are the geometric and dynamical requirements for the lubrication approximation. Then, at leading order, i.e., neglecting $O(\varepsilon)$-terms, the Navier-Stokes equation can be simplified, yielding,

$$\frac{dP}{dx} = \mu \frac{\partial^2 v}{\partial y^2}. \tag{7.5.8}$$

with $v = v_x$. This is basically the same equation as (7.1.5), the only differences being that dP/dx is not a constant, but depends on x, and $v = v(x,y)$, instead of $v = v(y)$. In any way, the integration over y of Eq. (7.5.8) remains straightforward, yielding Eq. (7.1.7), i.e.,

$$v(x, y) = U\left(1 - \frac{y}{H}\right) - \frac{H^2}{2\mu}\frac{dP}{dx}\frac{y}{H}\left(1 - \frac{y}{H}\right). \tag{7.5.9}$$

Now, imposing that the volumetric flow rate per unit width (in m²/s) is constant and equal to q, we have:

$$q = \int_0^H v(x,y)dy = \frac{UH}{2} - \frac{H^3}{12\mu}\frac{dP}{dx} \quad \Rightarrow \quad \frac{dP}{dx} = \frac{6\mu}{H^2}\left(U - \frac{2q}{H}\right). \tag{7.5.10}$$

Integrating between $x = 0$ and $x = L$, imposing that $P(0) = P(L)$ and applying the expression (7.5.5) for $H(x)$, we obtain a relation between q and U, i.e.,

$$q = UH_0 \frac{1 + \kappa}{2 + \kappa}, \quad \text{with} \quad \kappa = \frac{d}{H_0}. \tag{7.5.11}$$

Substituting this result into Eq. (7.1.10) we see that the pressure reaches a maximum, P_{max}, at a point x_{max} such that $H(x_{max}) = 2q/U$, finding:

$$x_{max} = L \frac{1 + \kappa}{2 + \kappa}; \quad P_{max} = \frac{3 \mu UL}{2} \frac{\kappa}{H_0^2 \, (1 + \kappa)(2 + \kappa)}. \tag{7.5.12}$$

Since $\frac{1}{2} < x_{max} < 1$, we see that the pressure maximum, P_{max}, always occurs in the downstream half of the gap. For example, when $d = H_0$, i.e., $\kappa = 1$, we obtain: $x_{max} = 2L/3$, while the two limiting cases correspond to having either $\kappa \ll 1$, i.e., $d \ll H_0$, so that $x_{max} = L/2$, or $\kappa \gg 1$, i.e., $d \gg H_0$, so that $x_{max} = L$.

From Eqs. (7.5.10) and (7.5.12) we see that $P = O(\mu UKL/H_0^2)$; on the other hand, the shear stress is $\tau = \mu \partial v/\partial y = O(\mu U/H_0)$, so that $\tau/P = O(\varepsilon)$. This is particularly important when we evaluate the force applied on the upper surface. On one hand, there is a pressure force, directed upward at an angle $\theta = \text{arctg}(d/L) \approx \varepsilon$ with respect to the vertical, y-direction Therefore, at leading order, considering that $\sin\varepsilon \approx \varepsilon$, and $\cos\varepsilon \approx 1$, the pressure force per unit area in the y-direction is P, while that in the x-direction is εP. Then, there is the shear force at the wall, which is directed along the quasi-horizontal direction; this force, as we have seen, is $\tau_w = O(\varepsilon P)$, and thus is of the same magnitude as the pressure force in the x-direction. Therefore, at leading order, considering the velocity profile (7.5.9), the forces (per unit width) along the x and y-directions are as follows:

$$F_x = -\int_0^L (\varepsilon P + \tau_w)dx, \quad \text{with} \quad \tau_w = \mu \frac{\partial v}{\partial y}\bigg|_{y=H} = \frac{2\mu}{H}\left(U - 3\frac{q}{H}\right). \tag{7.5.13}$$

and

$$F_y = \int_0^L Pdx. \tag{7.5.14}$$

Finally, integrating, we obtain:

$$F_x = 6\mu U \frac{L}{d}\left[\frac{2}{3}\ln(1 + \kappa) - \frac{\kappa}{2 + \kappa}\right]. \tag{7.5.15}$$

and

$$F_y = 6\mu U \left(\frac{L}{d}\right)^2 \left[\ln(1+\kappa) - 2\frac{\kappa}{2+\kappa}\right]. \qquad (7.5.16)$$

As expected, we have:

$$\frac{F_x}{F_y} \approx \frac{d}{L} = \varepsilon \ll 1. \qquad (7.5.17)$$

For example, when $\kappa = 1$, we find $F_x/F_y = 0.2\, d/L$. This shows that a thin gap filled with fluid can provide a very favorable ratio between its load capacity and the drag.

All bearings are based on this simple property, allowing the journal (i.e., the part of the shaft in contact with the bearing) to slide over the bearing surface.

7.6 Drainage of a Liquid Film from a Vertical Plate

Consider a thin film of a viscous liquid that is spread uniformly on a horizontal plane. Then, suddenly, the plane is rotated vertically, so that the film begins to drain from the, now vertical, surface, under the influence of gravity (see Fig. 7.8). In this Section, we will study this problem, trying to estimate the time required for the liquid to drain. Note that here, unlike the case studied in Sect. 4.1, the liquid is not continuously resupplied from the top of the plate so that, at the end, the liquid will drain completely off the plate.

Assume that (a) the film thickness, $h(x, t)$, varies slowly along x; (b) the fluid is viscous enough that inertial forces can be neglected; (c) surface tension forces can also be neglected. Then, proceeding as in the previous Section, considering that the pressure drops are hydrostatic, the Navier-Stokes equation reduces to

$$\frac{\partial v}{\partial t} = \nu \frac{\partial^2 v}{\partial y^2} + g, \qquad (7.6.1)$$

Fig. 7.8 Drainage of a liquid film from a vertical plane

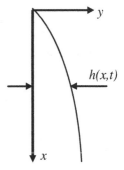

where v is the fluid velocity in the x-direction, and we have assumed that $\partial^2 v/\partial x^2 \ll \partial^2 v/\partial y^2$. The boundary conditions state that the fluid velocity is zero at the wall and the shear stress is zero at the free surface, $y = h(x, t)$. Note that, in general, it is this latter condition that determines an implicit dependence of v on x, so that $v = v$ (x, y, t), although Eq. (7.6.1) indicates that v depends explicitly on y and t.

Mathematically this problem is extremely complex, and it must be simplified if we want to solve it analytically. Then, let us assume to be in a quasi-stationary regime, so that the flow field at each position x, corresponding to a certain film thickness $h(x,t)$, is given by the steady-state solution of Sect. 7.1, obtaining:

$$v = \frac{g}{v}\left(hz - \frac{1}{2}z^2\right). \tag{7.6.2}$$

The volumetric flux per unit width, q, expressed in m^2/s, can be easily found as,

$$q(x,t) = \int_0^h v \, dy = \frac{gh^3(x,t)}{3v}. \tag{7.6.3}$$

At this point, let us consider a volume balance (coinciding with a mass balance, since density is constant), imposing that the temporal variation of the volume of the fluid enclosed between x and $x + dx$ is equal to the difference of the volumetric flux entering and the one exiting, i.e.,

$$\frac{\partial h}{\partial t}dx = q_x - q_{x+dx} \quad \Rightarrow \quad \frac{\partial h}{\partial t} = -\frac{\partial q}{\partial x} = -\frac{gh^2}{v}\frac{\partial h}{\partial x}. \tag{7.6.4}$$

This is a homogeneous, nonlinear partial differential equation, that can be solved assuming:

$$h(x,t) = Ax^\alpha t^\beta, \tag{7.6.5}$$

where A, α and β are constants to be determined. Substituting (7.6.5) into (7.6.4) we obtain:

$$\beta Ax^\alpha t^{\beta-1} = -\frac{g}{v}\left(Ax^\alpha t^\beta\right)^2 \alpha Ax^{\alpha-1}t^\beta, \tag{7.6.6}$$

and, upon equating the exponents of the x and t-terms,

$$\begin{array}{llll} x: & \alpha = 2\alpha + \alpha - 1 & \Rightarrow & \alpha = 1/2; \\ t: & \beta - 1 = 2\beta + \beta & \Rightarrow & \beta = -1/2. \end{array} \tag{7.6.7}$$

Finally, equating the coefficients (not the exponents) we find: $A = (v/g)^{1/2}$, and so,

$$h(x,t) = \left(\frac{vx}{gt}\right)^{1/2}. \tag{7.6.8}$$

Obviously, this solution is valid only when the time is not too short (and, in fact, the solution diverges when $t = 0$), so that the hypothesis of quasi steady state can be satisfied. Within this constraints, the solution (7.6.8) is in excellent agreement with experimental results.

7.7 Integral Methods

For complex problems, where analytical solutions are unavailable, the analysis proceeds using the *integral methods*. Here, the Navier-Stokes is averaged across an appropriate cross section and then, instead of solving for the velocity profile, we *assume* a certain form for it, as a function of certain parameter, i.e., a typical thickness δ, that is finally determined by manipulating the resulting equations.

Let us consider, for example, the fluid flow due to the rapid movement of a wall, that we have studied in Sect. 7.4. Integrating the two members of Eq. (7.4.1) along y, we obtain:

$$\int_0^\infty \frac{\partial v}{\partial t} dy = \frac{d}{dt} \int_0^\infty v dy = \frac{1}{A\rho} \frac{dQ}{dt}, \quad \text{with} \quad Q = A\rho \int v dy \tag{7.7.1}$$

$$v \int_0^\infty \frac{\partial^2 v}{\partial y^2} dy = -v \frac{\partial v}{\partial y}(0) = \frac{1}{\rho} \tau_w = \frac{1}{A\rho} F_w, \tag{7.7.2}$$

where Q is the fluid momentum, A the area of the wall, ρ the fluid density, while τ_w and F_w are, respectively, the shear stress and the friction force at the wall. Thus, as expected, we find Newton's equation of motion, stating that any difference in the x-directed momentum flow is equal to the net x-directed force acting on the solid surface,

$$F_w = \frac{dQ}{dt}. \tag{7.7.3}$$

Now, we take the next step in the integral method, *assuming* a specific form for the velocity profile after applying some physical reasoning. We know that the fluid is gradually set into motion, that is the fluid is quiescent outside of a region of thickness $\delta(t)$, growing monotonically with time. Therefore, assume that:

$$v(y,t) = \begin{cases} V[1 - f(\eta)] & \text{per} \quad 0 < y < \delta(t) \\ 0 & \text{per} \quad y > \delta(t) \end{cases} \qquad \eta = \frac{y}{\delta(t)}, \qquad (7.7.4)$$

where $f(\eta)$ is a dimensionless function to be determined, subjected to be bounded between 0 and 1. So, basically, we trade our lack of knowledge of the velocity field for another unknown function, δ, that is obviously much easier to determine.

Substituting Eq. (7.7.4) into (7.7.1) and (7.7.2), we obtain:

$$\frac{Q}{A\rho} = \int_0^\delta v\,dy = \alpha_1 V \delta \quad \text{and} \quad \frac{\tau_w}{\rho} = -v\frac{\partial v}{\partial y}(0) = \beta\frac{\nu V}{\delta}, \qquad (7.7.5)$$

where,

$$\alpha_1 = \int_0^1 [1 - f(\eta)]d\eta \quad \text{and} \quad \beta = f'(0) \qquad (7.7.6)$$

are two dimensionless constants. Therefore, Eq. (7.7.3) becomes,

$$\alpha_1 \delta \frac{d\delta}{dt} = \beta\nu. \qquad (7.7.7)$$

Hence:

$$\delta = \sqrt{2(\beta/\alpha_1)\nu t}, \qquad (7.7.8)$$

showing that the thickness of the moving fluid layer grows proportionally to the square root of time.

Now, let us specify the type of function f that we want to have. Often, $f(\eta)$ is chosen to be a polynomial of order n, that is

$$f(\eta) = a_0 + a_1\eta + a_2\eta^2 + \cdots + a_n\eta^n, \qquad (7.7.9)$$

where the $(n + 1)$ constants are chosen so as to satisfy the following conditions:

- $f(0) = 0$; $f(1) = 1$, to satisfy the boundary conditions.
- $f'(1) = f''(1) = \cdots = f^{(n-1)}(1) = 0$, to have a smooth transition between the two regions of flow field.

In this case, though, since pressure is constant, the velocity at the wall should be linear, so that its second derivative should be zero. Accordingly, we choose the following odd function,

$$f(\eta) = \frac{3}{2}\eta - \frac{1}{2}\eta^3,$$ (7.7.10)

that satisfies the conditions $f(0) = 0; f''(0) = 0; f(1) = 1$ and $f'(1) = 0$. Therefore, we find: $\alpha_1 = 3/8$ and $\beta = 3/2$, so that at the end we obtain: $\delta(t) = 2\sqrt{(2\nu t)}$. Thus, we conclude that the velocity profile is the following:

$$v = U\left[1 - \frac{3}{2}\eta + \frac{1}{2}\eta^3\right], \quad \text{with} \quad \eta = \frac{y}{2\sqrt{2\nu t}},$$ (7.7.11)

in excellent agreement with the exact result (7.4.8) and (7.4.8a). In addition, from Eqs. (7.7.5)–(7.7.8) we obtain:

$$\tau_w = -\mu\frac{\partial v}{\partial y}(0) = kV\sqrt{\frac{\mu\rho}{t}}, \quad \text{where} \quad k = \sqrt{\beta\alpha_1/2} = 3/\left(4\sqrt{2}\right) \cong 0.56,$$

that is within 5 % from the exact result (7.4.10), where $k = \sqrt{1/\pi} \cong 0.53$.

It is instructive to see what we obtain if we choose a "less smart" velocity profile, such as:

$$f(\eta) = 2\eta - \eta^2,$$ (7.7.12)

satisfying the conditions $f(0) = 0$, $f(1) = 1$ and $f'(1) = 0$ (naturally, here $f''(0) \neq 0$). Therefore, we find: $\alpha_1 = 1/3$ and $\beta = 2$, so that at the end we obtain: $\delta(t) = 2\sqrt{(3\nu t)}$. Although the velocity profile is clearly rather different than its exact value, we see that

$$\tau_w = kV\sqrt{\frac{\mu\rho}{t}}, \quad \text{where} \quad k = \sqrt{\beta\alpha_1/2} = 1/\sqrt{3} \cong 0.58,$$

that is not too far from its exact value, 0.53.

Clearly, the integral method is quite impressive, as it provides an approximate solution with very little mathematical effort. However, unlike the perturbation methods analyzed in Sects. 10.2 and 15.3, it is hard to predict the accuracy of the results, as they strongly depend on the form of the chosen trial function. This is why, nowadays, with the advent of numerical computation, the practical usefulness of the integral method has greatly declined, although it can still provide a powerful way to understand the underlying physical mechanism of the process.

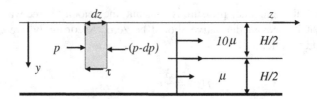

Fig. 7.9 Flow of two layered immiscible liquids

7.8 Problems

7.1 Assuming that the upper disk of Fig. 7.4 has mass M, find the distance $2H$ between the two disks for a given volumetric flow rate.

7.2 Consider the two incompressible and immiscible fluids represented in Fig. 7.9, that are separated by a plane interface. The heavier fluid has viscosity μ and occupies the lower half of the canal, with $H/2 < y < H$, with $y = H$ denoting the position of the rigid wall, while the lighter fluid has viscosity $10\ \mu$ and occupies the higher half, with $0 < y < H/2$, with $y = 0$ denoting the position of the free surface. The fluids are flowing, due to a given constant pressure drop per unit length, $\Delta p/L$, that is imposed by a pump or by gravity (imagine, for example, that the canal lays on an inclined plane). Assuming that at the free surface the air resistance is negligible,

- Determine the shear stress $\tau = \tau(y)$, and in particular its value at the wall, $\tau_w = \tau(H)$.
- Find the fluid velocity at the interface, $v(H/2)$, and the velocity at the free surface, $v(0)$.
- Sketch qualitatively the velocity field.

7.3 A Newtonian fluid with viscosity μ flows in a canal of height H. Using the same Cartesian axes as in Fig. 7.9, assume that the air resistance at the free surface, $y = 0$, can be neglected, while at the wall, $y = H$, the fluid moves with a, so called, *slip velocity.*, equal to $v_w = a|dv/dz|_w$, where a is a characteristic length, e.g., the wall roughness. Find the velocity profile and, in particular, its maximum value.

Chapter 8
Laminar Boundary Layer

Abstract In Sect. 3.3 we have seen that when a viscous fluid flows past a sub-merged object at high Reynolds number, $Re \gg 1$ then its convective momentum flux greatly exceeds its diffuse counterpart, i.e., inertial forces are much larger than viscous forces. This is true, however, provided that the fluid points that we are considering are not too close to the outer surface of the object, where the fluid velocity is null, due to the no-slip boundary condition. Accordingly, near the surface of the object, we define a small region of thickness δ, denoted *boundary layer*, such that, when the distance from the wall, y, is larger than δ, i.e. when $y > \delta$ inertial forces prevail while when $y < \delta$ viscous forces are dominant. Then, as we saw in Sect. 3.3, the boundary layer thickness can be determined by imposing that at the edge of this region, when $y \approx \delta$, inertial forces balance viscous forces. At the end, we found that the boundary layer thickness decreases, proportionally to the size, L, of the object, as the inverse of the square root of the Reynolds number, i.e., $\delta/L \simeq Re^{-1/2}$. This relation reveals why boundary layers are so important, although they occupy only a small region of space: the drag of the object depends on the shear stress at the wall, τ_w, which is determined by the velocity profile at the wall, and so it is inversely proportional to the boundary layer thickness; therefore, as Re increase, δ decreases and τ_w increases. In this chapter, after analyzing the scaling of the problem in Sect. 8.1, in Sect. 8.2 we will study the classical Blasius self-similar solution of the flow past a flat plate, and then in Sect. 8.3 consider more general cases, leading to flow separation. Finally, in Sect. 8.4, we analyze the approximate von Karman–Pohlhausen integral method.

8.1 Scaling of the Problem

Consider the simple case of a two-dimensional flow field, with velocity $\mathbf{v} = (v_x, v_y)$ and pressure P, near a flat plate, located at $y = 0$ (see Fig. 8.1). The upstream unperturbed velocity is uniform, equal to $(V, 0)$, with uniform pressure P_0, while at the wall the velocity is zero. Finally, assume that the Reynolds number, $Re = VL/\nu$ is

R. Mauri, *Transport Phenomena in Multiphase Flows*,
Fluid Mechanics and Its Applications 112, DOI 10.1007/978-3-319-15793-1_8

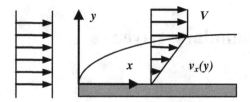

Fig. 8.1 Laminar boundary layer on a flat plate

large, with L denoting the plate size. The three unknowns, v_x, v_y and P, can be determined solving a system of differential equation, consisting of the Navier-Stokes (2 scalar equations) and the continuity equations,

$$\rho\left(v_x\frac{\partial v_x}{\partial x}+v_y\frac{\partial v_x}{\partial y}\right)=-\frac{\partial P}{\partial x}+\mu\left(\frac{\partial^2 v_x}{\partial x^2}+\frac{\partial^2 v_x}{\partial y^2}\right),\qquad(8.1.1)$$

$$\rho\left(v_x\frac{\partial v_y}{\partial x}+v_y\frac{\partial v_y}{\partial y}\right)=-\frac{\partial P}{\partial y}+\mu\left(\frac{\partial^2 v_y}{\partial x^2}+\frac{\partial^2 v_y}{\partial y^2}\right),\qquad(8.1.2)$$

$$\frac{\partial v_x}{\partial x}+\frac{\partial v_y}{\partial y}=0,\qquad(8.1.3)$$

with boundary conditions,

$$v_x=v_y=0 \text{ per } y=0; \quad v_x=V;$$
$$v_y=0 \text{ and } P=P_0 \text{ per } y\to\infty \text{ and per } x=0.\qquad(8.1.4)$$

Now, based on the fact that $Re \gg 1$, so that inertial forces are much larger than viscous forces, we could say that the last two terms in Eqs. (8.1.1) and (8.1.2) could be neglected. In doing so, however, the problem reduces mathematically from a second to a first-order boundary value problem, and therefore it becomes impossible to satisfy both boundary conditions, at the wall and at infinity.[1] In particular, since the condition at infinity must be satisfied, that means that the no-slip condition at the wall cannot be applied. The physical counterpart of this fact is that, as we saw, viscous forces can be neglected almost everywhere, with the exception of a narrow region very near the wall. That region, however, is where the fluid exchanges momentum with the submerged object and therefore it is responsible for the drag force. So, it is logically consistent that, neglecting viscous forces, the fluid velocity slips at the wall, so that there will be no shear stress and no viscous drag.

Based on the previous reasoning, as we know that viscous forces are relevant only within a so-called boundary layer of thickness δ, with $\delta \ll L$, let us study the

---

[1]In fact, a differential equation of order n is solved integrating n times and so introducing n constants, which in turn can be determined by imposing n (boundary or initial) conditions.

scaling of the problem (the analysis presented in Sect. 2.3 is clearly hand-waving) at the edge of the boundary layer, when $y \approx \delta$, (while $x \approx L$).

(a) Considering that at the edge of the boundary layer the longitudinal velocity is almost equal to its unperturbed value, $v_x \approx V$, the continuity Eq. (8.1.3) reveals that $V/L \approx v_y/\delta$, that is $v_y \approx V\delta/L$, meaning that the transversal velocity is much smaller than the longitudinal velocity.

(b) The two inertial terms at the LHS of Eq. (8.1.1) have the same magnitude. In fact, $\rho v_y \partial v_x / \partial y \approx \rho v_x \partial v_x / \partial x \approx \rho V^2 / L$, and the same holds for the inertial terms of Eq. (8.1.2).

(c) As for the viscous terms at the RHS of Eq. (8.1.1), $\mu \partial^2 v_x / \partial x^2 \ll \mu \partial^2 v_x / \partial y^2 \approx \mu V / \delta^2$, so that the first term can be neglected. The same happens for the viscous terms in Eq. (8.1.2).

Now, imposing that at the edge of the boundary layer the inertial term and the viscous term balance each other, we obtain:

$$\rho \frac{V^2}{L} \approx \mu \frac{V}{\delta^2} \quad \Rightarrow \quad \delta \approx \sqrt{vL/V} \quad \Leftrightarrow \quad \frac{\delta}{L} \approx \frac{1}{Re^{1/2}}, \qquad (8.1.5)$$

with $Re = VL/v$, where $v = \mu/\rho$ is the kinematic viscosity.

The other important information that we can infer from Eq. (8.1.1) concerns the role of pressure. In fact, considering that the pressure term of the Navier-Stokes equation must balance the inertial term, comparing Eqs. (8.1.1) and (8.1.2) term by term we see that, as $v_x \gg v_y$,

$$\frac{\partial P}{\partial y} \approx \rho \frac{Vv}{L} \ll \frac{\partial P}{\partial x} \approx \rho \frac{V^2}{L}. \qquad (8.1.6)$$

So, the pressure changes along the transversal direction can be neglected, i.e., $P = P(x)$, meaning that the pressure at a point (x, y) inside the boundary layer is equal to the pressure at a point (x, Y), with $Y \gg \delta$, outside the boundary layer. In the case of a flow past a flat plate, the unperturbed pressure is uniform, and therefore the velocity field within the boundary layer is linear, i.e., it consists of a Couette flow. In the more general case of a fluid flowing past a submerged object, since the wall curvature is much larger than the boundary layer thickness, the above scaling is still valid, when x represents a curvilinear longitudinal coordinate and y is the distance from the wall (see Fig. 8.2). In this case, though, outside the boundary layer the unperturbed velocity, $\mathbf{V} = (V_x, V_y)$, and pressure, P, are not constant, but solve the Euler equation, i.e., the Navier-Stokes equation where the viscous term is identically null,

$$\rho \mathbf{V} \cdot \nabla \mathbf{V} = -\nabla P, \qquad (8.1.7)$$

with boundary conditions, $\mathbf{V} = \mathbf{V}_\infty$ and $P = P_\infty$ at infinity.

Fig. 8.2 Boundary layer on a
curved surface

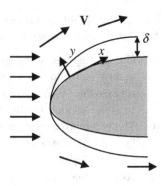

Let us go back to the case of a flow past a flat plate, with $\mathbf{V} = (U, 0)$, where, as we saw, there is no pressure term, as $P = P_\infty = $ constant, and the velocity profile within the boundary layer is linear. In this case, however, the problem has no characteristic dimension, apart from the distance from the edge of the plate, x. Therefore, Eq. (8.1.5) becomes:

$$\delta \approx \sqrt{vx/V} \Leftrightarrow \frac{\delta}{x} \approx \frac{1}{Re_x^{1/2}}, \quad \text{with} \quad Re_x = \frac{Vx}{v}, \tag{8.1.8}$$

where the Reynolds number has been defined in terms of x. Therefore, we see that the boundary layer thickness increases like the square root of the distance from the edge of the plate. Now, we can estimate the shear stress,

$$\tau_{xy} = -\mu \left(\frac{\partial v_x}{\partial y} \right)_{y=0} \approx \mu \frac{V}{\delta} = \rho V^2 \sqrt{\frac{v}{Vx}} = \rho V^2 Re_x^{-1/2}, \tag{8.1.9}$$

where we have considered that, as the velocity profile is linear, the velocity gradient is equal to the ratio between the velocity difference, V, and the thickness of the boundary layer, δ. Finally, we obtain the following expression for the friction factor,

$$f = \frac{\tau_{xy}}{1/2\rho U^2} \approx Re_x^{-1/2}. \tag{8.1.10}$$

An alternative way to obtain the scaling described in this section is to recognize the need to rescale the transversal variable, y, by stretching it using a small parameter, $\varepsilon = Re^{-1} \ll 1$, thus defining a new variable $\zeta = y\, \varepsilon^{-a}$. Then, maintaining a balance between inertial and viscous terms, requires that $a = \frac{1}{2}$, and we arrive again the scaling (8.1.8). This is the so-called *singular perturbation* method, that is briefly described in Sect. 10.2.2.

8.2 Blasius Self-similar Solution

In 1908 H. Blasius solved exactly the boundary layer problem on a flat plate[2] that we have analyzed in the previous section. First of all he realized that, within $O(Re^{-1/2})$-terms, pressure is uniform, the two inertial forces have the same magnitude, and the viscous contribution consists only of the y-derivative term. Then the problem has only 2 unknowns, v_x and v_y, satisfying the following equations,

$$v_x \frac{\partial v_x}{\partial x} + v_y \frac{\partial v_x}{\partial y} = v \frac{\partial^2 v_x}{\partial y^2}, \tag{8.2.1}$$

$$\frac{\partial v_x}{\partial x} + \frac{\partial v_y}{\partial y} = 0, \tag{8.2.2}$$

Subjected to the 4 boundary conditions

$$v_x = v_y = 0 \text{ when } y = 0; \quad v_x = V \text{ when } y \to \infty \text{ and when } x = 0. \tag{8.2.3}$$

Since the problem does not have any intrinsic characteristic length, we expect to find a self-similar solution. In particular, using the dimensional analysis of last section, we expect that the solution will depend on the self-similar variable,

$$\eta = \frac{y}{\delta(x)} = y \left(\frac{V}{xv} \right)^{1/2} \tag{8.2.4}$$

Now, all two-dimensional problems can be solved in terms of a *stream function* ψ, so that,

$$v_x = \frac{\partial \psi}{\partial y}; \quad v_y = -\frac{\partial \psi}{\partial x}. \tag{8.2.5a, b}$$

In this way, the continuity equation becomes: $\partial^2 \psi / \partial x \partial y - \partial^2 \psi / \partial x \partial y = 0$, and therefore it is identically satisfied.

At this point, let us try to guess how ψ should look like in order to find a self-similar solution of the type $v_x = VG(\eta)$. Obviously, from Eq. (8.2.5a, b), we see that if we choose a steam function of the type,

$$\psi = g(\eta) \sqrt{vVx}, \tag{8.2.6}$$

---

[2]It was the Ph.D. thesis of Paul Richard Heinrich Blasius (1883–1970), a German fluid dynamics physicist, who was one of the first student of L. Prandtl. For more than 50 years H. Blasius taught at the University of Hamburg.

we obtain: $G(\eta) = dg/d\eta \equiv g'(\eta)$, that is,

$$v_x(\eta) = Vg'(\eta). \tag{8.2.7}$$

Let us verify that this choice leads indeed to a self-similar solution. First of all, considering that,

$$\frac{\partial}{\partial x} = -\left(\frac{\eta}{2x}\right)\frac{d}{d\eta}; \quad \frac{\partial}{\partial y} = \left(\frac{V}{vx}\right)^{1/2}\frac{d}{d\eta}, \tag{8.2.8}$$

we find $v_y = \frac{1}{2}\left(\frac{vV}{x}\right)^{1/2}(\eta g' - g)$, and therefore Eq. (8.2.1) reduces to the following equation,

$$2g''' + gg'' = 0, \tag{8.2.9}$$

subjected to the three boundary conditions,

$$g(0) = g'(0) = 0; \quad g'(\infty) = 1. \tag{8.2.10}$$

Note that the two boundary conditions, $v_x(x = 0) = v_x(x \rightarrow \infty) = V$, reduce to the condition $g'(\infty) = 1$. Thus, this problem is well posed, as it consists of solving a third order ordinary differential equation with three boundary conditions.

As we have seen, the "complicated" part of this problem is not about solving a differential equation, but instead it is Blasius' intuition about the most convenient form of the stream function that allows to obtain a well-posed problem in η.

Solving Eqs. (8.2.9) and (8.2.10) numerically, we obtain the curve $g'(\eta) = v_x/V$ represented in Fig. 8.3. Therefore, considering that $v_x/V = 0.99$ when $\eta = 5$, it is generally agreed that the thickness of the boundary layer δ_V corresponds to

$$\delta_V = 5\sqrt{\frac{vx}{V}}. \tag{8.2.11}$$

Fig. 8.3 The velocity field within the boundary layer

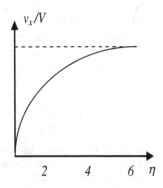

In addition, expanding the exact solution when $\eta \ll 1$, we obtain:

$$v_x(\eta) = V[0.332\eta - 0.0023\eta^4 + O(\eta^7)];$$
$$v_y(\eta) = \sqrt{\frac{\nu V}{x}}[0.083\eta^2 + O(\eta^5)], \tag{8.2.12}$$

showing that the velocity field in the vicinity of the wall can be considered to be linear, with excellent approximation, i.e., $v_x \propto y$. This result confirms the initial hypothesis of having a uniform pressure field, so that the velocity profile is a Couette flow.

At this point, we can calculate the shear stress at the wall,

$$\tau_{xy}(x) = -\mu\left(\frac{\partial v_x}{\partial y}\right)_{y=0} = \rho V^2 \sqrt{\frac{\nu}{Vx}}g''(0) = 0.332\rho V^2 Re_x^{-1/2}, \tag{8.2.13}$$

where we have considered that the numerical solution yields: $g''(0) = 0.332$. Obviously, Eq. (8.2.13) shows that the shear stress varies with x, together with the boundary layer thickness. Finally, from the definition of friction factor we obtain:

$$f(x) = \frac{\tau_{xy}(x)}{1/2\rho V^2} = 0.664 Re_x^{-1/2}. \tag{8.2.14}$$

At first, Eqs. (8.2.13) and (8.2.14) look wrong, since at $x = 0$ the shear stress and the friction factor diverge. This is due to the fact that at the edge of the plate, when $x = 0$, the boundary layer has zero thickness, so that the velocity gradient diverges. Accordingly, at $x = 0$, the starting assumption, $\partial v_x/\partial x \ll \partial v_y/\partial y$, is not valid anymore and the obtained solution is not applicable. However, from a practical viewpoint, we are interested in the shear force, F, that a fluid exerts on a plate as it flows past it. Denoting by L the length of the plate a by W its width, the shear force is:

$$F = W\int_0^L \tau_{xy}(x)dx = WL\overline{\tau_{xy}}, \tag{8.2.15}$$

where

$$\overline{\tau_{xy}} = \frac{1}{L}\int_0^L \tau_{xy}(x)dx = 0.332\rho V^2\sqrt{\frac{\nu}{V}}\left[\frac{1}{L}\int_0^L \frac{dx}{\sqrt{x}}\right] = 0.664\rho V^2 Re_L^{-1/2} = 2\tau_{xy}(L),$$

is the mean shear stress, and $Re_L = VL/\nu$ is the Reynolds number evaluated in terms of L. Therefore, we can define a mean friction factor, as,

Fig. 8.4 The streamlines of a
fluid flowing past a cylinder

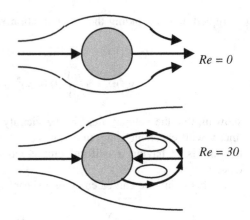

$$\bar{f} = \frac{\overline{\tau_{xy}}}{1/2\rho V^2} = \frac{1.328}{Re_L^{1/2}} = 2f(L), \qquad (8.2.16)$$

showing that the average friction does not diverge.

8.3 Flow Separation

As we saw, the inviscid approximation cannot account for the fluid drag on a solid
object. In fact, as in that case the fluid motion is necessarily time reversible, the
streamlines around a sphere must have a fore-and-aft symmetry. The same symmetry,
although with completely different fluid trajectories, is observed at vanishingly
small Reynolds number, as in the streamlines of a fluid flowing past a cylinder
represented in Fig. 8.4. In general, though, for not too small Reynolds numbers, the
actual picture tends to be more complicated, as the streamlines passing near the
forward face of the body break away at the side of the body and enclose fluid in
slow and sometimes unsteady motion. This is what we denote as *boundary-layer
separation*, where there are *separation lines* (or *separation points*, in 2D) where the
surface streamlines divide, with one branch departing from the surface, as shown in
Fig. 8.4. Physically, downstream from the separation points, there is a region where
the flow near the surface is reversed; then, the separated streamlines reattach to the
surface at some point, thereby enclosing the eddy. In the wake of a blunt body, that
produces a couple of counter-rotating, so-called *von Karman vortices*.[3] For a circular cylinder, as *Re* increases from 6 to 44, the separation points move forward and

---

[3]Named after Theodore von Kármán (1881–1963), a Hungarian-American mathematician, aerospace engineer and physicist. He was a student of L. Prandtl and taught first at the Aeronautical
Institute at Aachen until 1936, when he moved to Caltech, where in 1944 he founded the JPL
laboratories, and where he continued to teach until his death.

Fig. 8.5 The wake past a cylinder due to the vortex shedding of a very viscous oil

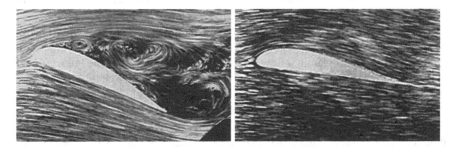

Fig. 8.6 Boundary layer separation from a wing at a high angle of attack

consequently the eddies, that are initially symmetric, grow in size. Then, as $Re > 44$, the eddies begin to shed periodically from either side of the body, forming a *von Karman vortex sheet*, that is a repeating pattern of swirling vortices caused by this unsteady flow separation. From experiments and numerical simulations it appears that the vortex shedding frequency f is such that the Strouhal number,[4] $St = fD/V$ is about 0.2 for cylinders, and about 0.3 for spheres, with D and V denoting, respectively, a typical dimension of the object (i.e., its diameter) and the unperturbed velocity. In Fig. 8.5 we see the wake past a cylinder due to the vortex shedding of a very viscous oil, with $Re = 73$.

Breakaway of the surface streamlines may occur even at $Re \approx 10$, as we have seen; however, the phenomenon becomes very important at large Re, in the presence of a boundary layer, because it can carry vorticity of large magnitude away from the walls. In fact, separation of a steady boundary layer on a smooth surface is observed to occur whenever the unperturbed velocity (i.e., the inviscid flow field outside the boundary layer) decreases in the flow direction sufficiently rapidly to induce an adverse pressure gradient at the wall, tending to decelerate the adjacent fluid. So, flow separation occurs when the kinetic energy of the fluid is no longer sufficient to overcome these forces. The importance of the boundary layer separation is quite evident in the pictures of Fig. 8.6, performed by L. Prandtl himself, showing the flow from left to right past an aerofoil. On the right side we see that, when the aerofoil is roughly aligned with the flow, there is no flow separation; therefore, as the unperturbed fluid velocity at the upper surface is larger than that on

---

[4]Named after Vincent Strouhal (1850–1922), a Czech physicist who first investigated the steady humming or singing of telegraph wires.

the lower surface, the upward pressure is larger than the downward pressure, resulting in a lift force that keeps the airplane up. On the other hand, the right-side picture shows that, when the aerofoil is not well aligned, i.e. when the *angle of attack* s larger than a critical value, the boundary layer separates from the upper wall, so that the fluid velocity suddenly decreases, with consequent increase of the pressure. The aerofoil is said to be *stalled*, a term that is associated with a severe lift reduction, often with tragic consequences. Similar considerations can also be applied to the motion of vehicles, where a "negative lift" improves their road holding.

Boundary layer separation is well illustrated in Fig. 8.7, where the velocity profiles in several cross sections of the boundary layer at the surface of an object are represented. Assuming laminar flow and stationary conditions, i.e. the velocity field is constant and does not fluctuate in time, we know that in the front of the body (i.e. the left part of the figure) the unperturbed velocity, $|\mathbf{V}|$, increases along the longitudinal coordinate, x. Therefore, $V_x(\partial V_v/\partial x) > 0$, and applying Eq. (8.1.7) we see that $\partial P/\partial x < 0$, corresponding to a forward pressure gradient that pushes the fluid along the positive x axis. Now, considering that, from Eq. (8.1.6) the pressure inside the boundary layer is equal to the pressure outside, the same inequality is valid also inside the boundary layer. That means that within the boundary layer the flow field is the sum of two terms: (a) a Couette-type linear velocity profile, due to the fact that the velocity is zero at the wall, while it is equal to V_x at the edge of the boundary layer; (b) a Poiseuille-type velocity profile, due to the forward pressure gradient. When the two flow fields are added to each other, the resulting velocity profile, as shown in the right part of Fig. 8.7, presents at the wall a higher velocity gradient as compared to the case of a flat plate, resulting in a larger friction. On the other hand, as we move downward to the back of the object, we find a decreasing unperturbed velocity and a concomitant adverse pressure gradient, with $\partial P/\partial x > 0$, so that the resulting flow field is now the difference between the Couette and the Poiseuille velocity profiles. At the end, the separation point is reached, where the velocity gradient at the wall, and consequently the wall shear stress as well, is null. At the separation point (actually, in 3D it is a line) there is no more adhesion of the fluid to the surface of the object, so that the boundary layer becomes detached or, equivalently, suddenly thickens. If we move further downstream, due to the increasing adverse pressure gradient, the fluid flow near the surface inverts its direction, forming eddies and vortices.

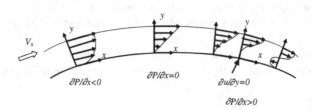

Fig. 8.7 Boundary layer separation

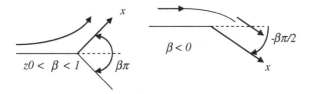

Fig. 8.8 Flow past a wedge

The shape of an object affects the boundary layer mainly by determining the pressure gradient. For example, the flow past a wedge shown in Fig. 8.8 yields a different pressure profile for each value of β. In this case, a similarity solution can be found,[5] showing that for concave wedges, with $\beta < 0$, the velocity profile near the wall has an inflection point and, when $\beta = -0.199$, we observe the detachment of the boundary layer (see the approximate solution in Sect. 8.4).

We should stress that the boundary layer theory can predict the velocity profile and the location of the separation point with great accuracy. However, the theory cannot be applied further downstream, since it assumes that the thickness of the boundary layer is much smaller than any macroscopic dimension, and that outside of that region the velocity corresponds to the inviscid fluid flow. On the other hand, downstream from the separation point, the region occupied by vortices is not narrow (see Fig. 8.6) and, above all, the velocity field is unstable even at relatively low Re. For that reason, flow separation at large Re is one of the most difficult phenomenon to simulate.

8.4   Von Karman–Pohlhausen Method

In Sect. 7.7 we have seen that sometimes it is convenient to use approximate, integral methods, when we lack all the necessary information to find the exact result. In the case of the boundary layer theory, integrate the x-component of the momentum equation over the y-direction, Eq. (8.1.1), neglecting the penultimate term, and obtain:

$$v \int_0^\infty \frac{\partial^2 v_x}{\partial y^2}\, dy = -v\frac{\partial v_x}{\partial y}(0) = -\frac{1}{\rho}\tau_w, \qquad (8.4.1)$$

$$\frac{1}{\rho}\int_0^\infty \frac{\partial P}{\partial x}\, dy = \frac{1}{\rho}\int_0^\infty \frac{dP}{dx}\, dy = -\int_0^\infty V_x\frac{dV_x}{dx}\, dy, \qquad (8.4.2)$$

[5]It was found by Falkner and Skan in 1931, and then extended by Hartree in 1937.

$$\int_0^\infty v_y \frac{\partial v_x}{\partial y} dy = [v_x v_y]_0^\infty - \int_0^\infty v_x \frac{\partial v_y}{\partial y} dy = V_x v_\infty + \int_0^\infty v_x \frac{\partial v_x}{\partial x} dy, \qquad (8.4.3)$$

$$\int_0^\infty v_x \frac{\partial v_x}{\partial x} dy = \frac{1}{2} \frac{d}{dx} \int_0^\infty v_x^2 dy. \qquad (8.4.4)$$

In Eq. (8.4.2) we have considered that the pressure depends only on x and satisfies the Euler Eq. (8.1.7), $dP/dx = -\rho V_x(dV_x/dx)$, where V_x is the longitudinal unperturbed velocity, while in Eq. (8.4.3) we have substituted the continuity equation, denoting by v_∞ the transversal velocity far from the wall, i.e., $v_\infty = v_y(y \to \infty)$. This latter can be obtained from the continuity equation, imposing the conservation of mass, as follows,

$$V_x v_\infty = V_x \int_0^\infty \frac{\partial v_y}{\partial y} dy = -\int_0^\infty V_x \frac{\partial v_x}{\partial x} dy = \int_0^\infty \frac{dV_x}{dx} v_x dy - \frac{d}{dx} \int_0^\infty V_x v_x dy. \qquad (8.4.5)$$

Finally, we obtain the following integral momentum equation,

$$\frac{d}{dx} \int_0^\infty v_x(V_x - v_x) dy + \frac{dV_x}{dx} \int_0^\infty (V_x - v_x) dy = \frac{\tau_w}{\rho}. \qquad (8.4.6)$$

Generally, it is preferable to write this equation in terms of the displacement thickness, δ_1, and the momentum thickness, δ_2,[6]

$$\delta_1 = \int_0^\infty \left(1 - \frac{v_x}{V_x}\right) dy \quad \text{and} \quad \delta_2 = \int_0^\infty \frac{v_x}{V_x} \left(1 - \frac{v_x}{V_x}\right) dy. \qquad (8.4.7a, b)$$

Physically, δ_1 represents the distance by which the external potential (i.e., inviscid) flow is displaced outwards, due to the decrease in velocity in the boundary layer. In fact, with respect to Fig. 8.9, denoting by $Y \gg \delta$ a distance much larger than the boundary layer thickness, from a mass balance between the two cross sections indicated in the figure, we obtain:

$$\int_0^Y V_x dy = \int_0^{Y+\delta_1} v_x dy = \int_0^Y v_x dy + V_x \delta_1,$$

---

[6]Sometimes, they are indicated as δ^* and θ, respectively.

Fig. 8.9 Streamlines of a
flow past a flat plate

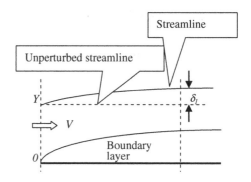

finding then Eq. (8.4.7a), since $v_x = V_x$ when $\delta < y < Y$. Proceeding in the same way, we see that δ_2 represents the loss of momentum in the boundary layer as compared with that of potential flow. In fact, from a momentum balance between the two cross sections indicated in Fig. 8.9, we obtain:

$$\delta_2 \rho V_x^2 = \rho \int_0^Y V_x^2 dy - \int_0^{Y+\delta_1} v_x^2 dy = \int_0^Y \left(V_x^2 - v_x^2\right) dy - V_x^2 \delta_1,$$

and therefore, using the definition (8.4.7a) of δ_1, we obtain Eq. (8.4.7b).

Substituting Eqs. (8.4.7a, b) into (8.4.6), we obtain the following, so-called, *von Karman–Pohlhausen* equation,

$$(\delta_1 + 2\delta_2)V_x \frac{dV_x}{dx} + V_x^2 \frac{d\delta_2}{dx} = \frac{\tau_w}{\rho}. \tag{8.4.8}$$

Now, as in Eq. (7.7.4), we assume that the velocity profile has a specific form, depending on a single parameter, δ, thus greatly simplifying the problem at the price of obtaining an approximate solution. We know that the fluid is unperturbed outside a boundary layer $\delta(x)$, that increases monotonically with x. Therefore, assume that:

$$\frac{v_x(x, y)}{V_x} = \begin{cases} f(\eta) & \text{at} \quad 0 < \eta < 1 \\ 0 & \text{per} \quad \eta > 1 \end{cases} \quad \text{with} \quad \eta = \frac{y}{\delta(x)}, \tag{8.4.9}$$

where $f(\eta)$ is any reasonable function. As we saw in Chap. 6, $f(\eta)$ is chosen to be a polynomial of order n, where the constants satisfy the following conditions:

- $f(0) = 0; f(1) = 1$, to satisfy the boundary conditions.
- $f'(1) = f''(1) = \cdots = f^{(n-1)}(1) = 0$, to have a smooth transition between the two regions of flow field.

After checking that $0 < f(\eta) < 1$, we have:

$$\delta_1 = \alpha_1\delta; \quad \delta_2 = \alpha_2\delta \quad \text{and} \quad \frac{\tau_w}{\rho} = -v\frac{\partial v_x}{\partial y}(0) = \beta\frac{vV_x}{\delta}, \qquad (8.4.10\text{a, b, c})$$

where:

$$\alpha_1 = \int_0^1 [1 - f(\eta)]d\eta,$$

$$\alpha_2 = \int_0^1 f(\eta)[1 - f(\eta)]d\eta \quad \text{and} \quad \beta = f'(0).$$

$$(8.4.11\text{a, b, c})$$

Finally, the von Karman–Pohlhausen Eq. (8.4.8) becomes:

$$\frac{1}{2}\alpha_2 V_x \frac{d\delta^2}{dx} + (\alpha_1 + 2\alpha_2)\frac{dV_x}{dx}\delta^2 = \beta v. \qquad (8.4.12)$$

- **Flow past a flat plate**

The simplest application of the integral methods is the boundary layer formed by a fluid flowing past a flat plate. In this case, as V_x is constant (i.e., there is no fluid acceleration), the momentum loss is absorbed by the friction at the wall, so that,

$$\rho V_x^2 \delta_2(x) = \int_0^x \tau_w(x)dx \quad \Rightarrow \quad \tau_w = \rho V_x^2 \frac{d\delta_2}{dx}, \qquad (8.4.13)$$

which coincides with Eq. (8.4.8) with constant V_x. Note that, based on the definition of friction factor, $f_F = \tau_w/(\frac{1}{2}\rho V_x^2)$, and considering that, as we saw from the exact Blasius solution, $\tau_w \propto x^{-1/2}$, here we obtain:

$$\delta_2(x) = xf_F(x). \qquad (8.4.14)$$

Integrating Eq. (8.4.12) with constant V_x, we find:

$$\delta(x) = \sqrt{\left(\frac{2\beta}{\alpha_2}\right)}\sqrt{\frac{vx}{V_x}} \quad \Rightarrow \quad \frac{\delta(x)}{x} = \sqrt{\left(\frac{2\beta}{\alpha_2}\right)}Re_x^{-1/2}. \qquad (8.4.15)$$

Now we must choose a reasonable function $f(\eta)$. As in the case studied in Sect. 6.6, pressure is constant, and therefore the velocity at the wall should be linear, so that its second derivative should be zero. Accordingly, we choose the following odd function,

$$f(\eta) = \frac{3}{2}\eta - \frac{1}{2}\eta^3, \tag{8.4.16}$$

that satisfies the conditions $f(0) = 0$; $f''(0) = 0$; $f(1) = 1$ and $f'(1) = 0$. Then, from (8.4.11a, b, c) we find:

$$\alpha_1 = \frac{3}{8}; \quad \alpha_2 = \frac{39}{280}, \quad \text{and} \quad \beta = \frac{3}{2}. \tag{8.4.16a}$$

Consequently, $\delta(x) = 4.641(vx/V_x)^{1/2}$, and we see that near the wall, that is when $\eta \ll 1$, the velocity profile is the following:

$$\frac{v_x}{V_x} = \frac{3}{2}\frac{y}{\delta} \cong 0.323\frac{y}{\sqrt{vx/V_x}}, \tag{8.4.17}$$

in excellent agreement with the exact velocity profile (8.2.12), where the numerical coefficient is 0.332, instead of 0.323. The shear stress can be determined by substituting into Eq. (8.4.10c) the values of β and $\delta(x)$:

$$\tau_w = \frac{3}{2}\frac{\mu V_x}{\delta} = f_F\left(\frac{1}{2}\rho V_x^2\right), \quad \text{where} \quad f_F = 0.646 Re_x^{-1/2}, \tag{8.4.18}$$

again in excellent agreement with the exact result (8.2.15), where the numerical coefficient is 0.664. So, we see that if we use a reasonable velocity profile, we obtain a value of the shear stress that is very close to the exact one.

Finally, for sake of completeness, let us compare the approximate values of the displacement thickness, δ_1, and of the momentum thickness, δ_2, with their exact values. First, we obtain: $\delta_1 = 3\delta/8 = 1.74(vx/V_x)^{1/2}$, in good agreement with its exact value,

$$\delta_1(x) = \int_0^\infty \left(1 - \frac{v_x}{V_x}\right)dy = \sqrt{\frac{vx}{V_x}}\int_0^\infty (1 - g')d\eta = 1.721\sqrt{\frac{vx}{V_x}}, \tag{8.4.19}$$

where g' is Blasius' function of Fig. 8.3 and the numerical value has been evaluated numerically. As for δ_2, from Eq. (8.4.14) we saw that it is proportional to the friction factor, f_F; in fact, $\delta_2 = 39\delta/280 = 0.646(vx/V_x)^{1/2}$.

- **Flow past a wedge**

The potential flow solution, i.e. the solution of the Euler Equation (8.1.7) for the flow past a wedge is:

$$V_x = V_0 \left(\frac{x}{L}\right)^m, \quad \text{with} \quad m = \frac{\beta}{2 - \beta}, \tag{8.4.20}$$

where β is indicated in Fig. 8.8, while L is a characteristic distance along the surface. Therefore, the von Karman–Pohlhausen Eq. (8.4.12) becomes:

$$\frac{d\delta^2}{dx} + 2\frac{(\alpha_1 + 2\alpha_2)}{\alpha_2} \frac{1}{V_x} \frac{dV_x}{dx} \delta^2 = \frac{2\beta}{\alpha_2} \frac{v}{V_x}, \tag{8.4.21}$$

and, substituting Eq. (8.4.20), we find:

$$\frac{d\delta^2}{dx} + 2\left(\frac{\alpha_1 + 2\alpha_2}{\alpha_2}\right) \frac{m}{x} \delta^2 = \frac{2\beta}{\alpha_2} \frac{vL^m}{V_0 x^m}.$$

Now look for solutions of the type

$$\delta^2 = K \frac{vL^m}{V_0} x^{(1-m)},$$

obtaining,

$$K = \frac{2\beta}{2m\alpha_1 + (3m + 1)\alpha_2}.$$

The boundary layer separation corresponds to a singular point of δ. Consequently, it occurs when,

$$m = \frac{-\alpha_2}{2\alpha_1 + 3\alpha_2}. \tag{8.4.22}$$

Now, choosing

$$f(\eta) = \eta, \tag{8.4.23}$$

from (8.4.11a) we find:

$$\alpha_1 = \frac{1}{2}; \quad \alpha_2 = \frac{1}{6}, \quad \text{and} \quad \beta = 1. \tag{8.4.23a}$$

Consequently, we see that flow separation occurs at $m = -1/9 = -0.\bar{1}$, corresponding to an angle $\beta = -1/5 = -0.2$, in excellent agreement with the exact result, $\beta = -0.199$.

8.5 Problems

8.1 Comment about the fact that the maximum speed and the road holding of a 3-volume car are better than those of a 2-volume car, at the same conditions.

8.2 Evaluate the wall shear stress of a fluid flowing past a flat plate using the von Karman–Pohlhausen integral equation and assuming a linear velocity profile.

8.3 Repeat the previous problem assuming sinusoidal velocity profile.

Chapter 9
Heat Conduction

Abstract In this chapter we start the second part of the book, which is devoted to heat and mass transport, by studying the diffusive heat transport, called *heat conduction*. First, in Sect. 9.1, we introduce some general concepts of heat transport, including the heat transfer coefficient. Then, in Sect. 9.2, heat conduction problems are solved in Cartesian, cylindrical and spherical coordinate. An important application of these concepts is the composite solid, that we considered in the following Sect. 9.3. Finally, in Sect. 9.4, we apply to heat transport the quasi steady state approximation that was studied in Sect. 5.5.

9.1 Introduction to Heat Transport

In heat transport, the uncoherent form of energy that we associate with internal energy can move from one region to another by one of the following modalities:

- **Diffusion**. The diffusion of energy, commonly denoted as **heat conduction**, consists of the transfer of internal energy from hotter to colder regions, without any net macroscopic movement, as it happens, for example, in solids.
- **Convection**. In moving fluids, the internal energy of a material element can also be transported by convection. Therefore, the study of heat convection requires that the flow field is determined beforehand.
- **Radiation**. Energy can also be transported as electromagnetic radiation, which travels in space even through a vacuum, i.e., in the absence of a medium.[1]

In general, if we isolate a volume element, at steady state and assuming that energy can be exchanged only as heat, we can write the following energy (per unit time) balance:

---

[1]This fact that today is taken for granted, is not so obvious and, in fact, during the second half of the 19th century, many important physicists kept looking for a fluid medium, called ether, whose vibrations should produce the light waves, just in the same way as the vibrations of the air produce the sound waves.

© Springer International Publishing Switzerland 2015   155
R. Mauri, *Transport Phenomena in Multiphase Flows*,
Fluid Mechanics and Its Applications 112, DOI 10.1007/978-3-319-15793-1_9

(heat power input) − (heat power output) + (heat power produced) = 0   (9.1.1)

Note that, since energy is conserved, heat can be "produced" only by degradation from other, "more noble" forms of energy, through processes by which energy becomes less available for doing work. In fact, we can have degradation of mechanical energy (i.e., viscous dissipation, as seen in Sect. 6.6) or electrical energy (i.e., the Joule heating), the conversion of chemical or nuclear energy into heat, etc.

Energy can enter or exit a control volume by conduction, convection or radiation. In this text, the theory of radiation is described in Chap. 19; elsewhere, we consider only convection and conduction. As we have seen in Sect. 1.4, the heat flux (that is the heat crossing a surface of unit area per unit time) due to convection is given by the following deterministic expression,

$$\mathbf{J}_{Uc} = \rho c \mathbf{v} T, \qquad (9.1.2a)$$

where temperature is referred to a convenient reference value. On the other hand, diffusive heat transport (i.e., conduction) is described by a constitutive equation that cannot be derived deterministically from continuum mechanics, as it reflects the microscopic properties of the medium; the simplest and most commonly valid constitutive relation is the Fourier law, stating that the diffusive heat flux $\mathbf{J}_{Ud}$ is proportional to the temperature gradient,[2]

$$\mathbf{J}_{Ud} = -k\nabla T, \qquad (9.1.2b)$$

where k is the coefficient of thermal conductivity of the medium. Note the negative sign, indicating that heat moves spontaneously from hot to cold regions, in agreement with the second law of thermodynamics (see Sect. 1.6).

In solids, thermal conductivity varies enormously from medium to medium, as can be seen in Appendix A. For example, $k \approx 100$ W/mK for metals, while $k \approx 0.01$ W/mK for polystyrene.[3] In liquids, thermal conductivity of water at ambient conditions is $k \approx 0.6$ W/mK, corresponding to a thermal diffusivity $\alpha = k/\rho c \approx 0.15 \times 10^{-3}$ cm^2/s, a value that is comparable to that of kinematic viscosity, $v \approx 10^{-2}$ cm^2/s. Finally in gases, as predicted in the kinetic theory of Sects. 1.5–1.7, α and v are not very different from each other.

A very important conclusion that we can draw from Eqs. (9.1.2a, b) is that the total heat flux (i.e., the sum of its convective and diffusive parts) is proportional to

---

[2]In the following, sometimes we will denote the heat flux by J_Q, stressing that we consider only the part of internal energy that can be associated with heat.

[3]As a curiosity, note that $k \approx 1$ W/mK for bricks and $k \approx 0.1$ W/mK for wood, which means that a house with wooden walls is 10 times better thermally insulated than a house with brick walls of the same thickness.

Fig. 9.1 Heat flux and temperature profile at an interface

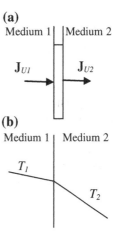

the temperature difference, which means that heat transfer is a linear process.[4] Therefore, there is a linear relation between cause (a temperature difference) and effect (a heat flux). In other words, if we cool a hot body using a cold fluid, the heat transfer between body and fluid is proportional to their temperature difference: if we double the temperature difference, the heat exchanges will also double.

When the energy (i.e., heat) balance (9.1.1) is applied to a volume that includes an interface, we must specify the appropriate **boundary conditions**. In general, at an interface there cannot be any jump in temperature nor in heat flux. The existence of a temperature jump is ruled out by the fact that, by Eq. (9.1.2b), it would imply an infinite heat flux; as for the continuity of the heat flux, it derives from a simple heat balance conducted on the thin slab of Fig. 9.1a, letting the thickness go to zero. So at the end, referring to Fig. 9.1b, we have the following boundary conditions at the interface, w,

$$T_1|_w = T_2|_w; \quad \mathbf{e}_n \cdot \mathbf{J}_{U1}|_w = \mathbf{e}_n \cdot \mathbf{J}_{U2}|_w \Rightarrow k_1 \frac{dT_1}{dz}\bigg|_w = k_2 \frac{dT_2}{dz}\bigg|_w. \qquad (9.1.3)$$

where $\mathbf{e}_n$ denotes a unit vector perpendicular to the interface, while z is a coordinate along $\mathbf{e}_n$. Note that, due to the fact that $k_1 \neq k_2$, the slope of the temperature distribution changes abruptly at the interface.

Sometimes, we know what happens on one side of the interface, so that the problem is greatly simplified. Consider the two following cases:

1. The temperature at the interface is fixed, with $T_w = T_0$. This applies, for example, when one of the two medium has a much larger thermal inertia than the other, i.e., $T_1 = T_0$ uniform, and therefore $T_2 = T_0$ at the interface.

---

[4]Naturally, this is true only approximately, that is neglecting the temperature dependence of ρ, c and k.

2. The flux at a solid interface is fixed, with $\mathbf{e}_n \cdot \mathbf{J}_U|_w = J_0$. Applying the Fourier law (9.1.2b), that means fixing the temperature gradient at the wall. This applies, for example, when $k_1 \ll k_2$, so that we obtain the no-flux boundary condition, $\mathbf{e}_n \cdot \nabla T_2|_w = 0$.

Sometimes, above all in the case of convective heat transfer, we do not know exactly what happens at the interface. For example, when a slab is cooled by a flowing fluid, the heat flux at the wall depends on the fluid flow regime and therefore it is very difficult to evaluate. Therefore, at the liquid-solid interface we apply the so-called *Newton's law of cooling*, simply stating that the heat flux at the wall is proportional to the difference between the temperature of the surface and that of the ambient fluid (i.e., the temperature of the fluid far from the wall), i.e.,

$$J_{Uw} = \mathbf{e}_n \cdot \mathbf{J}_U|_w = h(T_w - T_0). \qquad (9.1.4)$$

Here $\mathbf{e}_n$ is the unit vector perpendicular to the surface and directed outside the volume element, that is towards the fluid, while h is denoted as *heat transfer coefficient* and is (approximately) independent of temperature.

This condition states that when T_w is larger than T_0 the heat flux will be directed along $\mathbf{e}_n$ (that is, $\mathbf{e}_n \cdot \mathbf{J}_U > 0$) and therefore will exit the solid, as it should by the second law of thermodynamics. Note that, when $h = 0$, Newton's law of cooling reduces to $J_{Uw} = 0$, corresponding to an adiabatic wall, with no heat flux, while as $h \rightarrow \infty$, considering that the flux at the wall must remain finite, it becomes $T_w = T_0$, indicating that the temperature of the wall is fixed and equal to the bulk fluid temperature.

Clearly, Newton's law of cooling is an empirical rule, and therefore h can be seen as a coefficient of ignorance. In fact, it predicts a temperature jump at the wall, while in reality the temperature profile near the wall consists of a continuous line connecting the wall temperature, T_w, with the fluid bulk temperature, T_0, as represented in Fig. 9.2. However, although the physically correct boundary conditions consist of Eqs. (9.1.3), Newton's law (9.1.4) is not at all trivial, as it states that, since h is (approximately) independent of temperature, heat transfer is a linear process. As we saw, this linear dependence stems from the fact that both convective and diffusive heat fluxes (9.1.2a, b) are linear with the temperature difference, T. This is fundamentally different than in momentum transport where, for example,

Fig. 9.2 Temperature distribution at a solid-liquid interface

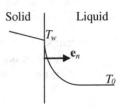

the drag force (i.e., the momentum flux at the wall) of a body moving in a quiescent fluid is proportional to the square of its speed at large Reynolds number, Re, while the dependence is linear at low Re. In fact, the convective momentum flux (1.4.1), $J_{Qc} = \rho V^2$, is quadratic in V, while its diffusive counterpart (1.4.4), $J_{Qd} = \mu V/L$, is linear. Naturally, had we considered radiative heat exchange, which is a highly non-linear process, introducing the heat transfer coefficient would bring no simplification.

9.2 Unidirectional Heat Conduction

9.2.1 Plane Geometry

Consider a steel slab with thermal conductivity k, thickness L, height $H \gg L$, and width $W \gg L$ (see Fig. 9.3). On one side, at $x = 0$, the wall is exposed to water in a well-stirred reservoir at temperature T_1, while on the other side, at $x = L$, the wall is lapped by air at a lower temperature, $T_0 < T_1$. Now, we want to find the heat exchanged between water and air, assuming that: (a) the water-slab heat transfer coefficient is large enough that the temperature of the slab on the left-side wall is T_1; (b) the air-slab heat transfer coefficient h is known; (c) edge effect can be neglected, so that the heat flow is unidirectional along x.

Apply the heat balance (9.1.1) to the volume element of Fig. 9.3, with thickness Δx, height H and width W, considering that there is no heat production term,

$$J_U(x)(HW) - J_U(x + \Delta x)(HW) = 0.$$

So, considering that x and Δx are arbitrary, we may conclude that the heat flux, J_U, is uniform,

$$J_U(x) = \text{constant} = J_U \quad \text{when} \quad 0 < x < L, \tag{9.2.1}$$

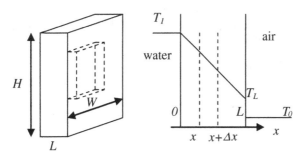

Fig. 9.3 Heat conduction in a slab

where the constant J_U is not known at this point. Now, applying the Fourier law at any point, we have $J_U = -k(dT/dx)$, where k is a constant, so we see that the temperature profile is linear and therefore $J_U = k(T_1 - T_L)/L$, where we have considered that $T(0) = T_1$ and $T(L) = T_L$. Finally, applying the boundary condition, $J(L) = h[T(L) - T_0]$, we obtain:

$$T_L = T(L) = T_1 - \frac{Bi}{1 + Bi}(T_1 - T_0) = T_0 + \frac{1}{1 + Bi}(T_1 - T_0), \qquad (9.2.2)$$

and

$$J_U = \frac{Bi}{1 + Bi}\frac{k(T_1 - T_0)}{L}; \quad Bi = \frac{hL}{k}, \qquad (9.2.3)$$

where Bi is the Biot number,[5] representing the ratio between the heat exchange at the surface and that within the slab (Note that $J_U > 0$ when $T_1 > T_0$, as it should).

Sometimes (actually, not very often), in addition to finding the heat flux, we are required to determine the temperature distribution as well. Here, considering that the profile is a straight line with a known slope and a known intercept with the $x = 0$ axis, we find:

$$\frac{T_1 - T(x)}{T_1 - T_0} = \frac{Bi}{1 + Bi}\frac{x}{L}. \qquad (9.2.4)$$

Now consider the two limit cases, corresponding to having the Biot number either very small or very large.

- $Bi \ll 1$. In this case, the heat exchange at the wall $x = L$ is very slow compared to the conduction within the slab, indicating that as soon as some heat is transferred by convection from the air to the slab, it is immediately transmitted by conduction to the water, so that the temperature within the slab is uniform. In fact, when $Bi = 0$, we find that $T_L = T_1$ (that is the temperature of the wall is equal to that of the water) and the heat flux is null, i.e., $J_U = 0$.
- $Bi \gg 1$. In this case, the heat exchange at the wall $x = L$ is very fast with respect to conduction: heat exits (or enters) very easily, but then it is very slow in propagating within the slab. In fact, when $Bi \to \infty$, we find that $T_L = T_0$ (that is the temperature of the wall is equal to that of the air) and heat flux is maximum, i.e., $J_U = k(T_1 - T_0)/L$.

The result (9.2.3) can be rewritten as:

$$J_U = k_{eff}\frac{(T_1 - T_0)}{L}, \quad \text{where: } k_{eff} = k\frac{Bi}{1 + Bi}, \quad \text{i.e., } \frac{1}{k_{eff}} = \frac{1}{k} + \frac{1}{hL}. \qquad (9.2.5)$$

---

[5]Named after Jean-Baptiste Biot (1774–1862), a French physicist, astronomer, and mathematician.

Fig. 9.4 Cylindrical shell

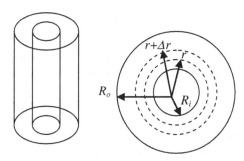

Here, considering that the inverse of thermal conductivity (and of the heat transfer coefficient as well) is a measure of the heat resistance, we see that there are two mechanisms that resist against the transport of heat, *in series* after one another: the former is due to heat conduction within the slab, while the latter is due to heat convection at the wall. In fact, we know that the characteristic relaxation time that is needed to reach steady state (that is, the time that it takes to the system to adapt to a sudden change of the boundary conditions) is $\tau_{ss} \cong L^2/\alpha_{eff}$, where $\alpha_{eff} = k_{eff}/\rho c$. Thus we obtain:

$$\tau_{ss} \cong \tau_d + \tau_c; \quad \tau_c = \frac{\rho c L}{h}; \quad \tau_d = \frac{\rho c L^2}{k} = \frac{L^2}{\alpha}, \tag{9.2.6}$$

indicating that the global relaxation time τ_{ss} is the sum of the two characteristic times regulating the transport of heat: one, τ_d, represents the time that takes to the heat to propagate by diffusion through the slab; the other, τ_c, indicates how long it takes to the heat to be exchanged by convection with the surrounding fluid. Since the two processes, i.e., diffusion within the slab and convection at the wall, occurs in series, the two times must be added to one another in order to obtain the global relaxation time.[6] Obviously, considering that the Biot number is the ratio between the two characteristic times,

$$Bi = \frac{\tau_d}{\tau_c} = \frac{hL}{k}, \tag{9.2.7}$$

we see that $\tau_{ss} = \tau_c$ when $Bi \ll 1$, while $\tau_{ss} = \tau_d$ when $Bi \gg 1$, indicating that, as usual, the slowest process controls the global kinetics (see Problem 9.11).

---

[6]Trivially, if the two processes were in parallel, we should add the inverses of the two characteristic times.

9.2.2 Cylindrical Geometry

Consider the circular pipe of a heat exchanger represented in Fig. 9.4. It is made of a steel having thermal conductivity k, with inner and outer radii, R_i and R_o, respectively, and height $H \gg R_o$. A liquid with temperature T_1 flows inside the tube, while the outer wall exchanges heat with a vapor, having temperature $T_0 < T_1$, with heat transfer coefficient h. We want to determine the heat power exchanged between the two fluids and the temperature profile.

As in the previous example, we assume that the heat exchange between the inner liquid and the pipe wall is very efficient, so that we may assume that the inner wall temperature is T_1, while at the outer wall we apply Newton's law of cooling. Thus, considering a heat balance in a cylindrical crown of radii r and $r + \Delta r$, assuming that the heat flux is radial (i.e., neglecting all edge effects), we obtain:

$$J_U(r)(2\pi rH) - J_U(r + \Delta r)(2\pi(r + \Delta r)H) = 0.$$

This relation indicates that the heat power, $\dot{Q} = J_U(r)2\pi rH$, is constant at each cross section, so that, unlike the planar case, where the heat flux is constant and the temperature profile is linear, here the product $J_U(r)r$ is constant, so that the heat flux is proportional to $1/r$ and therefore the temperature profile is logarithmic. In fact, applying the Fourier constitutive equation, integrating between R_i and r and imposing the boundary condition, $T(R_i) = T_1$, we have:

$$J_U = \frac{C}{r} = -k\frac{dT}{dr} \Rightarrow T = T_1 - \frac{C}{k}\ln\frac{r}{R_i}. \tag{9.2.8}$$

The constant C can be determined applying the boundary condition $J_U(R_o) = h[T(R_o) - T_0]$ at the outer wall, obtaining,

$$C = \frac{Bi}{1 + Bi \ln(R_o/R_i)}k(T_1 - T_0); \quad Bi = \frac{hR_o}{k},$$

where Bi is the Biot number. Note that C is positive, as it must be because the heat flux is directed outward and is therefore positive. So, since $J_U = C/r$ and $\dot{Q} = 2\pi rHJ_U = 2\pi HC$, we obtain:

$$\dot{Q} = \frac{2\pi kH(T_1 - T_0)Bi}{1 + Bi \ln(R_o/R_i)}. \tag{9.2.8a}$$

Finally, substituting C into (9.2.8), we obtain the temperature distribution,

$$\frac{T_1 - T}{T_1 - T_0} = \frac{Bi \ln(r/R_i)}{1 + Bi \ln(R_o/R_i)}, \tag{9.2.9a}$$

and, in particular, the temperature of the outer wall, $T_{ext} = T(R_o)$,

$$\frac{T_1 - T_{ext}}{T_1 - T_0} = \frac{Bi \ln(R_o/R_i)}{1 + Bi \ln(R_o/R_i)}. \tag{9.2.9b}$$

Now we can apply the same considerations of the planar case. When $Bi = 0$, then $T_{ext} = T_1$ and the heat flux vanishes, while when $Bi \gg 1$, then $T_{ext} = T_0$ and the heat power is maximum and equal to

$$\dot{Q}_{Bi\gg1} = \frac{2\pi kH(T_1 - T_0)}{\ln(R_o/R_i)}. \tag{9.2.10}$$

Note that when $R_i = 0$ the problem collapses, as $C = 0$ in (9.2.8), showing that $J_U = 0$ and $T = T_0$, irrespectively of the value of h. This is true for all full objects, with any geometry: if we put a potato in a hot oven and wait, at steady state its temperature will be uniform, equal to the temperature of the air in the oven, irrespective of the shape of the potato; the value of the heat transfer coefficient, h, will not influence the temperature profile at steady state, but it will determine how long it takes to reach it.

Finally, we leave as an exercise to show that when $(R_o - R_i) \ll R_o$ the above solution reduces to the planar case.

9.2.3 Spherical Geometry

Consider a circular crown, composed of a material having thermal conductivity k, with inner and outer radii, R_i and R_o, respectively. The inside wall of the crown is kept at a constant temperature, T_1, while the outer wall exchanges heat with a gas having temperature $T_0 < T_1$, with heat transfer coefficient h. We want to determine the heat power and the temperature profile.

Proceeding as in the previous case, we see that at steady state the heat power crossing any section, $\dot{Q} = 4\pi r^2 J_U$, is constant, i.e., independent of r. Therefore, integrating $J_U = C/r^2$, where C is a constant, applying the Fourier constitutive equation, and considering $T(R_i) = T_1$, we obtain:

$$T_1 - T = \frac{C}{k}\left(\frac{1}{R_i} - \frac{1}{r}\right). \tag{9.2.11}$$

Imposing the other boundary condition, $J_U(R_o) = h[T(R_o) - T_0]$, we finally find C and hence the temperature of the outer wall, the heat flux at the inner wall and the total heat power:

$$\frac{T_1 - T(R_o)}{T_1 - T_0} = \frac{Bi'}{1 + Bi'}, \tag{9.2.12}$$

$$J_{Ui} = \frac{k(T_1 - T_0)}{L} \frac{Bi'}{1 + Bi'} \frac{R_o}{R_i}; \quad \dot{Q} = 4\pi R_i^2 J_{Ui} = 4\pi R_i R_o \frac{k(T_1 - T_0)}{L} \frac{Bi'}{1 + Bi'},$$

$$(9.2.13)$$

where $Bi' = (hL/k)(R_o/R_i)$, with $L = R_o - R_i$.

Note that, even in this case, when the sphere is full, i.e., when $R_i = 0$, we find $T = T_0$ and $J_U = 0$, independent of h.

The general solution (9.2.11) shows that the steady temperature distribution in spherical coordinates varies like $1/r$, i.e., $T(r) = A + B/r$, where A and B are constants to be determined applying two boundary conditions. Therefore, considering the temperature profile *outside* a sphere of radius R with boundary conditions $T(R) = T_0$, and $T(r \to \infty) = T_\infty$, we obtain:

$$T - T_\infty = \frac{R}{r}(T_0 - T_\infty); \quad \dot{Q} = 4\pi R k (T_0 - T_\infty). \quad (9.2.14)$$

Now, if we try to do the same for cylindrical (or planar) coordinates, we run into problems. In fact, the general solution $T = A + B \ln r$ diverges as $r \to \infty$ and therefore the solution must be $T = A$ constant, which obviously cannot satisfy both boundary conditions. The same problem arises when we study the flow past an infinite cylinder at low Reynolds number, and can be resolved only by considering the edge effects: after all, there are no real two dimensional objects!

9.3 The Composite Solid

A solid consisting of two layers made of different materials of thickness L_1 and L_2, exchanges heat with two fluids having temperature T_{0L} and T_{0R}, with heat transfer coefficients h_L and h_R (see Fig. 9.5). Since at steady state each cross section is crossed by the same heat flux, we have:

$$J_U = h_L(T_{0L} - T_1) = \frac{k_1}{L_1}(T_1 - T_{12}) = \frac{k_2}{L_2}(T_{12} - T_2) = h_R(T_2 - T_{0R}) \quad (9.3.1)$$

Fig. 9.5 Heat conduction through a composit wall

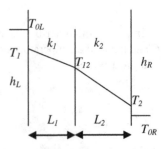

Hence, we can calculate the intermediate temperatures, i.e., T_1, T_{12} and T_2. However, since in general we are interested to find only the heat power exchanged, it is convenient to rewrite Eq. (9.3.1) as follows,

$$(T_{0L} - T_1) = \frac{\dot{Q}}{h_L A}; \quad (T_1 - T_{12}) = \frac{\dot{Q}L_1}{k_1 A}; \quad (T_{12} - T_2) = \frac{\dot{Q}L_2}{k_2 A};$$

$$(T_{0R} - T_1) = \frac{\dot{Q}}{h_R A}, \tag{9.3.2}$$

where $\dot{Q} = J_U A$ is the heat power, with A indicating the area of any cross section. Summing all terms of Eq. (9.3.2), we obtain:

$$\Delta T_{tot} = R_{th}\dot{Q}, \quad L \tag{9.3.3}$$

where $\Delta T_{tot} = T_{0L} - T_{0R}$ is the total temperature difference, while

$$R_{th} = \frac{1}{A}\left(\frac{1}{h_L} + \frac{L_1}{k_1} + \frac{L_2}{k_2} + \frac{1}{h_R}\right) \tag{9.3.4}$$

is the *thermal resistance*, which is therefore defined in (9.3.3) as the ratio between the total temperature difference and the exchanged heat power. So, the thermal resistance in (9.3.4) is the sum of the thermal resistances associated with each heat transfer process in series, that is,

$$R_{th} = \sum_i R_{th,i}; \quad R_{th,i} = \begin{cases} \dfrac{L_i}{Ak_i} & \text{(distributed conductive resistance)} \\ \dfrac{1}{Ah_i} & \text{(concentrated convective resistance)} \end{cases} \tag{9.3.5}$$

where L_i and k_i are the thickness and thermal conductivity of the material layer where heat is transported by conduction, while h is the heat transfer coefficient at the wall which, in general, is due to conduction.

It is quite evident that transport of heat (i.e., of internal energy) is very similar to the transport of electrical energy: heat flux and temperature difference take the place of electrical current and voltage, so that thermal resistance substitutes electrical resistance. Therefore, the total resistance is the sum of all partial resistances when, as it is the case here, the partial resistances are in series, that is they are crossed by the same heat flux, while the total temperature difference is the sum of all temperature drops. On the other hand, the inverse of the total resistance is the sum of the inverse of all partial resistances when the different sections are in parallel, that is each section is subjected to the same temperature difference, while the total flux is the sum of the fluxes crossing each section (see Problem 9.9).

Generalizing the analysis of Sect. 9.2.1 [see Eq. (9.2.6)], we see that the global relaxation time (that is, how long it takes for the system to adapt to a sudden variation of the boundary conditions) is the sum of all relaxation times associated

Fig. 9.6 Heat conduction through a composite circular pipe

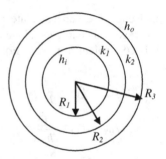

with each resistance, these latter being either diffusive, $\tau_d = \rho c L^2/k$, or convective, $\tau_c = \rho c L/h$, and whose ratio is the Biot number, as in Eq. (9.2.7).

9.3.1 Cylindrical Geometry

The conductive resistance of a circular tube can be determined using Eq. (9.2.10), while the convective resistance is concentrated and therefore it does not depend on the surface curvature. Therefore, in the case of a composite cylindrical pipe composed of different layers, Eq. (9.3.3) is still valid, with,

$$
R_{th} = \sum_i R_{th,i}; \ R_{th,i} = \begin{cases} \dfrac{\ln(R_o/R_i)}{2\pi k_i H} & \text{(distributed conductive resistance)} \\[2ex] \dfrac{1}{2\pi R h_i H} & \text{(concentrated convective resistance)} \end{cases}
$$

$$(9.3.6)$$

where we have considered that the heat transfer surface area is $A = 2\pi R H$. For example, in the case represented in Fig. 9.6, we have:

$$
R_{th} = \frac{1}{2\pi H} \left(\frac{1}{R_1 h_i} + \frac{\ln(R_2/R_1)}{k_1} + \frac{\ln(R_3/R_2)}{k_2} + \frac{1}{R_3 h_o} \right). \tag{9.3.7}
$$

It is convenient to write the thermal resistance as inversely proportional to the exchange area, as in Eq. (9.3.4). However, unlike the plane case, where this area is constant, in the cylindrical case it is proportional to the radius. For example, in the case of a tube exchanging heat with a fluid inside and another fluid outside, the thermal resistance can be written in terms of either the inner tube surface area, $S_i = 2\pi R_i H$ or the outer area, $S_o = 2\pi R_o H$,

$$
R_{th} = \frac{1}{S_i} \left(\frac{1}{h_i} + \frac{R_i \ln(R_o/R_i)}{k} + \frac{R_i}{R_o h_o} \right) = \frac{1}{S_0} \left(\frac{R_o}{R_i h_i} + \frac{R_o \ln(R_o/R_i)}{k} + \frac{1}{h_o} \right). \tag{9.3.8}
$$

9.4 Quasi Steady State Approximation

In agreement with Sect. 5.5, the Quasi Steady State (QSS) approximation consists of assuming that the characteristic relaxation time τ_{ss} that the temperature profile (and consequently the heat flux as well) takes to reach its steady state is much shorter than the characteristic time of variations of the boundary conditions.

The simplest example of QSS approximation is about finding the temperature profile of a metallic object with thermal conductivity k, linear dimension L, and initial temperature T_i, which, at time $t = 0$, is immersed in a liquid at temperature T_0, with which it exchanges heat with heat transfer coefficient h. In general, the temperature of the object will depend on both position and time[7]; therefore, as $T = T(r, t)$, the full solution of this problem is rather complex, as we will see in Sect. 12.4. However, when $Bi = hL/k \ll 1$, the problem becomes much simpler. Let us see why.

From a simple energy balance, we see that the speed with which the internal energy of the object, U, changes with time is equal to the total entering heat power, i.e. the net heat flux inlet at the wall, $(-J_{Uw})$, multiplied by the outer surface area of the object, S, i.e.,

$$\frac{dU}{dt} = \dot{Q}_{in} = -J_{Uw}S.$$

Then, consider that,

$$U = \int_V \rho c T dV = Mc\bar{T}, \quad \text{with} \quad \bar{T}(t) = \frac{1}{V}\int_V T(\mathbf{r}, t)dV,$$

where $M = \rho V$ is the mass of the object, V is its volume, ρ and c are its density and specific heat (that we assume to be constant), while $\bar{T}$ is its mean temperature. Thus, we obtain:

$$\rho c V \frac{d\bar{T}}{dt} = -SJ_{Uw} = -Sh(T_w - T_0), \tag{9.4.1}$$

where J_{Uw} is the heat flux at the wall exiting the object which, in turn, can be expressed in terms of the heat transfer coefficient h and the wall temperature, T_w, with the subscript w indicating "at the wall". Now, at steady state, the temperature of the object is uniform, with $\bar{T} = T_w = T(t)$, and therefore depending only on time. In this case, Eq. (9.4.1) becomes,

$$\rho c L \frac{dT}{dt} = -h(T - T_0), \tag{9.4.2}$$

---

[7]For example, when we take a potato out of the oven, as it cools down, the inside temperature is much higher than that of its skin.

where $L = V/S$ is a typical linear dimension of the object. Denoting by $\Delta T = T - T_0$ the temperature difference, we obtain:

$$\frac{d\Delta T}{\Delta T} = -\frac{dt}{\tau}; \qquad \tau = \frac{\rho c L}{h}, \qquad (9.4.3)$$

to be solved with initial condition $\Delta T(t = 0) = \Delta T_i = T_i - T_0$, finding:

$$\Delta T = \Delta T_i \, e^{-t/\tau}. \qquad (9.4.4)$$

Therefore, we see that τ is the characteristic time that the system takes to reach thermal equilibrium, provided that the Quasi Steady State hypothesis is satisfied.

Now we verify if and when the QSS hypothesis is satisfied, imposing that τ is much longer than the time that the object needs to reach steady state, i.e. in this case, a uniform temperature distribution. In order to understand this point, suppose that the object has initially a uniform temperature and then, suddenly, its wall temperature changes to a new value: the inside temperature will rearrange and reach a new distribution (uniform again, in this case), within a timescale τ_{ss}. Therefore, if the characteristic time of variation of the wall temperature, τ, is much larger than τ_{ss}, the temperature distribution within the object will always be equal to its steady value, corresponding here to a uniform profile.[8] Now, here $\tau_{ss} = L^2/\alpha$, where $\alpha = k/\rho c$ is the thermal diffusivity (see Sect. 1.4), so that, imposing $\tau_{ss} \ll \tau$, we obtain: $Bi = hL/k \ll 1$. We may conclude that the QSS hypothesis is verified, provided that the Biot number is very small.

As we mentioned above, for arbitrary values of Bi the problem becomes much more complicated and finding the temperature distribution as a function of both time and position is mathematically rather challenging. However, evaluating the magnitude of the characteristic cooling time is something that we have already seen in Sect. 9.2.1, finding Eq. (9.2.6), that is,

$$\tau \cong \tau_d + \tau_c; \quad \tau_c = \frac{\rho c L}{h}; \quad \tau_d = \frac{\rho c L^2}{k} = \frac{L^2}{\alpha}, \quad \text{with} \quad Bi = \frac{\tau_d}{\tau_c} = \frac{hL}{k}.$$

In the limit case that we have considered here, with $Bi \ll 1$, we find $\tau \cong \tau_c$, i.e., Eq. (9.4.3). On the other hand, when $Bi \gg 1$, we see that $\tau \cong \tau_d$. These limit cases correspond to a small metallic object, like a coin, and a poorly conducting object, like a potato, respectively.

Now, let us consider a more complex case. A thermally insulated reservoir is divided in two parts by a thin impermeable membrane of thickness d (see Fig. 9.7, where L and R indicate "left" and "right"). Assuming that the two parts have the

---

[8]It is equivalent to the quasi-static process in thermodynamics.

Fig. 9.7 Thermally insulated reservoir divided into two parts by a heat conducting membrane

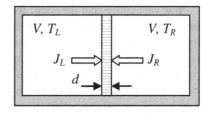

same volume V and are filled with a well-mixed liquid with different initial temperatures, T_{L0} and T_{R0}, we want to determine how the liquid temperatures vary with time.

First of all, we observe that the total internal energy of the reservoir must remain constant. Therefore, as for long times the system must be at thermal equilibrium, with a uniform temperature, i.e., with $T_L = T_R = T_f$, we obtain:

$$\rho c V T_{L0} + \rho c V T_{R0} = \rho c V T_L + \rho c V T_R = 2\rho c V T_f \Rightarrow T_L + T_R = 2T_f \qquad (9.4.5)$$

where ρ and c are the liquid density and specific heat, that we assume to be constant.

Here, we have neglected the heat capacity of the membrane, that is we have assumed that $\rho_s c_s V_s \ll \rho c V$, where ρ_s, c_s and V_s are the density, specific heat and volume of the membrane, respectively. In our case, assuming that $V = L^3$ and $V_s = dL^2$, this assumption becomes:

$$\frac{\rho_s c_s d}{\rho c L} = \frac{M_s c_s}{M c} \ll 1, \qquad (9.4.6)$$

where $M = \rho V$ and $M_s = \rho_s V_s$ are the masses of the fluids and the membrane, respectively, indicating that the heat stored within the membrane is negligible with respect to that stored in the fluids.

From Eq. (9.4.5) we see that the sum $(T_L + T_R)$ remains invariant in time and is equal to its final value, $2T_f$, with,

$$T_f = T_L(t \to \infty) = T_R(t \to \infty) = \frac{1}{2}(T_{L0} + T_{R0}). \qquad (9.4.5a)$$

Now, in order to find the transient solution, we must consider the two sides of the reservoir separately, obtaining,

$$\rho c V \frac{dT_L}{dt} = -A J_L \quad \text{and} \quad \rho c V \frac{dT_R}{dt} = -A J_R, \qquad (9.4.7)$$

where $A = L^2$ is the area of the membrane, while J_L and J_R are the heat fluxes leaving the L and R side, respectively (see Fig. 9.7).

Now we assume that the Quasi Steady State approximation can be applied within the membrane (note that the QSS approximation is always valid within the liquid volumes, since the liquid is well mixed, i.e., T_L and T_R are uniform). Accordingly, we assume that the temperature of the two liquid volumes, T_L and T_R, varies so slowly that within the membrane we have a steady temperature distribution. That means (a) the heat flux leaving, say, the left side is equal to that entering the right side (i.e., there is no heat accumulation within the membrane); (b) the temperature profile within the membrane is linear. In particular, assuming that the wall temperatures at the two sides of the membrane are equal to the liquid temperatures, we may conclude:

$$J_L = -J_R = k_s \frac{T_L - T_R}{d}, \tag{9.4.8}$$

where k_s is the thermal conductivity of the membrane. Now, summing the two balance Eqs. (9.4.7) we obtain: $d(T_L + T_R)/dt = 0$, expressing again that the total internal energy within the reservoir is constant. Instead, subtracting Eqs. (9.4.7) we find:

$$\rho c L \frac{d\Delta T}{dt} = -2J_L = -2\frac{k_s \Delta T}{d}, \tag{9.4.9}$$

where $\Delta T = T_L - T_R$ and $L = V/A$. Therefore we obtain the first order differential equation:

$$\frac{d\Delta T}{\Delta T} = -\frac{dt}{\tau}; \quad \tau = \frac{\rho c L d}{2k_s}, \tag{9.4.10}$$

to be solved with initial condition, $\Delta T(t = 0) = \Delta T_0$, obtaining:

$$\Delta T = \Delta T_0 e^{-t/\tau}. \tag{9.4.11}$$

Substituting Eq. (9.4.11) into (9.4.7) we find:

$$T_L = \frac{1}{2}T_{L0}\left(1 + e^{-t/\tau}\right) + \frac{1}{2}T_{R0}\left(1 - e^{-t/\tau}\right);$$
$$T_R = \frac{1}{2}T_{R0}\left(1 + e^{-t/\tau}\right) + \frac{1}{2}T_{L0}\left(1 - e^{-t/\tau}\right). \tag{9.4.12}$$

This solution shows that ΔT, T_L and T_R, tend exponentially to their equilibrium value, with a characteristic relaxation time τ given by Eq. (9.4.10).

Now, let us see if and when the QSS hypothesis is satisfied. Imposing[9] that $\tau \gg \tau_{ss} = d^2/\alpha_s$, where $\alpha_s = k_s/\rho_s c_s$ is the membrane thermal diffusivity, at the end we find again the condition (9.4.6). Thus, we see that the QSS hypothesis is valid when the thermal capacity of the membrane is much smaller than that of the liquid.

---

[9]The QSS approximation here concerns the steady state within the membrane, and so its characteristic distance is d.

9.5 Problems

9.1 A well-insulated little room loses heat through a window having a 1 m^2 area, with a glass of thickness $L = 0.5$ cm. The mean outer temperature is $T_0 = 10\,°C$, the inner temperature is $T_1 = 25\,°C$, while the inner and outer heat transfer coefficient is $h = 20$ W/m^{2}K. Calculate, in kWh, the annual heat consumption, knowing that the thermal conductivity of glass is $k_s = 0.75$ W/mK. Calculate the heat loss reduction deriving from using a double glazing window constituted by two glasses, of thickness $L/2 = 0.25$ cm, separated by a $l = 1$ mm thick air gap (assume that air is stagnant, with thermal conductivity $k_a = 0.024$ W/mK).

9.2 Calculate the heat losses (in W) through the walls of a house. The walls are $L_1 = 10$ cm thick and have an area $S = 100$ m^2. The inside and outside temperatures are 20 and 0 °C, respectively, while the inner and outer heat transfer coefficient is $h = 20$ W/m^{2}K, while the mean thermal conductivity of the wall is $k_1 = 0.1$ W/mK. Calculate the heat losses reduction that occurs when we add to the wall an layer of thickness $L_2 = 0.6$ cm, made of an insulating material with thermal conductivity $k_2 = 0.002$ W/mK (note that an insulator has a thermal conductivity that is about 10 times smaller than air's; unfortunately, they are not transparent and cannot be used in windows).

9.3 A wall is composed of an inner layer of bricks ($k_2 = 0.72$ W/mK) of thickness $L_2 = 10$ cm, and an inner layer of an insulating material ($k_1 = 0.06$ W/mK) of thickness $L_1 = 2$ cm. Assuming a $\Delta T = 20\,°C$ temperature difference, and inner and outer heat transfer coefficients $h_1 = 25$ and $h_2 = 8$ W/m^{2}K, respectively, find the heat flux.

9.4 Saturated vapor at $T_i = 180\,°C$ flows in a circular pipe ($R_1 = 2.1$ cm; $R_2 = 2.5$ cm) made of steel ($k_1 = 43$ W/mK), crossing an ambient at $T_0 = 25\,°C$. The pipe is covered with a layer of an insulating material ($k_2 = 0.04$ W/mK) of thickness $d = 1.5$ cm. The inner and outer heat transfer coefficient are, respectively, $h_i = 12{,}000$ and $h_o = 6$ W/m^{2}K. Calculate the heat losses per meter.

9.5 Consider the insulation around a pipe, assuming for sake of simplicity that the inside temperature of the insulator, at $r = R_i$, is kept constant and that the insulator exchanges heat with the surroundings, at $r = R_o$, with a heat transfer coefficient h. Determine the critical outer radius, $R_o = R_c$, corresponding to a maximum heat exchange.

9.6 A 5 cm diameter sphere, made of cupper and with uniform 800 °C temperature, is suddenly placed in a pool of water at 20 °C. Calculate the mean temperature of the sphere after 45 s, knowing that the heat transfer coefficient is $h = 450$ W/m^{2}K.

9.7 A metal object (density ρ_1, specific heat c_1, thermal conductivity k_1) of volume V_1 and external surface area S_1 is immersed in a liquid (density ρ_2, specific heat c_2, thermal conductivity k_2) of volume V_2, initially at a T_{20}

temperature and confined in an adiabatic container. Assuming that the liquid is well mixed and that the heat transfer coefficient h is such that the Biot number, Bi, is small, determine $(T_1 - T_2)$ and T_1 as functions of time.

9.8 Determine the characteristic time that it takes to cool the liquid contained in the hollow sphere shown in Fig. 9.8, with inner radius R_i and outer radius R_o, exchanging heat with air at temperature T_o, with a heat transfer coefficient h. Assume that the inner wall has the same temperature of the liquid, T, and that density, specific heat and thermal conductivity of the liquid, ρ, c and k and of the spherical crown, ρ_s, c_s and k_s, are known.

9.9 In a composite medium, the effective thermal conductivity, k_{eff}, is defined as the ratio between the mean heat flux $J_Q = \dot{Q}/S$ (i.e., the ratio between the heat power and the area of the cross section) and the mean temperature gradient, $\varDelta T/L$. Calculate k_{eff} in the two-phase medium shown in Fig. 9.9, where the two phases have thermal conductivities k_1 and k_2, and occupy fractions ε_1 and ε_2 of the total volume, with $\varepsilon_1 + \varepsilon_2 = 1$. Assume that the two phases are placed (a) parallel, or (b) perpendicular to the heat flux.

9.10 A composite medium consists of two layers of different materials, of thickness L_1 and L_2. Assuming that the two outer surfaces are kept at temperatures T_1 and T_1, determine the interface temperature T_{12}.

9.11 Solve the second problem in Sect. 9.4 assuming that fluids and membrane exchange heat with heat transfer coefficient h.

Fig. 9.8 Hollow sphere containing a liquid

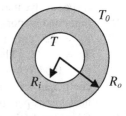

Fig. 9.9 Composite layered medium

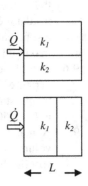

9.12 Consider a liquid contained in a square reservoir, where all walls are insulated, with the exception of one, whose outer temperature T_1 is kept constant.
Assuming to know the heat transfer coefficient h between the inner wall and
the fluid, find the temperature of the fluid as a function of time.

Chapter 10
Conduction with Heat Sources

Abstract In this chapter we study problems of conduction combined with internal heat generation, that is the conversion of mechanical energy into internal energy. Here we list some examples of heat generation, specifying the energy "dissipated" per unit volume and unit time, $\dot{q}$.

- Dissipation of electrical energy $\dot{q} = I^2/k_e$, where I is the current density, in amps cm^{-2}, while k_e is the electric conductivity, in Ω^{-1} cm^{-1}.
- Dissipation of mechanical energy (viscous dissipation), $\dot{q} = \mu|\nabla v|^2$, where μ is viscosity and ∇v the velocity gradient.
- Dissipation of chemical or nuclear energy.

First, in Sect. 10.1, we consider the general solution of this problem in plane, cylindrical and spherical geometries in the case of a uniform heat production. Then, in Sect. 10.2, we examine the case of a linear source term, using either regular or singular perturbation expansion methods to solve the limit cases corresponding to small and large heat generation, respectively.

10.1 Uniform Heat Generation

10.1.1 Plane Geometry

Consider a fluid confined between the two parallel plates shown in Fig. 10.1, that are placed at a distance $2L$, are kept at a fixed temperature, T_0, and move with respect to one another with a constant velocity (Couette flow). At steady state, a simple heat balance (9.1.1) [i.e., (heat power in) + (heat power generated) = (heat power out)] in a volume element having area $S = HW$ and thickness Δx, yields,

$$J_U(x)S - J_U(x + \Delta x)S + \dot{q}S\Delta x = 0, \qquad (10.1.1)$$

© Springer International Publishing Switzerland 2015
R. Mauri, *Transport Phenomena in Multiphase Flows*,
Fluid Mechanics and Its Applications 112, DOI 10.1007/978-3-319-15793-1_10

Fig. 10.1 Fluid flow between
the two parallel plates

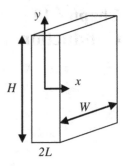

where J_U is the heat flux (defined as positive when it is directed along the positive
x direction) and $\dot{q}$ is the energy "dissipated" per unit volume and per unit time (here
it is constant, with $\dot{q} = \mu|\nabla v|^2$, as the velocity gradient is uniform). Dividing by Δx,
taking the limit as $\Delta x \to 0$, and applying the Fourier constitutive relation,
$J_U = -k\,dT/dx$, where k is the fluid thermal conductivity, we obtain:

$$\frac{dJ_U}{dx} = -k\frac{d^2T}{dx^2} = \dot{q} \quad -L \leq x \leq L, \tag{10.1.2}$$

This is a second order differential equation, requiring two boundary conditions.
First, let us assume that the fluid temperature at the wall coincides with the wall
temperature, i.e., $T(-L) = T(L) = T_0$. Then, Eq. (10.1.2) can be easily solved,
obtaining the following parabolic temperature profile,

$$\Delta T(x) = T(x) - T_0 = \frac{\dot{q}L^2}{2k}\left(1 - \frac{x^2}{L^2}\right). \tag{10.1.3}$$

Therefore, as expected, the maximum temperature is reached at the centerline, with

$$\Delta T_{\max} = \Delta T(0) = \dot{q}L^2/2k. \tag{10.1.3a}$$

An important check of this result consists in verifying that the total heat power
exiting the system, $J_U(\pm L)S$, is equal to the total heat power produced, $2\dot{q}LS$ (see
Eq. (10.1.7a) below).

Note that the solution (10.1.3) is identical to a Poiseuille profile, with $\Delta T = T - T_0$
and $\dot{q}$ replacing v and ΔP (v, in fact, is the difference between the velocity of the fluid
and that of the wall, that is assumed to be zero).

The mean temperature can be easily evaluated considering that,

$$\overline{T} = \frac{1}{V}\int_V T(\mathbf{r})\,d^3\mathbf{r}. \tag{10.1.4}$$

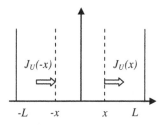

Fig. 10.2 Heat balance between the sections $-x$ and $+x$

In the plane case that we are considering, $V = 2LHW$ and $d^3\mathbf{r} = HWdx$, where H and W are the height and the thickness of the slab. Therefore, we find,

$$\overline{\Delta T} = \frac{1}{2L} \int_{-L}^{L} \Delta T(x)dx = \frac{\dot{q}L^2}{3k}. \tag{10.1.5}$$

Now, assume that the fluid (that is still assumed to be kept at a constant temperature, T_0) exchanges heat with the walls with a given heat transfer coefficient, h, so that boundary conditions become:

$$J_U(\pm L) = \pm h[T(\pm L) - T_0]. \tag{10.1.6}$$

Instead of solving the problem by integrating Eq. (10.1.2), let us consider a macroscopic heat balance in the volume element located between the sections $-x$ and $+x$ (see Fig. 10.2), imposing that the heat exiting from the outer boundaries equals the heat produced within the volume, obtaining:

$$-J_U(-x)S + J_U(x)S = 2\dot{q}Sx,$$
$$\text{i.e., } J_U(x) = \dot{q}x, \tag{10.1.7}$$

where we have considered that $J_U(-x) = -J_U(x)$ out of symmetry.[1] In particular, at the wall we have:

$$J_L \equiv J_U(L) = \dot{q}L. \tag{10.1.7a}$$

Now, imposing that the heat flux $J_U(L)$ is expressed through Newton's boundary condition (10.1.6), i.e., $J_L = \dot{q}L = h(T_L - T_0)$, we find the fluid temperature at the wall,

$$T_L \equiv T(L) = T_0 + \dot{q}L/h. \tag{10.1.8}$$

---

[1]Remind that J_U is positive when it is directed along the positive x direction, so that $-J_U(-x)$ is the heat flux exiting the control volume at section at $-x$.

Fig. 10.3 Parabolic
temperature profile

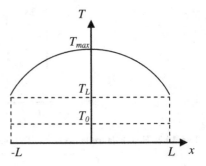

Inside the fluid, instead, heat propagates by conduction and the temperature distribution is determined by applying the Fourier constitutive relation, obtaining,

$$T(x) = T_L + \frac{\dot{q}L^2}{2k}\left(1 - \frac{x^2}{L^2}\right).$$

$$\text{i.e., } \Delta T(x) = \frac{\dot{q}L}{h} + \frac{\dot{q}L^2}{2k}\left(1 - \frac{x^2}{L^2}\right).$$

(10.1.9)

Therefore,

$$\Delta T_{\max} = \Delta T(0) = \frac{\dot{q}L^2}{2k}\left(1 + \frac{2}{Bi}\right),$$

(10.1.9a)

where $Bi = hL/k$ is the Biot number. The solution (10.1.9), represented in Fig. 10.3, shows that the parabolic shape of the temperature profile, $T(x) - T_L$, depends only on k and is independent of h, while the temperature drop at the wall, $T_L - T_0$, depends only on h and is independent of k. As we have seen in Sects. 9.2 and 9.3, this always occurs whenever the conductive and convective heat resistances are in series. These features can be seen starting from Eq. (10.1.7), showing that the heat flux at the wall, $J_U(L) = \dot{q}L$, is a fixed quantity, independent of h and k, and depending only on the heat power produced. Therefore, (a) applying the Fourier constitutive relation we see that, by decreasing k, the temperature gradient at the wall increases and thus the temperature profile becomes more elongated, independently of h; (b) applying Newton's boundary condition we see that, by decreasing h, the temperature drop at the wall increases, independently of k.

10.1.2 Cylindrical Geometry

Consider an electric wire of circular cross section with radius R and length L, through which an electric current is flowing, dissipating a uniform power per unit volume, $\dot{q}$. Now, if our objective consists of determining the heat flux, $J_R \equiv J_U(R)$, exiting the wire, we do not need to solve the problem. In fact, imposing, as in the plane case, that the total heat power exiting the system, $J_R(2\pi RL)$, is equal to the total heat power produced, $\dot{q}(\pi R^2 L)$, we find:

$$J_R = \dot{q}R/2. \tag{10.1.10a}$$

This result is very similar Eq. (10.1.7a) for plane geometry, i.e., $J_L = \dot{q}L$. Again, we see that the heat leaving the system is determined by the dissipation rate and therefore does not depend on any heat transport quantity.

Instead, if we want to determine the maximum temperature, we consider a heat balance on a cylindrical crown volume of thickness Δr and height L (see Fig. 10.4), obtaining:

$$[J_U(2\pi rL)]_r - [J_U(2\pi rL)]_{r+\Delta r} + \dot{q}(2\pi rL\Delta r) = 0.$$

Dividing by Δr and taking the limit as $\Delta r \to 0$ find:

$$\frac{d}{dr}(J_U r) = \dot{q}r,$$

with a general solution,

$$J_U(r) = \frac{1}{2}\dot{q}r + \frac{C_1}{r}.$$

Now, since on the centerline, $r = 0$, the solution must remain finite (and actually, out of symmetry, the flux must be zero there), it must be that $C_1 = 0$, so that now we obtain:

$$J_U(r) = \frac{1}{2}\dot{q}r. \tag{10.1.10}$$

This equation is very similar to Eq. (10.1.7) for the plane case, i.e., $J_U(x) = \dot{q}x$. In particular, at the wall, for $r = R$, Eq. (10.1.10) reduces to (10.1.10a), as it must.

Now, applying the Fourier constitutive relation, $J_U(r) = -k\, dT/dr$, and imposing a boundary condition at the wall, we find the temperature profile.

Fig. 10.4 Heat balance on a
cylindrical crown

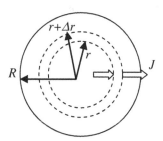

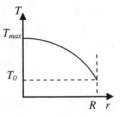

In particular, assuming that the wall temperature is fixed, i.e., $T(R) = T_0$, we obtain
again a parabolic profile (see Fig. 10.5),

$$\Delta T(r) = \Delta T_{\text{max}}\left(1 - \frac{r^2}{R^2}\right); \quad \Delta T_{\text{max}} = \Delta T(0) = \frac{\dot{q}R^2}{4k}. \tag{10.1.11}$$

It is easy to verify that Eq. (10.1.10a) is identically verified by explicitly determining the heat flux at the wall, $J_R = -k(dT/dr)_{r=R}$.

Note that here we have only one explicit boundary condition, although, in
reality, we have used another condition on the centerline that has allowed us to find
the other integration constant, i.e. $C_1 = 0$. Thus, in the case of a hollow cylinder, the
$1/r$ term in the flux does not vanish, so that by integration we obtain a logarithmic
term in the temperature profile, as we have seen in Sect. 9.2.2. In that case, though,
we have two boundary conditions at the inner and the outer walls, so that the
problem is well posed.

The mean temperature can be determined by applying the general definition
(10.1.4), where $V = \pi R^2 L$ and $d^3\mathbf{r} = 2\pi r dr L$, obtaining:

$$\overline{\Delta T} = \frac{1}{\pi R^2}\int_0^R \Delta T(r)2\pi r dr = \frac{\dot{q}R^2}{8k}. \tag{10.1.12}$$

Now, assume that the boundary condition is $J_U(R) = h[T(R) - T_0]$. Clearly, this
does not affect the heat balance, leading to Eq. (10.1.10), i.e., $J_U(r) = \dot{q}r/2$.
Then, applying this expression at the wall, when $r = R$, we find:
$T_R \equiv T(R) = T_0 + \dot{q}L/2h$. Inside the cylinder, applying the Fourier constitutive
relation as in the plane case, we find the parabolic profile (10.1.11), with T_R
replacing T_0, thus obtaining:

$$T(r) = T_R + \frac{\dot{q}R^2}{4k}\left(1 - \frac{r^2}{R^2}\right). \tag{10.1.13}$$

Thus:

$$\Delta T_{max} = \Delta T(0) = \frac{\dot{q}R^2}{4k}\left(1 + \frac{2}{Bi}\right), \quad \text{with } Bi = \frac{hR}{k}. \tag{10.1.13a}$$

10.1.3 Spherical Geometry

Consider a spherical nuclear pellet of radius R_i, surrounded by an aluminum cover whose external surface, at radius R_o is kept at temperature T_0. Knowing the heat generated per unit volume and per unit time, $\dot{q}$, calculate the maximum temperature, which is obviously reached at the center of the sphere.

Let us consider a heat balance on a spherical crown of thickness Δr:

$$\left[J_U\left(\pi r^2\right)\right]_r - \left[J_U\left(\pi r^2\right)\right]_{r+\Delta r} + \dot{q}\left(4\pi r^2 \Delta r\right) = 0.$$

Dividing by Δr and taking the limit as $\Delta r \to 0$ we obtain:

$$\frac{d}{dr}\left(J_U r^2\right) = \dot{q} r^2, \tag{10.1.14}$$

with a general solution,

$$J_U = \frac{1}{3}\dot{q}r + \frac{C}{r^2}. \tag{10.1.15}$$

In the region I, defined as that occupied by the nuclear fuel, i.e., when $0 < r < R_i$, imposing that at the center, $r = 0$, the solution must remain finite (and actually, out of symmetry, the flux must be zero there), it must be that $C = 0$, so that we obtain: $J_U^{(I)}(r) = \dot{q}r/3$. In region II, instead, defined as the aluminum covering, for $R_i < r < R_o$, $\dot{q} = 0$ and therefore Eq. (10.1.15) becomes: $J_U^{(II)}(r) = C/r^2$. Imposing the continuity of the flux at the interface, i.e., $J_U^{(I)}(R_i) = J_U^{(II)}(R_i)$, we obtain: $C = \dot{q}R_i^3/3$, and so we may conclude:

$$J_U^{(I)} = \frac{\dot{q}}{3}r \quad 0 < r < R_i; \quad J_U^{(II)} = \frac{\dot{q}R_i^3}{3}\frac{1}{r^2}. \tag{10.1.16a, b}$$

As a check, it is easy to verify that the total heat power leaving the nuclear element (and the aluminum sheet as well), $\dot{Q} = (\dot{q}R_i/3)(4\pi R_i^2)$, equals the heat power generated, $\dot{q}(4\pi R_i^3/3)$.

Now, applying the Fourier constitutive relation, $J_U(r) = -k\,dT/dr$ to Eq. (10.1.16b) and integrating with boundary condition $T^{(II)}(R_o) = T_0$, we obtain:

$$\Delta T^{(II)}(r) = \frac{\dot{q}R_i^2}{3k^{II}}\left(\frac{R_i}{r} - \frac{R_i}{R_o}\right), \quad \text{as } R_i < r < R_o, \tag{10.1.17}$$

where $\Delta T = T - T_0$, while k^{II} is the aluminum thermal conductivity. Then, applying the Fourier constitutive relation to Eq. (10.1.6) and integrating with boundary condition $T^{(I)}(R_i) = T^{(II)}(R_i)$, we obtain:

$$\Delta T^{(I)}(r) = \frac{\dot{q}R_i^2}{3k^{II}}\left(1 - \frac{R_i}{R_o}\right) + \frac{\dot{q}R_i^2}{6k^I}\left(1 - \frac{r^2}{R_i^2}\right), \tag{10.1.18}$$

where k^I is the thermal conductivity of the nuclear fuel. The maximum temperature is:

$$\Delta T_{\max} = \Delta T^{(I)}(0) = \frac{\dot{q}R_i^2}{3k^{II}}\left(1 - \frac{R_i}{R_o}\right) + \frac{\dot{q}R_i^2}{6k^I}. \tag{10.1.19}$$

Note that in the absence of the aluminum sheet, i.e. when $R_o = R_i$, the solution becomes:

$$\Delta T(r) = \Delta T_{\max}\left(1 - \frac{r^2}{R^2}\right); \quad \Delta T_{\max} = \Delta T(0) = \frac{\dot{q}R^2}{6k}. \tag{10.1.20}$$

In addition, the mean temperature can be determined by applying the general definition (10.1.4), with $V = 4\pi R^3/3$ and $d^3\mathbf{r} = 4\pi r^2 dr$, obtaining:

$$\overline{\Delta T} = \frac{3}{4\pi R^3}\int_0^R \Delta T(r)4\pi r^2 dr = \frac{\dot{q}R^2}{15k}. \tag{10.1.21}$$

Comparing Eq. (10.1.20) with (10.1.3) and (10.1.11) we see that the temperature profiles are always parabolic, as one would expect, since the heat fluxes are linear in all cases. In addition, the maximum and the mean temperatures depend on the dimension, d, of the problem (i.e., $d = 1$ in the plane case, $d = 2$ in the cylindrical case and $d = 3$ in the spherical case) and can be written as,

$$\Delta T_{\max} = \Delta T(0) = \frac{\dot{q}L^2}{(2d)k} \quad \text{and} \quad \overline{\Delta T} = \frac{\dot{q}R^2}{d(d+2)k}, \tag{10.1.22}$$

where L is a characteristic size (half-width or radius) of the problem. Therefore, we see that the temperature profile becomes flatter as we move from plane to cylindrical and from cylindrical to spherical geometry, as the surfaces where the cooling takes place become proportionally larger.

10.2 Heat Conduction with Chemical Reaction

Consider a slab of thickness L composed of a polymer-monomer mixture, where the polymerization reaction is at equilibrium when $T = T_0$, so that when $T > T_0$ the reaction is endothermic (i.e., it absorbs heat), while when $T < T_0$ the reaction is esothermic (i.e., it produces heat). For sake of simplicity, assume a linear relation, $\dot{q} = -r(T - T_0)$ for the heat source, where r is a positive constant. Now assume that one wall, at $x = L$, is kept at the equilibrium temperature, i.e., $T(L) = T_0$, while the other wall, at $x = 0$, is heated to a higher temperature $T_1 > T_0$, so that $T(0) = T_1$. We want to determine the heat flux, $J_U(0)$, entering at $x = 0$.

From a heat balance we obtain Eq. (10.1.2), i.e., $-k\,d^2T/dx^2 = \dot{q}$, which in our case we can rewrite as:

$$\frac{d^2\Theta}{d\xi^2} - Da\,\Theta = 0, \quad \text{with boundary conditions } \Theta(0) = 1 \text{ and } \Theta(1) = 0,$$

$$(10.2.1)$$

where:

$$\Theta = \frac{T - T_0}{T_1 - T_0}; \quad \xi = \frac{x}{L}; \quad Da = \frac{rL^2}{k}. \tag{10.2.2}$$

Here, Da denotes the *Damköhler* number, a dimensionless quantity expressing the ratio between the diffusion time, L^2/α, and the reaction time, $\rho c/r$, where $\alpha = k/\rho c$ is thermal diffusivity. Also, the flux,

$$J(0) = -k\left(\frac{dT}{dx}\right)_{x=0}, \tag{10.2.3}$$

can be expressed in dimensionless form, defining the Nusselt number,[2]

$$Nu = \left|\frac{J}{J_{cond}}\right| = -\left(\frac{d\Theta}{d\xi}\right)_{\xi=0}. \tag{10.2.4}$$

The Nusselt number is the ratio[3] between the flux at the wall and its conductive part, i.e., the flux that one would measure if conduction were the only heat transfer mechanism and in the absence of all the other phenomena, such as convection, heat generation, etc.

---

[2]Named after Ernst Kraft Wilhelm Nusselt (1882–1957), a German engineer. He held the chair of Theoretical Mechanics in München from 1925 to 1952, where was succeeded by Ernst Schmidt.
[3]The Nusselt number is often defined as twice the one defined in (10.2.4), where $Nu = 1$ when conduction is the only heat transfer mechanism.

Fig. 10.6 Temperature
profile as a function of Da

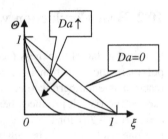

The solution of Eq. (10.2.1) is:

$$\Theta(\xi) = C_1\sinh\left(\sqrt{Da}\xi\right) + C_2\cosh\left(\sqrt{Da}\xi\right),$$

where C_1 and C_2 are constants to be determined by satisfying the boundary conditions. At the end we find the following solution, illustrated in Fig. 10.6,

$$\Theta(\xi) = \cosh\left(\sqrt{Da}\xi\right) - \frac{\cosh\sqrt{Da}}{\sinh\sqrt{Da}}\sinh\left(\sqrt{Da}\xi\right). \qquad (10.2.5)$$

and

$$Nu = \frac{\cosh\sqrt{Da}}{\sinh\sqrt{Da}}\sqrt{Da}. \qquad (10.2.6)$$

In the following, we consider separately the limit cases $Da \ll 1$ and $Da \gg 1$.

10.2.1 Asymptotic Expansion for Small Da

In this case, the heat generation (or consumption) term is small compared to the conduction term. In fact, when $Da = 0$, we find the usual linear profile, $\Theta = 1 - \xi$, and consequently $Nu = 1$. The correction to this solution can be found directly from (10.2.5) and (10.2.6), observing that, for any $\varepsilon \ll 1$, we have: $\text{Cosh}(\varepsilon) = 1+\varepsilon^2/2 + \cdots$ and $\text{Sinh}(\varepsilon) = \varepsilon + \varepsilon^3/6 + \cdots$, so that we obtain:

$$Nu \approx \frac{1 + Da/2 + \cdots}{1 + Da/6 + \cdots} \approx \left(1 + \frac{1}{2}Da + \cdots\right)\left(1 - \frac{1}{6}Da + \cdots\right)$$
$$\approx 1 + \frac{1}{3}Da + O(Da^2). \qquad (10.2.7)$$

Now, suppose that we want to find this first-order correction to the leading-order solution, without solving the whole problem. This is important in many cases, where the full solution is not known, but nevertheless it can be expressed as a power series of a small parameter ε (in our case, the Damköhler number). For example, this is what happens in fluid mechanics when we calculate the drag force exerted by a fluid on a rigid sphere at $Re \ll 1$.

In our case, denoting $\varepsilon = Da \ll 1$, let us assume that the solution can be expressed as the following expansion:

$$\Theta(\xi) = \Theta_0(\xi) + \varepsilon\Theta_1(\xi) + \varepsilon^2\Theta_2(\xi) + \cdots;$$
$$\text{where } 0 \leq \xi \leq 1, \quad \text{with } |\Theta_n| = O(1). \tag{10.2.8}$$

Note that here we assume that the nth term of the expansion is of $O(\varepsilon^n)$ and therefore it is much smaller than the preceding term. In particular, that means that Θ_1 must remain much smaller than $1/\varepsilon$ in the whole domain $0 \leq \xi \leq 1$, so that the leading-order term of this expansion, Θ_0, is everywhere much larger than $\varepsilon\Theta_1$, which therefore can be considered as the first-order correction. The same can be said about the successive terms. When this condition is satisfied, the power series (10.2.8) is denoted as *regular perturbation expansion*.

Now, let us substitute Eq. (10.2.8) into Eq. (10.2.1), obtaining:

$$\Theta_0''(\xi) + \varepsilon\left[\Theta_1''(\xi) - \Theta_0(\xi)\right] + \varepsilon^2\left[\Theta_2''(\xi) - \Theta_1(\xi)\right] + \cdots = 0 \tag{10.2.9}$$

with boundary conditions,

$$\Theta_0(0) + \varepsilon\Theta_1(0) + \varepsilon^2\Theta_2(0) + \cdots = 1; \quad \Theta_0(1) + \varepsilon\Theta_1(1) + \varepsilon^2\Theta_2(1) + \cdots = 0 \tag{10.2.10}$$

and Nusselt number,

$$Nu = Nu_0 + \varepsilon Nu_1 + \cdots = \left[-\Theta_0'(0)\right] + \varepsilon\left[-\Theta_1'(0)\right] + \cdots \tag{10.2.11}$$

Now, since the solution must be valid for any value of ε, terms of the same magnitude must balance each other. Consider first all $O(1)$-terms:

$$\Theta_0''(\xi) = 0; \quad \text{with } \Theta_0(0) = 1 \text{ and } \Theta_0(1) = 0. \tag{10.2.12}$$

Hence,

$$\Theta_0(\xi) = 1 - \xi, \tag{10.2.13}$$

and thus,

$$Nu_0 = 1. \tag{10.2.14}$$

This is the leading-order solution, corresponding to the pure conductive case with $Da = 0$.

Now consider all $O(\varepsilon)$-terms:

$$\Theta_1''(\xi) = \Theta_0(\xi) = 1 - \xi; \quad \text{with} \ \Theta_1(0) = \Theta_1(1) = 0, \tag{10.2.15}$$

where we have substituted Eq. (10.2.13). Integrating we obtain:

$$\Theta_1(\xi) = -\frac{1}{3}\xi + \frac{1}{2}\xi^2 - \frac{1}{6}\xi^3. \tag{10.2.16}$$

Thus,

$$Nu_1 = \frac{1}{3}, \tag{10.2.17}$$

which coincides with the first-order result (10.2.7) that was obtained from the exact solution. Proceeding further, we could determine the solution up to any degree of approximation. Thus, we see that, unlike the integral method described in Sect. 7.7, perturbation methods allow us to keep track of the magnitude of the neglected terms, together with a systematic procedure to increase accuracy. Obviously, however, perturbation methods can be applied only to problems where a very small (or a very large) parameter can be defined.

10.2.2 Asymptotic Expansion for Large Da

This case is more complicated. Let us see why. Here, the heat generation term is dominant with respect to the conductive term. That means that the system tends to have a uniform temperature T_0, that is $\Theta = 0$, in the whole domain. However, in the vicinity of the wall $\xi = 0$, the temperature Θ must shoot up from 0 to 1: this is the sign of the existence of a boundary layer at the wall. In fact, from the exact solution (10.2.6), considering that $\exp(-\sqrt{Da}) \cong 0$, we obtain:

$$\Theta(\xi) \cong \exp\left(-\sqrt{Da}\,\xi\right); \quad Nu \cong \sqrt{Da}. \tag{10.2.18}$$

Again, let us see if this solution can be determined without having to solve the whole problem. Denoting $\varepsilon = Da^{-1}$, the governing Eq. (10.2.1) becomes,

$$\varepsilon \frac{d^2\Theta}{d\xi^2} - \Theta = 0; \quad \Theta(0) = 1; \quad \Theta(1) = 0. \tag{10.2.19}$$

When $\varepsilon = 0$, this 2nd order differential equation is transformed into a 0th order equation, obtaining $\Theta = 0$, that is a solution that (by chance, in this case) satisfies the boundary condition at $\xi = 1$, but not that at $\xi = 0$. Therefore, we will find that the temperature profile $\Theta = 0$ is valid almost everywhere, with the exception of a small region near the wall at $\xi = 0$, called *thermal boundary layer*, where the profile shoots up to $\Theta = 1$ (see Fig. 10.6). This is the typical behavior of all types of boundary layers: a small parameter ε multiplies the highest-order derivative in the governing equation. Consequently, when $\varepsilon = 0$, the governing equation is transformed into an equation of lower order, that cannot satisfy all the imposed boundary conditions. In fluid mechanics' boundary layer, the small parameter is the inverse Reynolds number, $\varepsilon = Re^{-1}$, which multiplies the viscous, 2nd order term in the Navier-Stokes equation; therefore, when $\varepsilon = 0$, the governing equation is transformed into the Euler equation, i.e. a 1st order differential equation, that requires only one boundary condition. Accordingly, the inviscid solution is valid almost everywhere, with the exception of a small boundary layer, thus allowing the boundary condition at the wall to be satisfied.

The above discussion boils down to concluding that within the boundary layer the profile shoots up from $\Theta = 0$ to $\Theta = 1$, so that the temperature gradient is large and the diffusive term in Eq. (10.2.19) balances the heat generation term. That means imposing that the Damköhler number defined in terms of the boundary layer thickness δ is of $O(1)$, i.e.,

$$Da_\delta = \frac{r\delta^2}{k} = 1 \Rightarrow \delta = L/\sqrt{Da}. \tag{10.2.20}$$

In fact, the basic idea defining the boundary layer is that, at the edge of this region, the different terms of the governing equation balance each other.

More rigorously, we can define a *stretching coordinate*, $\zeta = \xi\varepsilon^{-\alpha}$, where α is a positive constant that can be defined by imposing that in the region where $\zeta = O(1)$ the conduction term and the heat generation term balance each other. Thus, rewriting Eq. (10.2.19) in terms of the stretching coordinate, we obtain:

$$\varepsilon^{1-2\alpha}\frac{d^2\Theta}{d\zeta^2} - \Theta = 0; \quad \Theta(0) = 1; \quad \Theta\left(\frac{1}{\varepsilon^\alpha} \to \infty\right) = 0. \tag{10.2.21}$$

Accordingly, the two terms in Eq. (10.2.21) balance each other when $\alpha = \frac{1}{2}$ and therefore $\zeta = \xi/\sqrt{\varepsilon}$. Then, the solution of Eq. (10.2.12) becomes:

$$\Theta(\zeta) = e^{-\zeta} \Rightarrow \Theta(\xi) = \exp(-\xi/\sqrt{\varepsilon}) = \exp(-\sqrt{Da}\xi), \tag{10.2.22}$$

and

$$Nu = \sqrt{Da}, \tag{10.2.23}$$

in perfect agreement with the exact result.

This technique was generalized in the years 1950–1960, leading to the development of the *singular perturbation expansion* method, often referred to as *matched asymptotic expansion*. It starts by observing that when $\varepsilon = 0$ the problem is ill-posed, and therefore there is no hope in finding a regular perturbation expansion (10.2.8) that is valid in the whole domain. Therefore, the domain is divided into two regions, that are described using two different coordinate, i.e., ξ and ζ in our case. Within the two regions, then, it is assumed that the solutions can be written as, so-called, inner and outer regular perturbation expansions. In our case, that means assuming the following expressions:

$$\Theta^{(o)}(\xi) = \Theta_0^{(o)}(\xi) + \varepsilon\Theta_1^{(o)}(\xi) + \varepsilon^2\Theta_2^{(o)}(\xi) + \cdots \quad 0 < \xi \le 1, \quad (10.2.24a)$$

$$\Theta^{(i)}(\zeta) = \Theta_0^{(i)}(\zeta) + \varepsilon\Theta_1^{(i)}(\zeta) + \varepsilon^2\Theta_2^{(i)}(\zeta) + \cdots \quad \zeta \ge 0. \quad (10.2.24b)$$

Now, each of these expansion can satisfy only one boundary condition, namely $\Theta^{(o)}(\xi = 1) = 0$ and $\Theta^{(i)}(\zeta = 0) = 1$, so that they can only be expressed in terms of some integration constants. At the end, these constants are determined by matching the two expansions, that is imposing the continuity of the temperature profile at the edge of he two domains, where the inner and outer regions merge into one another, i.e.,

$$\lim_{\xi \to 0} \Theta_0^{(o)}(\xi) = \lim_{\zeta \to \infty} \Theta_0^{(i)}(\zeta); \quad \lim_{\xi \to 0} \Theta_1^{(o)}(\xi) = \lim_{\zeta \to \infty} \Theta_1^{(i)}(\zeta); \quad \text{etc.} \quad (10.2.25)$$

In our case, the matching is trivial, since the outer solution is $\Theta^{(o)} = 0$, while the inner solution tends to zero exponentially when $\zeta \to \infty$.

10.3 Problems

10.1 Calculate the temperature profile and the heat flux at the walls in a slab of thickness L and heat conductivity k, whose surfaces $x = 0$ and $x = L$ are kept, respectively, at temperature T_0 and T_L, and in which there is a uniform heat source (heat power produced per unit volume) $\dot{q}$.

10.2 In a graphite moderated nuclear reactor, heat is uniformly generated within cylindrical uranium bars of radius $R = 2.5$ cm at a rate $\dot{q} = 8 \times 10^7$ W/m^3 per unit volume. The fuel bars are cooled using water at $T_0 = 120$ °C, with heat transfer coefficient $h = 30$ kW/m^2 K. Knowing that the uranium thermal conductivity is $k = 30$ W/m K, determine the temperatures at the center and at the surface of the bars.

10.3 A new type of graphite-moderated nuclear reactor is composed of cylindrical bars with radius $R = 2$ mm, made of enriched uranium, with thermal conductivity $k = 30$ W/m K, that are cooled using water at $T_0 = 100$ °C, with a heat transfer coefficient $h = 30$ kW/m^2 K. Since for safety reason the maximum temperature of the bars cannot exceed $T_{max} = 500$ °C, determine the maximum power per unit volume $\dot{q}_{max}$, so that the maximum enrichment degree of uranium can be found.

10.4 Calculate the average temperature in the three geometries (planar, cylindrical and spherical) examined in Sect. 10.1.

10.5 Consider a metallic wire of radius R and thermal conductivity k_1, where an imposed electric current produces a uniform heat power per unit volume, $\dot{q}$. The wire is covered with an insulating layer of thickness d and thermal conductivity k_2, whose outer surface is kept at a constant temperature, T_0. Determine the maximum temperature in the wire.

10.6 A heat power $\dot{q}$ per unit volume is generated within a sphere of radius R and thermal conductivity k, that is cooled by a fluid at temperature T_0, with heat transfer coefficient h. Find the total heat leaving the sphere per unit time (why doesn't it depend on k?) and the maximum temperature.

10.7 A fluid flows in laminar Poiseuille flow between two plates at a distance $2B$. Assuming that the fluid temperature at the walls, T_0, is fixed, determine the temperature profile considering that the fluid is heated by viscous dissipation.

10.8 Consider a hollow wire, with inner radius R, outer radius $2R$ and thermal conductivity k. At the outer surface, when $r = 2R$, the wire is insulated, while inside it exchanges heat with a coolant at temperature T_0, with heat transfer coefficient h.

   (a) Knowing that the electric current generates a uniform heat power per unit volume, $\dot{q}$, write the heat balance equation with boundary conditions in dimensionless form, defining $\Theta = (T - T_0)/(\dot{q}R^2/k)$.
   (b) Assuming that the heat exchange between wire and fluid is very efficient (what does that mean?), find $T(2R)$.

10.9 Consider a spherical electric heater of radius R. Water is heated through an electric resistance producing a heat power (energy per unit time) $\dot{Q}$, while at the outer surface heat is exchanged with the surrounding air at temperature T_0, with heat transfer coefficient h.

   (a) Write the differential equation describing the temporal evolution of the water temperature, $T = T(t)$.
   (b) Determine the final temperature.

10.10 Hemichordates are invertebrate marine worms living in shallow sea waters. They can be modeled as cylinders with thermal conductivity $k = 1$ W/m K, whose inner metabolism produces a heat power per unit volume, $\dot{q} = 1$ W/cm^3. Assuming that the waters where they live has a temperature $T_0 = 15$ °C and that the maximum temperature that can be reached in their bodies is $T_{max} = 40$ °C, determine the maximum radius that these creatures can reach.

Chapter 11
Macroscopic Energy Balance

Abstract This chapter is closely related to Chap. 4, as we analyze the macroscopic aspect of heat transport by performing energy balances over macroscopic slab or shells perpendicular to the direction of the heat flow. First, in the introductory Sect. 11.1, the Bernoulli equation is re-derived, stressing all its heat-related features, instead of the terms related to momentum transport, as we did in Chap. 3. In fact, we see that the heat exchanged equals the enthalpy variation and is the product between a temperature difference (i.e., the driving force) and a coefficient of heat transfer. This latter, as shown in Sect. 11.2, can be determined using experimental or analytical expressions, which are then applied in Sects. 11.3 and 11.4 to design shell&tube and finned heat exchangers, respectively.

11.1 Introduction

As we have seen in Chap. 4, energy conservation (that is, the first law of thermodynamics) applied to the control volume represented in Fig. 4.1 states that the energy accumulated within the volume equals the difference between the energy entering and the energy leaving through the closed surface enclosing the volume, i.e.,

$$\frac{d}{dt} \int_V \rho e dv = (\rho v e S)_{in} - (\rho v e S)_{out} + \dot{W} + \dot{Q}. \qquad (11.1.1)$$

Here, e represents the energy (kinetic, potential and internal) per unit mass,

$$e = \frac{energy}{fluid\ mass} = u + \frac{1}{2}v^2 + gz, \qquad (11.1.2)$$

where u is the specific internal energy. In Eq. (11.1.1), $\rho v e$ is the convective energy flux, which is entering and exiting through the cross sections S_1 and S_2, while $\dot{W}$ and $\dot{Q}$ are, respectively, the work and the heat exchanged per unit time through the

© Springer International Publishing Switzerland 2015
R. Mauri, *Transport Phenomena in Multiphase Flows*,
Fluid Mechanics and Its Applications 112, DOI 10.1007/978-3-319-15793-1_11

surface enclosing the control volume. In turn, the work can be considered as the sum of the work done by pressure forces on the fluid, PvS, and the work exerted by a pump on the fluid flow (the so called *shaft work*), that we indicate in terms of the mass flow rate, $\dot{m}$, as $\dot{W}_p = \dot{m}w_p$. Therefore, we obtain:

$$\frac{d}{dt}\int_V \rho e dv = -\Delta\left(\frac{1}{2}\rho v^3 S\right) - g\Delta(\rho v z S) - \Delta(pvS) - \Delta(\dot{m}u) + \dot{m}w_p + \dot{m}q,$$

$$(11.1.3)$$

where we have denoted $\Delta f = f_{out} - f_{in}$ and $\dot{m}q = \dot{Q}$. At steady state, since the mass flow rate is constant, considering that $pvS = \dot{m}p/\rho$, and dividing by $\dot{m}$ we obtain:

$$0 = \Delta\left(\frac{1}{2}v^2 + gz + u + \frac{p}{\rho}\right) - q - w_p. \qquad (11.1.4)$$

Now, proceeding as in Sect. 4.2, we obtain the Bernoulli equation. Instead, let us remind of the definition $h = u + p/\rho$ of enthalpy[1] per unit mass. In particular, for ideal gases, h depends only on temperature,

$$\Delta h = \int_{T_1}^{T_2} c_p dT \quad \text{for ideal gases,}$$

where c_p is the specific (i.e., per unit mass) heat at constant pressure, while for incompressible flows, with ρ being constant, the specific heat at constant pressure and that at constant volume are equal to each other, i.e., $c_p = c_v = c$, so that,

$$\Delta h = \int_{T_1}^{T_2} c dT + \frac{1}{\rho}(p_2 - p_1) \quad \text{for incompressible flows.}$$

In most cases, whenever there are heat exchanges, the enthalpy variations are much larger than any change in kinetic and potential energy; conversely, the exchanged heat power $\dot{Q}$ is much larger than the shaft work, so that in the end the balance equation (11.1.4) becomes,

$$\Delta h = q \Leftrightarrow \dot{m}\Delta h = \dot{Q}. \qquad (11.1.5)$$

---

[1]Not to be confused with the heat transfer coefficient!!

In this equation, enthalpy is a function of the temperature of the system,[2] while the exchanged heat power can be determined case by case, expressing it in terms of a heat transfer coefficient at the boundaries.

11.2 The Heat Transfer Coefficient

As we have seen in the previous chapters, the heat exchanged between an object whose wall has a temperature T_w, and a fluid at temperature T_a, is proportional to the product between a heat transfer coefficient, h, the area of the interface between the object and the fluid, and the temperature difference, $T_w - T_0$,

$$\dot{Q} = -hS(T_w - T_0). \qquad (11.2.1)$$

Note the sign: $\dot{Q}$ is positive when it represents the heat entering the system, i.e., when it indicates the heat flowing from a warmer fluid to a colder wall. Due to the linearity of both convective and diffusive heat fluxes with respect to the temperature difference, we know that there must be a linear dependence between cause (the temperature difference) and effect (the heat flux); therefore, h does not depend on temperature, although it depends, in general, on the fluid physical properties and its flow characteristics, e.g., the Reynolds number.[3] The fluid temperature, T_0, is the fluid bulk temperature, that is measured at a sufficiently large distance from the wall, so that it may be assumed to be uniform.

Let us consider the simplest case, consisting of a sphere of radius R and temperature T_w, that exchanges heat with an unbounded medium at temperature T_0. In the absence of any convection (e.g., assume that the medium is either solid or a very viscous liquid), we have seen in Sect. 9.2.3 that the temperature of the medium decays like $1/r$, obtaining Eq. (9.2.14),

$$(T - T_0)/(T_w - T_0) = R/r. \qquad (11.2.2)$$

Therefore, the heat flux at the wall is:

$$J_U = -\frac{\dot{Q}}{S} = -k\left(\frac{dT}{dr}\right)_{r=R} = \frac{k\Delta T}{R}, \qquad (11.2.3)$$

where $\Delta T = T_w - T_0$. Note that J_U is positive when it is directed along the unit vector perpendicular to the wall, pointing outward of the control volume; therefore

---

[2]Although this is rigorously true only for incompressible fluids, it is generally true that even for gases the dependence of enthalpy on pressure can be neglected.

[3]This is an approximation, since all fluid properties, and so h as well, depend on the temperature, although the dependence is rather weak.

it has an opposite sign with respect to $\dot{Q}$. Comparing Eq. (11.2.3) with (11.2.1), we have:

$$h = k/R, \qquad (11.2.4)$$

indicating that, in the case of purely conductive heat exchange, the heat transport coefficient is the ratio between the thermal conductivity of the medium and a characteristic linear dimension, L. Therefore, it comes natural to define the *Nusselt number*, as a dimensionless quantity representing the ratio between the heat transfer at the wall (due to both convection and diffusion), $J_U = h\Delta T$, and its conductive part, $J_{Ud} = k\Delta T/L$,[4]

$$Nu = \frac{J_U}{J_{Ud}} = \frac{hL}{k}. \qquad (11.2.5)$$

In general, the characteristic dimension L appearing in the definition (11.2.5) of Nu is either the hydraulic radius, $R_H = 2S/P$, where S is the section area and P the wetted volume, or the hydraulic diameter, $D_H = 2R_H$, depending on the circumstances. In fact, on one hand, in pipe flow it is customary to use the tube diameter as a characteristic dimension of the problem, probably because this is what one can measure most easily with a caliber. On the other hand, though, using the radius, R, as a characteristic dimension, is most convenient in the study of the heat exchange around a sphere, since from Eqs. (11.2.3) to (11.2.5) we see that the Nusselt number in this case is exactly the ratio between the total heat flux and its conductive part. The following examples will clarify the different conventions that can be found in the literature.

11.2.1 Forced Convection, Internal Flow

In pipe flows, heat transport takes place along the radial direction, perpendicular to the mean fluid velocity. Therefore, in case of laminar flow, heat flux is only conductive, while for turbulent flow its convective component is predominant by far. In all cases, conventionally, the Nusselt number is defined in terms of the hydraulic diameter, i.e., $Nu = hD_H/k$.

---

[4]Note the Nusselt number has an identical expression as the Biot number, Bi, defined in Eq. (9.2.3), as both dimensionless quantities represent the ratio between the total heat flux at the wall, J_U, and a conductive heat flux, J_{Ud}. However, in the definition of Bi, k and J_{Ud} denote the thermal conductivity and the heat flux inside the solid medium, which means that the two heat fluxes J_U, and J_{Ud} appearing in the definition of Bi take place in two different locations, i.e. one in the fluid (h) and the other in the solid medium (k). On the other hand, in the definition of Nu, J_{Ud} represents the heat flux at the wall that one would have if the heat exchange were exclusively conductive. Therefore, in this case, k denotes the thermal conductivity of the fluid, and the two heat fluxes J_U, and J_{Ud} appearing in the definition of Nu take place at the same location, i.e., within the fluid.

(a) Laminar flow
In this case, as expected, Nu is a constant that depends solely on the geometry of the problem (i.e., on the choice of the characteristic length L). When the wall temperature is constant, we find $Nu = 3.66$ for circular tubes (with $D_H = D$), and $Nu = 7.54$ for channels (with $D_H = 2H$, where H is the channel height). When, instead, the flux at the wall is fixed, we find $Nu = 4.36$ for circular tubes, and $Nu = 8.23$ for channels. These are exact results, that were obtained analytically by Graetz (1885).

(b) Turbulent flow
Here we use experimental correlations, relating the Nusselt number, $Nu = hD_H/k$, to the Reynolds number, $Re = VD_H/\nu$, and the Prandtl number, $Pr = \nu/\alpha$, where ν is the kinematic viscosity and α the thermal diffusivity. One of the most common and particularly simple expression is the Dittus-Boelter correlation,

$$Nu = \frac{hD}{k} = 0.023\,Re^{4/5}Pr^n, \tag{11.2.6}$$

where $n = 0.4$ for heating (i.e., the wall is warmer than the bulk fluid) and 0.33 for cooling (i.e., the wall is colder than the bulk fluid). The Dittus-Boelter correlation is valid when $Re > 10^4$ and $0.7 < Pr < 160$. More complex correlations can be obtained considering the tube roughness and applying the Colburn-Chilton analogy (see Sect. 13.3) to the Moody's curve (see Sect. 4.5).

11.2.2 Forced Convection, External Flow

In analyzing the heat transfer associated with the flow past the exterior surface of a solid, the situation is complicated by phenomena such as boundary layer separation. In most cases, a mean Nusselt number can be calculated using the Colburn analogy (see Sect. 13.3).

(a) Laminar flow around a sphere
In the absence of convection, we have seen that the Nusselt number (defined in terms of the sphere diameter, D) is 2. Then, for $Pr > 0.5$, we find experimentally,

$$Nu = \frac{hD}{k} = 2 + 0.60\,Re^{1/2}Pr^{1/3}. \tag{11.2.7}$$

This result is in excellent agreement with the theoretical analysis of Sect. 13.2.

(b)   Laminar flow around a cylindrical tube
The following correlation can be used:

$$Nu = \frac{hD}{k} = \left[0.35 + 0.56\, Re^{0.52}\right] Pr^{0.3}. \tag{11.2.8}$$

11.2.3 Laminar Convection Past a Flat Plate

(a)   When $Pr > 0.5$, we find analytically the mean value of the Nusselt number,

$$\overline{Nu} = \frac{\bar{h}L}{k} = \frac{2}{3}\, Re_L^{1/2} Pr^{1/3}, \tag{11.2.9}$$

where L is the distance from the plate edge, $Re_L = UL/v$ and $\bar{h}$ is the mean heat transfer coefficient,

$$\bar{h} = \frac{1}{L} \int_0^L h(x)\, dx.$$

(b)   For liquid metals, where $10^{-3} < Pr < 10^{-1}$, we find:

$$\overline{Nu} = \frac{\bar{h}L}{k} = 1.13\, Re_L^{1/2} Pr^{1/2}. \tag{11.2.10}$$

11.3   Heat Exchangers

Heat exchangers are pieces of equipment where heat is transferred in an efficient way from a hot to a cold fluid. The two fluids may be in direct contact with one another, or they may be separated by solid walls. Here we consider the latter case.

11.3.1 Simple Geometries

In Fig. 11.1 we can see the simplest geometry of heat exchangers, namely a double pipe heat exchanger composed of two concentric pipes. There are two different possible flow arrangement: in *co-current flows*, corresponding to *parallel-flow* heat exchangers, the two fluids enter the exchanger at the same end, and travel in parallel to the other side. In *counter-current flows*, corresponding to *counter-flow* heat

Fig. 11.1 Double pipe heat exchanger

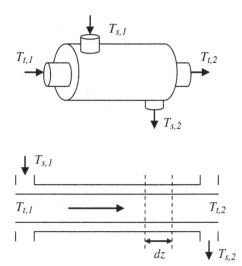

exchangers, the fluids enter the exchanger from opposite ends.[5] Although here we consider the parallel flow case, all the results can be extended to the counter-current case as well.

In the following, we will denote by the subscript "t" (i.e., tube-side) all quantities related to the flux inside the tubes, and by "s" (i.e., shell-side) those related to the flow outside the tubes. In addition, the subscript "1" and "2" indicate the sections 1 and 2 of the heat exchanger; in co-current flows, section 1 and section 2 correspond to inlet and outlet flows, respectively, for both tube- and shell-side flows; with counter-current flow arrangement, instead, section 1 corresponds to the inlet tube-side and the outlet shell-side flows, while section 2 corresponds to the inlet shell-side and the outlet tube-side flows. For sake of convenience, let us assume that the tube-side fluid is colder and is heated by the hotter shell-side fluid, which is consequently cooled. Anyway, the final result does not depend on this assumption.

First of all, let us consider a global heat balance, assuming that the system is adiabatic, that is neglecting all heat losses (this assumption is largely satisfied in all applications). Then imposing that heat lost by the warmer fluid $-\dot{Q}_s$, is absorbed by the colder fluid, $\dot{Q}_t$, we obtain:

$$\dot{Q}_t = \dot{m}_t c_t \left(T_{t,2} - T_{t,1}\right) = \dot{m}_s c_s \left(T_{s,1} - T_{s,2}\right), \quad (11.3.1)$$

where c_t and c_s are the specific heat of the two fluids, while $\dot{m}_t$ and $\dot{m}_s$ are their mass flow rates. Now, consider the heat balance on the volume of thickness dz represented in Fig. 11.1:

---

[5]There a third flow arrangement, namely *cross-flow*, where the fluids travel roughly perpendicular to one another through the heat exchanger (see Sect. 11.3.2).

$$d\dot{Q}_t = \dot{m}_t c_t dT_t = -d\dot{Q}_s = -\dot{m}_s c_s dT_s. \tag{11.3.2}$$

Finally, the heat exchanged between the two fluids can be determined through Eq. (11.1.1) in terms of the exchange area, dS, the temperature difference between the two fluids, $\Delta T = T_s - T_t$, and an overall heat transfer coefficient, h_{tot}, defined in Eq. (9.3.3), with $h_{tot} = (R_{th}S)^{-1}$,

$$d\dot{Q}_t = h_{tot}\Delta T \, dS. \tag{11.3.3}$$

Note that dS can indicate the inner or the outer tube surface, depending on how h_{tot} is defined, as in Eq. (9.3.9).[6] Now, deriving dT_t and dT_s from Eq. (11.3.2), and substituting Eq. (11.3.3), we obtain:

$$d(\Delta T) = -d\dot{Q}_t\left(\frac{1}{\dot{m}_t c_t} + \frac{1}{\dot{m}_s c_s}\right) = -h_{tot}\Delta T\left(\frac{1}{\dot{m}_t c_t} + \frac{1}{\dot{m}_s c_s}\right)dS. \tag{11.3.4}$$

Then, integrating:

$$\ln\left(\frac{\Delta T_1}{\Delta T_2}\right) = h_{tot}S\left(\frac{1}{\dot{m}_t c_t} + \frac{1}{\dot{m}_s c_s}\right), \tag{11.3.5}$$

where $\Delta T_1 = T_{s,1} - T_{t,1}$ and $\Delta T_2 = T_{s,2} - T_{t,2}$ are the temperature differences between the two fluids at section 1 and section 2, respectively. Equation (11.3.5) connects the four inlet and outlet temperatures of the two fluids quantities, with their respective mass flow rates and the exchange area. In general, however, instead of this equation we prefer to use another expression, that we can obtain by substituting $1/\dot{m}_t c_t$ and $1/\dot{m}_s c_s$ from Eq. (11.3.1) into Eq. (11.3.5):

$$\dot{Q}_t = h_{tot}S(\Delta T)_{lm}, \tag{11.3.6}$$

where,

$$(\Delta T)_{ln} = \frac{\Delta T_1 - \Delta T_2}{\ln(\Delta T_1/\Delta T_2)} = \frac{(T_{s,1} - T_{t,1}) - (T_{s,2} - T_{t,2})}{\ln\left[(T_{s,1} - T_{t,1})/(T_{s,2} - T_{t,2})\right]}, \tag{11.3.7}$$

is the logarithmic mean temperature difference (LMTD). In general, the LMTD is the temperature driving force for heat transfer in flow systems, as we see in Sect. 11.5 where another example is illustrated.

Obviously, if the shell-side fluid were the coldest, all signs should be inverted, leaving the result unchanged. Also, for counter-flow heat exchangers, the final

---

[6]Generally, by exchange surface we denote the outer tube surface, as tubes are measured by their outside diameter.

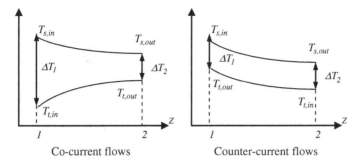

Fig. 11.2 Co- and counter-current temperature profiles

result (11.3.6)–(11.3.7) remains unchanged, provided that the subscripts are given their appropriate meaning, that is:

$$\Delta T_1 = T_{s,in} - T_{t,in}; \quad \Delta T_2 = T_{s,out} - T_{t,out} \quad \text{for co-current flows} \qquad (11.3.8)$$

while:

$$\Delta T_1 = T_{s,in} - T_{t,out}; \quad \Delta T_2 = T_{s,out} - T_{t,in} \quad \text{for counter-current flows} \qquad (11.3.9)$$

The temperatures of the two fluids are represented in Fig. 11.2 for both co-current and counter-current flow arrangements. In the former case, as we move from the inlet to the outlet, the temperature difference decreases, while in the latter case it remains almost constant so that the average temperature difference in the latter case is greater. For that reason, the counter-current arrangement is more efficient than the co-current flow arrangement, as it can transfer more heat for fixed inlet and outlet temperatures and for a given exchange area (see Problem 11.4). In addition, we see that the maximum or minimum temperature that can be reached within a parallel heat exchanger is an intermediate temperature between the two inlet temperatures, while for counter-current flow arrangements the maximum temperature that the tube-side fluid can reach is the inlet temperature of the shell-side fluid, and vice versa.

Note the following special conditions.

(a) When $\dot{m}_t c_t \ll \dot{m}_s c_s$, or $\dot{m}_t c_t \gg \dot{m}_s c_s$, we may assume that T_t is constant or T_s is constant, respectively.

(b) In the case of condensers or evaporators, the heat balance must be written in terms of the specific enthalpies, that is using the expression $\dot{Q} = \dot{m}\Delta h$.

11.3.2 Complex Geometries

Most heat exchangers have more complex geometries that a double tube. The most common geometry is the so-called *shell and tube*, consisting of a pressure vessel, i.e. the shell, with a bundle of tubes inside it. One fluid runs through the tubes and

Fig. 11.3 Shell and tubes
heat exchanger with U tubes

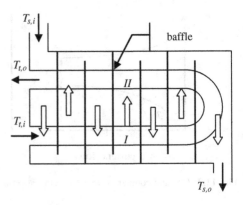

Fig. 11.4 The correction
factor F for U-tube heat
exchangers

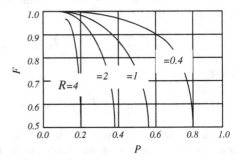

another fluid flows over the tubes (through the shell), as indicated in Fig. 11.3, where a U-tube heat exchanger is represented. In this case, the tube-side fluid presents two passages (I and II in the figure), while the shell-side fluid has only one passage. In addition, the shell side is partitioned by baffles, designed to support tube bundles and direct the fluid in cross-flow, that is perpendicularly to the tubes, for maximum efficiency.

In all these cases, assuming that the shell-side fluid temperature is a function of the axial coordinate,[7] Eq. (11.3.6) is modified as follows,

$$\dot{Q} = h_{tot} S (\Delta T)_{\ln} F, \qquad (11.3.10)$$

where $\Delta T_{\ln}$ is the LMTD (11.3.7) for a counter-current flow arrangement, while F is a geometric correction factor, indicated in Fig. 11.4.

---

[7]This hypothesis means assuming that the shell-side fluid is well mixed and there are very many baffles, so that the temperature is a continuous function of the axial coordinate, z.

11.4 Heat Exchanging Fins

Fins are used to promote the heat exchange from a surface by increasing the available exchange area. An automobile radiator or a domestic heater are good examples, where highly conductive surfaces are added that extend the area of the system out into the convective fluid. The simplest example, sketched in Fig. 11.5, consists of a rectangular fin on a planar surface at temperature T_w, exchanging heat with a surrounding fluid at temperature T_a. The fin has thickness $2B$, width W and length L. Generalization to more complex geometries is straightforward and will be illustrated at the end of this section.

First, we begin assuming that the temperature is uniform along the thin, x-direction. This hypothesis is verified when the transversal Biot number (i.e., the ratio between the heat flux at the wall and the transversal conductive heat flux) is very small, that is $Bi_B = hB/k \ll 1$, where k is the fin thermal conductivity and h the heat transfer coefficient at the wall. In addition, assuming that $B \ll W$, the edge area is much smaller than the upper and lower area, so that the heat losses from the fin edges can be neglected. Therefore, the temperature is only a function of z, and we obtain the following steady state heat balance across the small volume element $2BW\Delta z$ indicated in Fig. 11.5,

$$[J_U(z) - J_U(z + \Delta z)]2BW - h[T(z) - T_a]2W\Delta z = 0. \tag{11.4.1}$$

Here, heat propagates by conduction along the z-axis across the sections of area $2BW$, with heat flux $J_U = -kdT/dz$, while it is exchanged with the surrounding fluid on the upper and lower surface $W\Delta z$, with thermal flux $h(T - T_a)$. Dividing by Δz and taking the limit for $\Delta z \to 0$, we obtain:

$$\frac{d}{dz}\left(k\frac{dT}{dz}\right) = \frac{h}{B}[T(z) - T_a]. \tag{11.4.2}$$

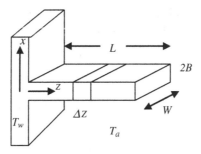

Fig. 11.5 Rectangular thin fin

This is a second-order differential equation, requiring two boundary conditions. In this case, we have:

$$T = T_w \quad \text{at} \quad z = 0, \tag{11.4.3}$$

$$-k\left(\frac{dT}{dz}\right) = h(T - T_a) \quad \text{at } z = L. \tag{11.4.4}$$

Now, define the dimensionless quantities:

$$\Theta = \frac{T - T_a}{T_w - T_a}; \quad \zeta = \frac{L - z}{L}; \quad N^2 = \frac{hL^2}{kB} = Bi\left(\frac{L}{B}\right); \quad Bi = \frac{hL}{k}, \tag{11.4.5}$$

where Bi is the longitudinal Biot number, (that is, the ratio between the heat flux at the wall and the longitudinal conductive heat flux), while N^2 is a modified Biot number. We obtain:

$$\frac{d^2\Theta}{d\zeta^2} - N^2\Theta = 0; \quad \Theta(1) = 1; \quad \frac{d\Theta}{d\zeta}(0) = -Bi\Theta(0). \tag{11.4.6}$$

Now, assume that $Bi \ll 1$, so that the second boundary condition becomes,

$$\frac{d\Theta}{d\zeta}(0) = 0. \tag{11.4.7}$$

Physically, this condition means assuming that the heat losses at the fin end can be neglected, that is the end area is much smaller than the upper and lower areas, i.e., $B \ll L$.

The general solution of Eq. (11.4.6) is:

$$\Theta(\zeta) = C_1 \cos h(N\zeta) + C_2 \sin h(N\zeta).$$

From the boundary condition $\Theta'(0) = 0$ we see that $C_2 = 0$, while applying the other condition $\Theta(1) = 1$ we finally obtain:

$$\Theta(\zeta) = \frac{\cos h(N\zeta)}{\cos hN}. \tag{11.4.8}$$

Therefore, the heat exchanged with the surrounding fluid per unit time is:

$$\dot{Q} = 2Wh \int_0^L (T - T_a)dz = 2WLh\Delta T \int_0^1 \Theta(\zeta)d\zeta = 2WLh\Delta T \frac{\tan hN}{N}, \tag{11.4.9}$$

where $\Delta T = T_w - T_a$. As a check, it is easy to verify that

$$\dot{Q} = 2BW\left[-k\left(\frac{dT}{dz}\right)_{z=0}\right], \qquad (11.4.10)$$

confirming that at steady state the total heat exchanged equals the heat flux entering (or leaving) from $z = 0$.

Now, consider the upper bounds of the heat exchanged, corresponding to the case where the temperature of the fin is uniform and equal to the wall temperature T_w, so that the heat exchanged across a surface $2LW$ with flux $h\Delta T$ is[8]

$$\dot{Q}_{max} = 2WLh\Delta T. \qquad (11.4.11)$$

Thus, we can define the *effectiveness* η of the fin as:

$$\eta = \frac{\dot{Q}}{\dot{Q}_{max}} = \frac{\tan hN}{N}, \quad \text{with} \quad N = \sqrt{\frac{hL^2}{kB}}, \qquad (11.4.12)$$

Note that, since $\tanh N \cong 1$ when $N \gg 1$, we see that

$$\eta = \frac{1}{N} \quad \text{when} \quad N \gg 1. \qquad (11.4.12a)$$

If we repeat all the previous analysis without assuming $B \ll L$, and therefore applying the full boundary conditions (11.4.6), we obtain the same result (11.4.12), where the fin length L is replaced by a corrected length L_c, with

$$L_c = L + B. \qquad (11.4.13)$$

For more complicated cases, such as fins with triangular or trapezoidal cross sections, the fin effectiveness presents quite similar expressions, provided that a properly corrected length is defined.[9]

11.5 Problems

11.1 Evaluate the heat dissipated by an aluminum fin ($k = 200$ W/mK) having $B = 1$ cm half thickness, $W = 1$ m width and $L = 1$ m length, assuming that the hot surface has temperature $T_w = 100$ °C, the ambient temperature is $T_a = 25$ °C, and the heat transfer coefficient is $h = 50$ W/m²K.

---

[8]We can also determine a lower bound, corresponding to the case where there is no fin at all, so that the heat exchanged across a surface area $2BW$ with heat flux $h\Delta T$ is $2WBh\Delta T$.

[9]See Perry's Chemical Engineers' Handbook.

11.2   A long steel bar, with diameter $D = 4.8$ cm and initially at temperature $T_i = 400\ °C$ exchanges heat with air at $T_0 = 20\ °C$ with heat transfer coefficient $h = 15\ W/m^2K$. After how long the temperature of the bar will be equal to $60\ °C$?

11.3   A steel plate of thickness $2L = 2.66$ mm is first heated in an oven at $T_i = 575\ °C$, then it is cooled with air at $T_a = 22\ °C$ with heat transfer coefficient $h = 14\ W/m^2K$. After how long the temperature at the center reaches $T = 55\ °C$?

11.4   Consider a double tube heat exchanger. The fluid inside the tube is a viscous oil with $c_p = 2091\ J/kgK$ and a mass flow rate $m_t = 0.1$ kg/s, having inlet and outlet temperatures $T_{t,in} = 127\ °C$ and $T_{t,out} = 67\ °C$, respectively. The fluid outside the tube is water $(c_p = 4179\ J/kgK)$ with mass flow rate $m_S = 0.15$ kg/s and inlet temperature $T_{s,in} = 30\ °C$. Determine the exchange area using both co-current and counter-current flow arrangements, considering that the overall heat transfer coefficient is $h_{tot} = 445\ W/m^2K$.

11.5   We want to heat 40 kg/s of glycerol from 20 to $34\ °C$, by cooling water from 80 to $48\ °C$. The shell-side heat transfer coefficient is $h_o = 300\ W/m^2K$, referred to the outer tube diameter. We use a shell and tube heat exchanger consisting of 75 U-tubes and 8 passages, with ¾ inch outer diameter. Determine: (a) the heat exchanged; (b) the water mass flow rate; (c) the tube length. Data: tube inner diameter: $D = 1.66$ cm; Specific heat of glycerol and water: $c_t = 2385\ J/kgK$ and $c_s = 4187\ J/kgK$.

Chapter 12
Time Dependent Heat Conduction

Abstract In Chaps. 9 and 10 we have considered heat conduction problems with only one independent variable. Then, in Chap. 11, we have shown how simple heat-flow problems can be solved using shell energy balances. In this Chapter, we discuss more complex cases, where the temperature distribution depends on more than one variable, generally one spatial coordinate and time. In Sect. 12.1 the energy equation for flow systems is derived using an Eulerian approach, showing that at the end we obtain the same result as in Chap. 6, where a Lagrangian approach was adopted. Then, we study heat conduction in a semi-infinite slab, due to either a sudden heating of the wall (Sect. 12.2) or to a heat pulse (Sect. 12.3). Unsteady heat conduction in a finite slab is considered in Sect. 12.4, using the separation of variable approach and obtaining at the end a solution in the form of an eigenvalue expansion. Finally, we consider the steady transport of heat in a pipe, where, as independent variable, time is replaced by the axial coordinate. This problem is studied either using an overall shell balance approach (Sect. 12.5) or solving the full energy equation (Sect. 12.6).

12.1 Heat Balance Equation

The heat transport in incompressible media is described by Eq. (6.7.8),

$$\frac{\partial T}{\partial t} + \mathbf{v} \cdot \nabla T = \alpha \nabla^2 T + \frac{\dot{q}}{\rho c_p}, \qquad (12.1.1)$$

where $\alpha = k/\rho c$ is the thermal diffusivity, $\dot{q}$ is the heat produced per unit volume and per unit time, and we have assumed that thermal conductivity, k, density, ρ, and specific heat, c, are constant. Note that for an incompressible medium the specific heat at constant volume and that at constant pressure are equal to each other, i.e., $c_v = c_p = c$. This equation can be derived directly from a balance of the internal

© Springer International Publishing Switzerland 2015
R. Mauri, *Transport Phenomena in Multiphase Flows*,
Fluid Mechanics and Its Applications 112, DOI 10.1007/978-3-319-15793-1_12

energy contained in a volume element $U = \int \rho c(T - T_0)dV$, where T_o is a reference temperature,

$$\frac{dU}{dt} = -\dot{U}_{out} + \dot{U}_{gen}, \quad \text{i.e.,} \quad \frac{d}{dt} \int_V \rho c_v(T - T_0)\, dV = -\int_S \mathbf{e}_n \cdot \mathbf{J}_U \, dS + \int_V \dot{q}\, dV,$$

$$(12.1.2)$$

where $\mathbf{J}_U$ is the heat flux, while $\mathbf{e}_n$ is the unit vector perpendicular to the surface S enclosing the volume V and directed outward. Now, applying the divergence theorem, we obtain:

$$\int_V \left[\rho c_v \frac{\partial T}{\partial t} + \nabla \cdot \mathbf{J}_U - \dot{q} \right] dV = 0. \qquad (12.1.3)$$

Considering that the heat flux is the sum of a convective and a diffusive component[1]

$$\mathbf{J}_U = \rho c_v(T - T_0)\mathbf{v} - k\nabla T, \qquad (12.1.4)$$

we obtain the differential equation:

$$\rho c \left[\frac{\partial T}{\partial t} + \nabla \cdot (\mathbf{v}T) \right] = k\nabla^2 T + \dot{q}. \qquad (12.1.5)$$

From here, considering that $\nabla \cdot \mathbf{v} = 0$ for incompressible flows, we find Eq. (12.1.1).

12.2 Heat Conduction in a Semi-infinite Slab

Consider a semi-infinite solid, initially at a uniform temperature T_0, whose wall is suddenly quenched to a temperature T_w. We want to determine the temperature profile, $T(y, t)$ as a function of position, y, and time, t. This is intended to model, for example, the cooling of a metal plate which, after undergoing a thermal treatment, is thrown in a cold water pool. With reference to Fig. 12.1, where $y = 0$ is one of the plate walls, we consider sufficiently short contact times or sufficiently large plate thicknesses, so that we presence of the other plate wall can be ignored. At the end of the Section this statement will become clearer.

---

[1]Note the difference between this, so-called Eulerian approach and the Lagrangian scheme adopted in Chap. 6. In the latter, the reference system moves together with the fluid, so that the heat flux presents only a diffusive component.

Fluid at
$T=T_w$

y

Fig. 12.1 Cooling of a metal plate

This process is described by the unsteady heat Eq. (12.1.1) in the absence of convection, i.e.,

$$\frac{\partial T}{\partial t} = \alpha \frac{\partial^2 T}{\partial y^2}, \tag{12.2.1}$$

To be solved with initial condition $T(y, t = 0) = T_0$ and boundary condition, $T(y = 0, t) = T_w$ and $T(y \rightarrow \infty, t) = T_0$, where we have assumed that if we move far enough from the wall the temperature is still equal to its unperturbed value.

Rewrite the problem using the dimensionless variables, $\Theta = (T - T_w)/(T_0 - T_w)$:

$$\frac{\partial \Theta}{\partial t} = \alpha \frac{\partial^2 \Theta}{\partial y^2}; \quad \Theta(y, 0) = \Theta(\infty, t) = 1; \quad \Theta(0, t) = 0. \tag{12.2.2}$$

As we have seen in Sect. 7.4, the solution of this problem is as follows,

$$\Theta(y, t) = erf(\eta); \quad \eta = \frac{y}{\sqrt{4\alpha t}}, \tag{12.2.3}$$

where erf is the error function, defined as the integral of a Gaussian,

$$erf(x) = \frac{2}{\sqrt{\pi}} \int_0^x e^{-\xi^2} d\xi \tag{12.2.4}$$

with $erf(0) = 0$ and $erf(\infty) = 1$.

Now, define the *penetration depth* of the heat as the distance δ_T corresponding to the temperature being 99 % of its maximum excursion, that is,

$$0.99 = erf\left(\frac{\delta_T}{\sqrt{4\alpha t}}\right). \tag{12.2.5}$$

Then, considering that $erf(2) = 0.99$, we obtain: $\delta_T = 4\sqrt{\alpha t}$. Thus, we see that the semi-infinite approximation is valid provided that, for a given contact time τ, the thickness L of the plate is much larger than the heat penetration depth, i.e.,

$L \gg \sqrt{\alpha\tau}$, or, conversely, for a given thickness L, the time interval is short enough, i.e., $\tau \ll L^2/\alpha$.

Finally, from this solution, we can determine the heat flux at the wall,

$$J_U = -k\left(\frac{\partial T}{\partial y}\right)_{y=0} = -k(T_0 - T_w)\left(\frac{d\Theta}{d\eta}\right)_{\eta=0}\frac{d\eta}{dy} = -\frac{k(T_0 - T_w)}{\sqrt{\pi\alpha t}}, \qquad (12.2.6)$$

showing that the flux decreases in time like $1/\sqrt{t}$. So, it seems that initially the flux is infinite, as one would expect because the temperature gradient is initially infinite. In reality, this condition does not occur, even if it is true that initially the heat flux is very large. In any event, though, the heat exchanged per unit area during a certain finite time interval t remains finite and grows like $\sqrt{t}$:

$$E = \left|\int_0^t J_U dt\right| = \frac{k(T_0 - T_w)}{\sqrt{\pi\alpha}}\int_0^t \frac{dt}{\sqrt{t}} = \frac{2k(T_0 - T_w)}{\sqrt{\pi\alpha}}\sqrt{t}. \qquad (12.2.7)$$

In a qualitative way, we could obtain this result by applying some of the fundamental properties of diffusion that we have seen in Sect. 1.9. In fact, considering that the penetration depth scales like $\delta_T \approx \sqrt{(\alpha t)}$, the flux can be determined as $J_U \approx -k\Delta T/\delta_T$, where $\Delta T = T_0 - T_w$, obtaining $J_U \approx -k\Delta T/\sqrt{(\alpha t)}$. This result coincides with the exact result (12.2.6), apart from an $O(1)$ constant factor.

12.2.1 Two Solids in Contact

We want to find the temperature distribution in two semi-infinite solid slabs, having the same thermal diffusivity, α, and with initial temperatures T_1 and T_2, that at time $t = 0$ are put in contact with one another (see Fig. 12.2). Using the dimensionless temperature, $\Theta = (2T - T_1 - T_2)/(T_2 - T_1)$, we find:

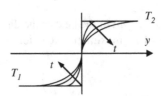

Fig. 12.2 Time dependent temperature distribution in two contacting semi-infinite solid slabs

$$\frac{\partial\Theta}{\partial t} = \alpha\frac{\partial^2\Theta}{\partial y^2}; \quad \Theta(y<0,0) = \Theta(-\infty,t) = -1; \quad \Theta(y>0,0) = \Theta(+\infty,t) = 1.$$

$$(12.2.8)$$

The solution is $\Theta(y,t) = erf(\eta)$, with $\eta = y/\sqrt{4\alpha t}$, where we have considered that $erf(\eta)$ is an odd function, that is $erf(\eta) = -erf(-\eta)$. Therefore, we see that at the interface $\Theta(y=0,t) = 0$, i.e., the temperature is constant in time and equal to the half sum of T_1 and T_2. The same result is obtained also when the two solids have different thermal diffusivities. In that case, stretching the coordinates for $y > 0$ and $y < 0$, Eq. (12.2.8) is obtained again.

12.2.2 Cooling of a Free Falling Film

A liquid film of thickness h and initially at temperature T_0 flows along an inclined plane of length L (see Fig. 12.3). The wall is isolated, while the liquid is cooled by air at temperature T_a. We want to find the temperature profile for short contact times.

Using the coordinate system of Fig. 12.3, the velocity profile has the form: $v(y) = V\left[1 - (y/h)^2\right]$. Therefore, at steady state the heat equation becomes:

$$V\left[1 - \left(\frac{y}{h}\right)^2\right]\frac{\partial T}{\partial z} = \alpha\left(\frac{\partial^2 T}{\partial y^2} + \frac{\partial^2 T}{\partial z^2}\right). \qquad (12.2.9)$$

For short contact times, the heat penetration depth is much smaller than the film thickness and the plane length i.e., $\delta_T \ll h$ and $\delta_T \ll L$, so that in the above equation $y \ll h$ and $\partial^2 T/\partial z^2 \ll \partial^2 T/\partial y^2$. More rigorously, since $y = O(\delta_T)$ and $z = O(L)$, with $\delta_T/h = \varepsilon_y \ll 1$ and $\delta_T/L = \varepsilon_z \ll 1$, define $\tilde{y} = y/\delta_T$ and $\tilde{z} = z/L$, obtaining:

$$\frac{V\delta_T^2}{\alpha L}\left[1 - \varepsilon_y^2\tilde{y}^2\right]\frac{\partial T}{\partial\tilde{z}} = \frac{\partial^2 T}{\partial\tilde{y}^2} + \varepsilon_z^2\frac{\partial^2 T}{\partial\tilde{z}^2}.$$

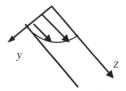

y

z

Fig. 12.3 A liquid film flowing along an inclined plane

Therefore, we see that at leading order Eq. (12.2.9) reduces to:

$$V\frac{\partial T}{\partial z} = \alpha\frac{\partial^2 T}{\partial y^2}.$$

(12.2.10)

Concluding, for short contact times heat penetration is very shallow so that (a) the fluid is perceived as if having a uniform, free surface velocity V; (b) the transversal temperature gradient is much larger than its longitudinal counterpart. Equation (12.2.10) requires, to be solved, that one boundary condition in z and two in y be specified. In fact, we have: $T(y, z = 0) = T_0$, $T(y = 0, z) = T_a$, together with the following third boundary condition: $\partial T/\partial y(y = h, z) = 0$. In our case, with short contact times, this condition can be replaced with the following: $T(y \gg \delta_T, z) = T_0$, i.e., $T(y \to \infty, z) = T_0$.

Defining $t = z/V$ as the time that takes the fluid to travel a distance z, Eq. (12.2.10) transforms into (12.2.2), with the same boundary conditions, obtaining the same solution,

$$\Theta(y, z) = \frac{T(y, z) - T_a}{T_0 - T_a} = erf\left(\frac{y}{\sqrt{4\alpha t}}\right) = erf\left(\frac{y}{\sqrt{4\alpha z/V}}\right).$$

(12.2.11)

Considering that at the end of the inclined plane, at $z = L$, we have $\delta_T = 4\sqrt{\alpha L/V}$, we see that in our approximation, as we have assumed $\delta_T \ll h$, it must be: $L/h \ll Vh/\alpha = Pe$.

12.3 Temperature Field Generated by a Heat Pulse

The surface $y = 0$ of a semi-infinite solid $y \geq 0$ is heated by a heat pulse, i.e., a finite energy per unit area, U, is radiated on the wall surface during a very short (infinitesimal, let's say) time interval. Assuming that the solid is initially at a uniform temperature, T_0, the heat equation is:

$$\frac{\partial \Phi}{\partial t} = \alpha\frac{\partial^2 \Phi}{\partial y^2}; \qquad \Phi = T - T_0,$$

(12.3.1)

to be solved with initial condition $\Phi(y, t = 0) = 0$. In addition, we assume that the thermal conductivity of the medium occupying the region $t < 0$ is much smaller than that of the solid, so that the heat can propagate by conduction only within the solid and does not diffuse outside of it. Therefore, imposing that the internal energy is conserved within the solid, we obtain:

$$\rho c \int_0^\infty \Phi dy = U. \tag{12.3.2}$$

Now, consider the spatial derivative, $\Theta'(y, t)$, of the temperature $\Theta(y, t)$ given by the self-similar expression (12.2.3). Obviously, Θ' satisfies the same Eq. (12.2.2) that is satisfied by Θ, with the same boundary condition at infinity and the same initial condition at $y > 0$. In addition, at the wall $y = 0$, initially the temperature distribution Θ is a discontinuous step function and therefore Θ' is not defined. For sake of simplicity, let us say (I hope that mathematicians will not get too angry at me) that initially Θ' is infinite at $y = 0$ and zero elsewhere, in such a way that its integral remains finite. Therefore, since Θ' satisfies the same equation and the same homogeneous conditions as Φ, the two functions must coincide with each other, within a multiplying constant C, obtaining:

$$\Phi(y, t) = C \frac{2}{\sqrt{\pi}} \frac{\partial}{\partial y} \left(\int_0^{y/\sqrt{4\alpha t}} e^{-\xi^2} d\xi \right) = \frac{C}{\sqrt{\pi \alpha t}} e^{-\frac{y^2}{4\alpha t}}.$$

Now, imposing that the condition (12.3.2) is satisfied, we find $C = U/\rho c$ and therefore we conclude:

$$\Phi(y, t) = \frac{U}{\rho c \sqrt{\pi \alpha t}} e^{-\frac{y^2}{4\alpha t}}. \tag{12.3.3}$$

This is a *Gaussian* (or *normal*) function, with zero mean and standard deviation, or variance, equal to $\sqrt{2\alpha t}$. Naturally, at $t = 0$, the normal distribution Φ tends to a so-called Dirac delta function,[2] that is a "thing" that is equal to zero for any $y \neq 0$, while it is strongly discontinuous at the origin $y = 0$, in such a way that its integral remains finite and equal to $2U/\rho c$, in agreement with Eq. (12.3.2), i.e.,

$$\lim_{t \to 0} \Phi(y, t) = 2 \frac{U}{\rho c} \delta(y), \text{ where } \int_0^\infty \delta(x) dx = \frac{1}{2}. \tag{12.3.4}$$

As we saw in Chap. 1, the normal distribution describes the probability that any real observation of a random variable will fall between any two real limits and, as such, it describes a diffusive process.

---

[2]This "thing" in reality is not a function, but a distribution, that is the limit of a series of functions.

12.4 Heat Conduction in a Finite Slab

Consider a slab of thickness $2L$, initially at temperature T_0, and assume that at time $t = 0$ the two walls at $x = -L$ and $x = L$ are suddenly quenched to a temperature T_1. We want to determine the temperature profile $T(x, t)$.

First, defining the dimensionless variables $\Theta(\tilde{x}, \tilde{t})$, with $\Theta = (T - T_1)/(T_0 - T_1)$, $\tilde{x} = x/L$ and $\tilde{t} = t/(L^2/\alpha)$, the problem can be formulated as:

$$\frac{\partial \Theta}{\partial \tilde{t}} = \frac{\partial^2 \Theta}{\partial \tilde{x}^2}; \quad \Theta(-1, \tilde{t}) = \Theta(1, \tilde{t}) = 0; \quad \Theta(\tilde{x}, 0) = 1. \qquad (12.4.1)$$

Often, the dimensionless time is denoted as the Fourier number, i.e., $Fo \equiv \tilde{t} = \alpha t/L^2$.

Note that the solution must satisfy the equality $T(-x, t) = T(x, t)$, indicating that the temperature profile for $-L < x < 0$ is equal to its mirror image for $0 < x < L$. In particular, that means that at the origin the temperature profile must be an extreme (maximum or minimum), that is $\partial T/\partial x(0, t) = 0$. Therefore, the problem can be formulated within the interval $0 < x < L$ by rewriting Eq. (12.4.1) as follows:

$$\frac{\partial \Theta}{\partial \tilde{t}} = \frac{\partial^2 \Theta}{\partial \tilde{x}^2}; \quad \frac{\partial \Theta}{\partial \tilde{x}}(0, \tilde{t}) = \Theta(1, \tilde{t}) = 0; \quad \Theta(\tilde{x}, 0) = 1. \qquad (12.4.2)$$

Considering that both the governing equation and its boundary conditions are homogeneous, we see that the steady state solution is trivially $\Theta_{ss}(\tilde{x}) = 0$. Therefore, the solution must satisfy the following condition:

$$\lim_{t \to \infty} \Theta(\tilde{x}, \tilde{t}) = \Theta_{ss}(\tilde{x}) = 0. \qquad (12.4.3)$$

Now we apply the so-called *separation of variables* method. It consists of assuming that the temperature Θ can be written as the product of two functions, depending separately on $\tilde{x}$ and $\tilde{t}$, that is: $\Theta(\tilde{x}, \tilde{t}) = \phi(\tilde{x})\, T(\tilde{t})$. Substituting this expression into the heat equation and dividing by ϕT, we obtain:

$$\frac{dT/d\tilde{t}}{T} = \frac{d^2\phi/d\tilde{x}^2}{\phi}.$$

Now, since the LHS is a function of $\tilde{t}$ only, while the RHS is a function of $\tilde{x}$ only, this equality can hold only when both members are equal to a constant, that we indicate as $-\lambda^2$ (the reason why it must be negative will become obvious in a while). Therefore, we obtain the following two equations:

$$\frac{dT}{d\tilde{t}} = -\lambda^2 T,$$

and:

$$\frac{d^2\phi}{d\tilde{x}^2} + \lambda^2\phi = 0.$$

The first of these equations can be solved, obtaining:

$$T(\tilde{t}) = Ae^{-\lambda^2\tilde{t}},$$

where A is an arbitrary constant. That's why the above constant must be negative: for long time the solution must tend to its asymptotic limit, as stated in Eq. (12.4.3). Had we assumed a positive constant, we would obtain instead a temperature that increases exponentially with time.

Now, let us turn to the second equation, $\phi'' + \lambda^2\phi = 0$, to be solved with homogeneous boundary conditions $\phi'(0) = \phi(1) = 0$. This is a typical eigenvalue problem, admitting an infinite number of solutions. In our case, these solutions are of the type

$$\phi(\tilde{x}) = B\cos(\lambda\tilde{x}) + C\sin(\lambda\tilde{x}),$$

where B and C are constants. In general, the functions $\phi(\tilde{x})$ are denoted as *eigenfunctions*, while the respective λ^2 are named *eigenvalues*. Here, the boundary condition $\phi'(0) = 0$ requires that $C = 0$, while imposing that $\phi(1) = 0$ we obtain: $\lambda_n = n\pi/2$, with $n = 2m + 1$ indicating an odd integer. Finally, we conclude that the eigenfunctions have the following form:

$$\phi(\tilde{x}) = B\cos(n\pi\tilde{x}/2), \quad \text{with } n \text{ odd}.$$

Thus, any function $\phi T = \cos(n\pi\tilde{x}/2)\exp(-n^2\pi^2t/4)$, with $n = 2m + 1$ is a solution of the problem. Now, since the problem is linear, with homogeneous boundary conditions,[3] we can apply the superposition principle, stating that if Θ_1 and Θ_2 are two solutions, any linear combination, $c_1\Theta_1 + c_2\Theta_2$, is a solution, too, where c_1 and c_2 are two arbitrary constant. Therefore, the general solution of this problem is the following:

$$\Theta(\tilde{x}, \tilde{t}) = \sum_{n=1}^{\infty} c_n\phi_n(\tilde{x})e^{-\lambda_n^2\tilde{t}}, \tag{12.4.4}$$

i.e.,

$$\Theta(\tilde{x}, \tilde{t}) = \sum_{m=0}^{\infty} c_m\cos\left[\left(m + \frac{1}{2}\right)\pi\tilde{x}\right]e^{-\left(m+\frac{1}{2}\right)^2\pi^2\tilde{t}}. \tag{12.4.4a}$$

---

[3]It means that they remain unchanged if temperature is multiplied by a constant.

Now we have to determine the constant c_m, imposing that the initial condition be satisfied, i.e.,

$$\Theta(\tilde{x},0) = 1 = \sum_{m=0}^{\infty} c_m \cos\left[\left(m+\frac{1}{2}\right)\pi\tilde{x}\right]. \qquad (12.4.5)$$

Consider the following equality:

$$\int_0^1 \cos\left[\left(m+\frac{1}{2}\right)\pi\tilde{x}\right] \cos\left[\left(n+\frac{1}{2}\right)\pi\tilde{x}\right] d\tilde{x} = \begin{cases} 0 & \text{if } m \neq n \\ 1/2 & \text{if } m = n \end{cases}$$

Then, multiplying Eq. (12.4.5) by $\cos\left[(n+\frac{1}{2})\pi\tilde{x}\right]$ and integrating between 0 and 1, we obtain:

$$\int_0^1 \Theta(\tilde{x},0) \cos\left[\left(n+\frac{1}{2}\right)\pi\tilde{x}\right] d\tilde{x} = \sum_{m=0}^{\infty} c_m \int_0^1 \cos\left[\left(m+\frac{1}{2}\right)\pi\tilde{x}\right] \cos\left[\left(n+\frac{1}{2}\right)\pi\tilde{x}\right] d\tilde{x},$$

i.e.,

$$c_n = \frac{2}{\left(n+\frac{1}{2}\right)\pi} \sin\left(n+\frac{1}{2}\right)\pi = \left\{ \begin{array}{ll} \frac{2}{\left(n+\frac{1}{2}\right)\pi} & \text{when } n \text{ is even} \\ -\frac{2}{\left(n+\frac{1}{2}\right)\pi} & \text{when } n \text{ is odd} \end{array} \right\} = \frac{2(-1)^n}{\left(n+\frac{1}{2}\right)\pi}.$$

Finally, we conclude:

$$\Theta(\tilde{x},\tilde{t}) = \sum_{n=0}^{\infty} \frac{2(-1)^n}{\left(n+\frac{1}{2}\right)\pi} \cos\left[\left(n+\frac{1}{2}\right)\pi\tilde{x}\right] e^{-\left(n+\frac{1}{2}\right)^2 \pi^2 \tilde{t}}, \qquad (12.4.6)$$

that is,

$$\Theta(\tilde{x},\tilde{t}) = \frac{4}{\pi}\cos(\pi\tilde{x}/2)e^{-\pi^2\tilde{t}/4} - \frac{4}{3\pi}\cos(3\pi\tilde{x}/2)e^{-9\pi^2\tilde{t}/4} + \cdots$$
$$= \frac{4}{\pi}\cos(\pi\tilde{x}/2)e^{-\pi^2\tilde{t}/4}\left[1 + O\left(e^{-8\pi^2\tilde{t}/4}\right)\right], \qquad (12.4.6a)$$

Obviously, for long times, $\tilde{t} \gg 1/\pi^2$, the dominant term of the infinite series (12.4.6) is the first one and the series converges very fast. Therefore, we see that $\tilde{t} \approx 4/\pi^2$ is a (dimensionless) relaxation time, i.e., the characteristic time needed to reach steady state, corresponding to a time $\tau \approx 4L^2/\alpha\pi^2$.

For short time, $\tilde{t} \ll 1/\pi^2$, the infinite series converges very slowly and therefore this solution is not very convenient. However, for short times, the temperature variations concern only the near-wall regions, where the influence of the opposite

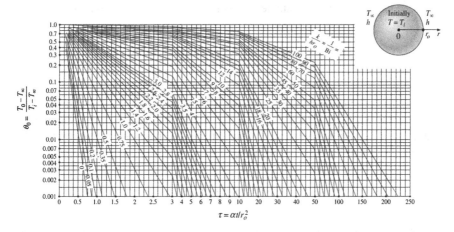

Fig. 12.4 Transient temperature at the center of a sphere of radius r_0, initially at uniform temperature T_i, exchanging heat with a fluid at temperature T_∞ with a heat transfer coefficient h, as a function of time and of the inverse Biot number

wall has not arrived yet. Accordingly, for short times the solution coincides with the self-similar temperature distribution for semi-infinite media,

$$\Theta(\tilde{x}, \tilde{t}) = erf(\eta'); \quad \eta' \frac{1 - \tilde{x}}{\sqrt{4t}}; \quad 0 < \tilde{x} < 1. \qquad (12.4.7)$$

Similar solutions can be obtained when the slab exchanges heat with a fluid at constant temperature, with a constant heat transfer coefficient, h. In this case, as expected, the solution depends on the Biot number, $Bi = hL/k$.

Analogous results are obtained with different geometries, as it is evident from Fig. 12.4, representing the temperature at the center of a sphere as a function of time and of the inverse Biot number. Note that, at long times, the curves continue to decrease exponentially, i.e., linearly in this semi-log plot.

12.5   Heat Exchange in a Pipe

Consider a fluid flowing along the axial direction, z, of a circular pipe of radius R, whose walls at $r = R$ are kept at a constant temperature T_w. Assuming a fully developed velocity field, $v = v(r)$, with a uniform fluid temperature at the inlet equal to T_i, we want to determine the mean temperature at a distance z from the entrance as a function of the mass flow rate and the tube diameter.

First, consider the steady state heat equation in terms of the fluid temperature, T (r,z):

$$\rho c v(r)\frac{\partial T}{\partial z} = k\left[\frac{1}{r}\frac{\partial}{\partial r}\left(r\frac{\partial T}{\partial r}\right) + \frac{\partial^2 T}{\partial z^2}\right].$$

The ratio between longitudinal convection and diffusion is the Peclet number, $Pe = \rho c U R/k$, where U is the mean velocity. Assuming that $Pe \gg 1$, longitudinal conduction is negligible compared to convection, and therefore the last term in the equation above can be omitted, obtaining,

$$\rho c v(r)\frac{\partial T}{\partial z} = k\frac{1}{r}\frac{\partial}{\partial r}\left(r\frac{\partial T}{\partial r}\right). \tag{12.5.1}$$

This equation must be solved with the following conditions (1 B.C. in z and 2 in r),

$$T(z = 0) = T_i; \quad T(r = R) = T_w; \quad \frac{dT}{dr}(r = 0) = 0. \tag{12.5.2}$$

Now, instead of solving the whole problem (see Sect. 12.6), here we want to determine the behavior of the average, so-called *bulk* fluid temperature within a cross-section, defined as,

$$\bar{T}_b(z) = \frac{\langle Tv \rangle}{\langle v \rangle}(z) = \frac{\int_0^R T(r,z)v(r)2\pi r dr}{\int_0^R v(r)2\pi r dr} = \frac{\int_0^R T(r,z)v(r)2\pi r dr}{\dot{V}}, \tag{12.5.3}$$

where $\dot{V}$ is the volumetric flow rate, while $\langle\,\rangle$ (also indicated with an overbar) represents the arithmetic mean within the cross section, i.e.,

$$\langle T \rangle(z) = \bar{T}(z) = \frac{1}{\pi R^2}\int_0^R T(r,z)2\pi r dr. \tag{12.5.4}$$

The *bulk temperature* $\bar{T}_b$ (also referred to as *cup mixing temperature or flow average temperature*) is the flow-weighted mean temperature and does not coincide with the arithmetic mean temperature within the cross section, $\bar{T}$. It represents the temperature that one would measure if the tube were chopped off and if the fluid issuing forth were collected in a container and thoroughly mixed. Note that, if the tube were adiabatic, $\bar{T}_b$ is, predictably, constant with z, while $\bar{T}$ is not and tends to $\bar{T}_b$ as $z \rightarrow \infty$, i.e., when the temperature profile is uniform. In general, in the bulk averaging (12.5.3) the faster fluid at the center of the tube "counts more" than that near the wall. So, for example, if the fluid is heated at the walls, i.e., $T_i < T_w$, we have $\bar{T}_b < \bar{T}$; then, as we proceed along the channel and the temperature distributes uniformly within the cross-section, both $\bar{T}_b$ and $\bar{T}$ will tend exponentially to T_w.

Multiplying Eq. (12.5.1) by r and integrating in r we have:

$$\rho c \dot{V} \frac{d\bar{T}_b}{dz} = 2\pi k \int_0^R \frac{\partial}{\partial r}\left(r \frac{\partial T}{\partial r}\right) dr = 2\pi k R \left(\frac{\partial T}{\partial r}\right)_{r=R}. \tag{12.5.5}$$

Equation (12.5.5) reflects the fact that the axial variation of the energy flux (i.e., the quantity on the LHS) is equal to the heat flux exchanged at the wall per unit length (i.e., the quantity on the RHS). In fact, Eq. (12.5.5) could be obtained from a macroscopic heat balance on a pipe segment of length dz.

Now, define the heat transfer coefficient h as the ratio between the heat flux at the wall and the difference between the wall temperature, T_w, and the bulk temperature, $\bar{T}_b$, i.e.,

$$-k \left(\frac{\partial T}{\partial r}\right)_{r=R} = h[\bar{T}_b - T_w]. \tag{12.5.6}$$

Equation (12.5.5) becomes:

$$\dot{m} c \left(\frac{d\bar{T}_b}{dz}\right) = -\pi D h[\bar{T}_b - T_w],$$

where $\dot{m} = \rho \dot{V}$ is the mass flow rate and $D = 2R$ is the tube diameter. Now we want to apply this equation to determine the bulk temperature at the outlet, assuming that h is known. Imposing an inlet and an outlet bulk temperatures, $T_i = \bar{T}_b(z = 0)$ and $T_o = \bar{T}_b(z = L)$, we can integrate the equation above, obtaining:

$$\frac{T_0 - T_w}{T_i - T_w} = \exp\left(-\frac{\pi h D L}{\dot{m} c}\right) = \exp\left(-4St \frac{L}{D}\right). \tag{12.5.7}$$

Here, we have defined the *Stanton* number, *St*, as

$$St = \frac{Nu}{Pe} = \frac{h}{\rho c U}; \quad \text{where} \quad Nu = \frac{hR}{k} \quad \text{and} \quad Pe = \frac{\rho c U R}{k}, \tag{12.5.8}$$

where U is the mean velocity and we have considered that $\dot{m} = \rho \dot{V} = \rho U(\pi D^2/4)$.

It is often more convenient to express this result in terms of a mean temperature difference between the fluid and the wall, defined as

$$\Delta T = -\dot{Q}/hA, \tag{12.5.9}$$

where $A = \pi D L$ is the exchange area, while $\dot{Q}$ is the total heat flux,

$$\dot{Q} = \dot{m} c(T_o - T_i), \tag{12.5.10}$$

which is defined as positive when it enters the system. From the last two equations we obtain:

$$\Delta T = [\dot{m}c(T_i - T_o)]/[\pi DLh],$$

while from Eq. (12.5.8) we have:

$$[\pi DLh]/[\dot{m}c] = \ln[(T_i - T_w)/(T_o - T_w)].$$

Therefore we conclude that:

$$\Delta T = \frac{T_i - T_0}{\ln\left(\frac{T_i - T_w}{T_0 - T_w}\right)} = \frac{T_0 - T_i}{\ln\left(\frac{T_0 - T_w}{T_i - T_w}\right)} = \frac{(T_i - T_w) - (T_0 - T_w)}{\ln\left(\frac{T_i - T_w}{T_0 - T_w}\right)} = \Delta T_{\ln}. \qquad (12.5.11)$$

This result shows that ΔT, as defined in (12.5.9), coincides with the logarithmic mean temperature difference between the fluid and the wall.

12.6 Heat Transfer Coefficient in Laminar Flow

In the previous Section we have assumed that the heat transfer coefficient, h, defined in (12.5.6), is a known quantity. In reality, h is rather complicated to be determined analytically: think, for example, to the case of turbulent flow, where the velocity field is time dependent. In the case of laminar regimes, though, the problem is solvable and it constitutes the so-called Graetz solution.[4] It amounts to solving Eqs. (12.5.1) and (12.5.2) by separating the variables, assuming a Poiseuille flow velocity profile,

$$v(r) = 2U\left(1 - r^2/R^2\right),$$

where U is the mean velocity. Following the same procedure illustrated in Sect. 12.4, we can define the dimensionless variable $\Theta(\tilde{r}, \tilde{z})$, with $\Theta = (T - T_w)/(T_i - T_w)$, $\tilde{r} = r/R$ and $\tilde{z} = z/(2RPe)$, where $Pe = UR/\alpha$ is the Peclet number and $\alpha = k/\rho c$ is thermal diffusivity. Thus, the problem can be formulated as:

$$(1 - \tilde{r}^2)\frac{\partial \Theta}{\partial \tilde{z}} = \frac{1}{\tilde{r}}\left[\frac{\partial}{\partial \tilde{r}}\left(\tilde{r}\frac{\partial \Theta}{\partial \tilde{r}}\right)\right]; \quad \Theta(1, \tilde{z}) = \frac{\partial \Theta}{\partial \tilde{r}}(0, \tilde{z}) = 0; \quad \Theta(\tilde{r}, 0) = 1.$$

$$(12.6.1)$$

---

[4]Named after Leo Graetz (1856–1941), a German physicist.

The solution is of the type (12.4.5), i.e.,

$$\Theta(\tilde{r}, \tilde{z}) = \sum_{n=1}^{\infty} c_n \phi_n(\tilde{r}) e^{-\lambda_n^2 \tilde{z}}, \qquad (12.6.2)$$

with

$$\frac{1}{\tilde{r}} \left[\frac{d}{d\tilde{r}} \left(\tilde{r} \frac{d\phi_n}{d\tilde{r}} \right) \right] + \lambda_n^2 (1 - \tilde{r}^2) \phi_n = 0; \quad \phi_n(1, \tilde{z}) = \frac{d\phi_n}{d\tilde{r}}(0, \tilde{z}) = 0. \qquad (12.6.3)$$

This is an eigenvalue problem, whose solutions are the confluent hypergeometric functions;[5] in particular, we have: $\lambda_1 = 2.7044$. The constant c_n can be determined by imposing that the boundary condition at the entrance, $\tilde{z} = 0$, is satisfied. Defining the Nusselt number in terms of the tube diameter, $Nu = 2hR/k$, substituting the expression (12.5.6) for the heat transfer coefficient, we obtain:

$$Nu(\tilde{z}) = -2 \frac{\partial \Theta / \partial \tilde{r}(1, \tilde{z})}{\Theta_b(\tilde{z})}, \qquad (12.6.4)$$

where Θ_b is the dimensionless bulk average temperature. Then, substituting the infinite series solution (12.6.2) into both the numerator and the denominator of (12.6.4), and using the properties of the eigenfunctions that can be derived from the governing differential equation (12.6.3), we finally obtain the following result for the Nusselt number:

$$Nu(\tilde{z}) = \frac{\sum_{n=1}^{\infty} c_n \phi_n'(1) e^{-\lambda_n^2 \tilde{z}}}{2 \sum_{n=1}^{\infty} c_n \frac{1}{\lambda_n^2} \phi_n'(1) e^{-\lambda_n^2 \tilde{z}}}, \quad \text{with} \quad \phi_n' \equiv \frac{d\phi_n}{d\tilde{r}}. \qquad (12.6.5)$$

Now, assuming to be at a large distance from the entrance, i.e. at $\tilde{z} \gg 1$, then only the first term in the infinite series in Eq. (12.6.5) is important. Therefore, we obtain:

$$Nu(\tilde{z}) \xrightarrow{\tilde{z} \gg 1} \frac{1}{2} \lambda_1^2 = 3.656. \qquad (12.6.6)$$

Similar exact results are obtained assuming that the heat flux at the wall is fixed, or for different laminar flow velocity profiles, e.g., Poiseuille 2D channel flow, or Couette flow.

Near the entrance, instead, the series (12.6.5) converges very slowly and therefore the so-called Leveque self-similar solution is used, as shown in Sect. 17.3.

---

[5]See Abramowitz and Stegun, *Handbook of Mathematical Functions*, Dover, Chap. 13.

12.7 Problems

12.1   The surface of a thick (explain what "thick" means) iron slab, initially at 0 °C, is heated (instantaneously, let us assume) to 60 °C. Calculate its temperature after 1 min at a 3 cm depth from the surface.

12.2   The surface of a thick (explain what "thick" means) nickel slab initially at 20 °C temperature is exposed to an instantaneous heat impulse of 12 MJ/m^2. Calculate the wall temperature after 5 s.

12.3   A potato, initially at a $T_i = 20$ °C temperature, is put in an oven at $T_a = 160$ °C, where it is heated with heat transfer coefficient $h = 13$ W/m^2 K. After how long can we take it out, knowing that it is cooked as soon as its temperature is everywhere above $T_0 = 100$ °C? Assume that the potato is a perfect sphere (miracles of genetic engineering...) with $R = 4$ cm radius.

12.4   Work out all the details of the exact solution of the problem in Sect. 12.4, assuming that the slab walls are located at $x = 0$ and $x = L$. Verify that the temperature at the center of the slab is identical to the infinite series (12.4.4).

12.5   A slab of thickness L, insulated on one side, at $x = 0$, has an initial temperature, T_i. Assuming that the other wall, at $x = L$, is instantaneously quenched to a temperature T_a, determine the time variation of the temperature at $x = 0$.

Chapter 13
Convective Heat Transport

Abstract In Sect. 3.2 we saw that when a hot (or cold) fluid flows past a cold (or hot) body of size L, with a characteristic velocity U such that the Peclet number is large, i.e., $Pe = UL/\alpha \gg 1$, then the convective heat flux is in general much larger than is diffusive counterpart. This is true everywhere, with the exception of a thin layer near the surface of the body of thickness δ_T, called thermal boundary layer. In fact, although far from the walls convection prevails, at the wall the fluid velocity is null and therefore heat is exchanged only by diffusion. Therefore, at a certain distance from the walls, convective and diffusive effects must balance each other: this is what defines the thickness δ_T of the thermal boundary layer. Although the amount of fluid contained in the thermal boundary layer is but a tiny fraction of the total fluid volume, it nevertheless plays a very important role, as it regulates the heat exchange between fluid and body, which is what concerns us the most. In this chapter, after investigating the scaling of the problem in Sect. 13.1, thermal boundary layer is studied in Sect. 13.2, determining its dependence on the fluid velocity and its physical properties. Finally, in Sect. 13.3, the Colburn-Chilton analogy between heat and momentum transport is analyzed, showing that the properties of one of the two transport phenomena can be determined, once the other is known.

13.1 Scaling of the Problem

Consider a fluid, with unperturbed homogeneous velocity **V**, pressure P_∞ and temperature T_∞, flowing past a still body, of size L and temperature T_0, as shown in Fig. 13.1. In principle, the 5 unknowns of this problem, i.e., the fluid velocity **v**, pressure P and temperature T, can be determines by solving the Navier-Stokes, continuity and heat equations,

$$\frac{D\mathbf{v}}{Dt} = \mathbf{v} \cdot \nabla \mathbf{v} = -\frac{1}{\rho}\nabla P + \nu \nabla^2 \mathbf{v}, \tag{13.1.1}$$

© Springer International Publishing Switzerland 2015
R. Mauri, *Transport Phenomena in Multiphase Flows*,
Fluid Mechanics and Its Applications 112, DOI 10.1007/978-3-319-15793-1_13

Fig. 13.1 Fluid flowing past
a still body having a different
temperature, T_0

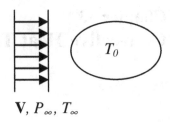

$$\mathbf{V}, P_\infty, T_\infty$$

$$\nabla \cdot \mathbf{v} = 0, \tag{13.1.2}$$

$$\frac{DT}{Dt} = \mathbf{v} \cdot \nabla T = \alpha \nabla^2 T, \tag{13.1.3}$$

with the following boundary conditions at the body surface,

$$\mathbf{v} = \mathbf{0}; \quad T = T_0 \text{ at the surface;}$$
$$\mathbf{v} = \mathbf{V}; \quad T = T_\infty \text{ far from the surface.} \tag{13.1.4}$$

At the end, as usual, we want to determine the heat flux at the wall, i.e., the body surface,

$$J_U = -k\mathbf{e}_n \cdot \nabla T \text{ at the wall,} \tag{13.1.5}$$

where $\mathbf{e}_n$ is the unit vector perpendicular to the body surface.[1] Note that, as the fluid velocity at the body surface is null, the flux of energy (as well as that of momentum and of mass) at the wall is only diffusive.

The problem (13.1.1)–(13.1.4) can also be formulated in terms of the dimensionless quantities,

$$\tilde{\mathbf{r}} = \frac{\mathbf{r}}{L}; \quad \tilde{t} = \frac{t}{(L/V)}; \quad \tilde{P} = \frac{P - P_\infty}{(\mu V/L)};$$
$$\tilde{\mathbf{v}} = \frac{\mathbf{v}}{V}; \quad \tilde{T} = \frac{T - T_0}{T_\infty - T_0}, \tag{13.1.6}$$

obtaining,

$$Re\frac{D\tilde{\mathbf{v}}}{D\tilde{t}} = Re\tilde{\mathbf{v}} \cdot \tilde{\nabla}\tilde{\mathbf{v}} = -\tilde{\nabla}\tilde{P} + \tilde{\nabla}^2\tilde{\mathbf{v}}, \tag{13.1.7}$$

$$\tilde{\nabla} \cdot \tilde{\mathbf{v}} = 0, \tag{13.1.8}$$

---

[1]In reality, we want to find the total heat flux, that is the integral of the heat flux (13.1.5) over the whole body surface.

$$Pe\frac{D\tilde{T}}{D\tilde{t}} = Pe\tilde{\mathbf{v}} \cdot \tilde{\nabla}\tilde{T} = \tilde{\nabla}^2\tilde{T}, \tag{13.1.9}$$

$$\tilde{\mathbf{v}} = \tilde{T} = 0 \text{ at the wall}; \quad \tilde{\mathbf{v}} = \tilde{T} = 1 \text{ far from the wall.} \tag{13.1.10}$$

From this dimensional analysis, two dimensionless numbers arise naturally, i.e., the Reynolds number, $Re = VL/v$ and the Peclet number, $Pe = VL/\alpha$, equal to the ratio between convective and diffusive fluxes in the transport of momentum and heat, respectively. The Peclet number can also be expressed as the product between the Reynolds number and the Prandtl number, $Pr = v/\alpha$. This latter represents the ratio between the efficiency of the diffusive momentum transport and that of the diffusive heat transport. Therefore,

$$Re = \frac{VL}{v}; \quad Pe = \frac{VL}{\alpha}; \quad Pr = \frac{v}{\alpha}. \tag{13.1.11}$$

The heat flux, too, can be expressed in a non-dimensional form, defining the Nusselt number as the ratio between the heat flux at the wall and its conductive part, that is,

$$Nu = \frac{|J_U|}{k\Delta T/L} = \left|\mathbf{e}_n \cdot \tilde{\nabla}\tilde{T}\right| \text{ at the wall}, \tag{13.1.12}$$

where $\Delta T = |T_\infty - T_0|$. In particular, in this Section we consider the cases where $Pe \gg 1$.

At first sight, we are tempted to say that, as convective fluxes are predominant over their diffusive counterpart, the RHS of Eq. (13.1.9) can be neglected altogether. However, in doing so, the problem would reduce from second to first order, and therefore it would be impossible to satisfy two boundary conditions.[2] Consequently, imposing that the boundary condition at infinity is satisfied, the solution of Eq. (13.1.3) with no diffusive term would be trivially $T = T_\infty$ and it would be impossible to impose also that the fluid temperature at the wall is T_0. Thus, we may conclude that, in the absence of diffusion, the resulting temperature field would be uniform, with no heat transfer at the wall: an unacceptable conclusion. The physical explanation of this unreasonable result is that at the wall convection is null, so that the heat flux is all diffusive; consequently, if the diffusive fluxes are neglected altogether, it is obvious that the heat transport at the wall results to be null.

The solution of this problem is to define the thermal boundary layer, just as in Chap. 8 we have defined the momentum boundary layer. Accordingly, the thermal boundary layer consists of a thin fluid layer of thickness δ_T (not necessarily uniform) in contact with the body. Outside this region of space, diffusive fluxes can be

---

[2]Remember that a differential equation of nth degree can be solved by integrating it n times, and therefore introducing n constant that can be determined by satisfying n boundary conditions.

effectively neglected so that, in our case, temperature is uniform with $T = T_\infty$, while inside it they become progressively more important as we near the wall. The basic idea is that the temperature gradients within the thermal boundary layer are of O $(\Delta T/\delta_T)$, where $\Delta T = T_\infty - T_0$, so that, if δ_T is sufficiently small, the effects of the diffusive fluxes will balance those of convection.

13.2 Laminar Thermal Boundary Layer

Since $\delta_T \ll L$, in studying the velocity and temperature profiles within the boundary layer we can neglect the curvature of the body surface, assuming that it is flat. Consider the simplest case of a fluid flowing past a plate, $y = 0$, with a 2D velocity field, $\mathbf{v} = (v_x, v_y)$, pressure P and temperature T (see Fig. 13.2). The unperturbed velocity is uniform, $\mathbf{V} = (V, 0)$, with uniform pressure P_∞, and temperature T_∞, while at the wall the velocity is null, $\mathbf{v} = 0$, and the temperature is T_o. In addition, let us assume that the Peclet number, $Pe = VL/\alpha$, is very large, where L is the plate size.

The four unknown functions, v_x, v_y, P and T, are determined through the x- and y-components of the Navier-Stokes equation, the continuity equation and the heat equation,

$$v_x \frac{\partial v_x}{\partial x} + v_y \frac{\partial v_x}{\partial y} = -\frac{1}{\rho}\frac{\partial P}{\partial x} + v\left(\frac{\partial^2 v_x}{\partial x^2} + \frac{\partial^2 v_x}{\partial y^2}\right), \tag{13.2.1}$$

$$v_x \frac{\partial v_y}{\partial x} + v_y \frac{\partial v_y}{\partial y} = -\frac{1}{\rho}\frac{\partial P}{\partial y} + v\left(\frac{\partial^2 v_y}{\partial x^2} + \frac{\partial^2 v_y}{\partial y^2}\right), \tag{13.2.2}$$

$$\frac{\partial v_x}{\partial x} + \frac{\partial v_y}{\partial y} = 0, \tag{13.2.3}$$

$$v_x \frac{\partial T}{\partial x} + v_y \frac{\partial T}{\partial y} = \alpha\left(\frac{\partial^2 T}{\partial x^2} + \frac{\partial^2 T}{\partial y^2}\right), \tag{13.2.4}$$

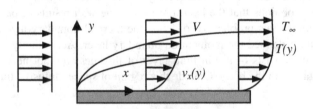

Fig. 13.2 Momentum and thermal boundary layers past a flat plate

with the following boundary conditions,

$$v_x = v_y = 0; \quad T = T_0 \text{ at } y = 0;$$
$$v_x = V; \quad v_y = 0; \quad P = P_\infty; \quad T = T_0 \text{ at } y \to \infty \text{ and at } x = 0. \tag{13.2.5}$$

Finally, the heat flux (13.1.5) is given by:

$$J_U = -k \left(\frac{dT}{dy} \right)_{y=0}. \tag{13.2.6}$$

Considering that convection is negligible within the boundary layer (and it is null at the wall), so that the temperature profile at the wall is linear, we can already estimate the magnitude of J_U as equal to $\kappa \Delta T / \delta_T$. The results of this analysis can be written in terms of the Nusselt number,

$$Nu = \frac{|J_U|}{k\Delta T / L} = \frac{L}{\delta_T}, \tag{13.2.7}$$

where $L \approx x$ is a macroscopic distance and $\Delta T = |T_\infty - T_0|$. We could also express it in terms of a heat transfer coefficient, defined in (8.1.4),

$$h = \frac{J_U}{\Delta T} = \frac{k}{L} Nu. \tag{13.2.8}$$

Now, two cases are considered, assuming either the presence or the absence of the momentum boundary layer. That means that we will considered separately the cases where, $Re \gg 1$ and that where $Re \ll 1$, provided anyway that $Pe \gg 1$.

13.2.1 Large Reynolds Number

In this case, two boundary layers are present, a momentum boundary layer, since $Re \gg 1$, and a thermal boundary layer, since $Pe \gg 1$. As we have seen in Chap. 7, the former is independent of the temperature field and has a thickness $\delta \approx L \, Re^{-1/2}$, where L is a macroscopic linear dimension, corresponding to the plate size. Now we have to guess whether the thermal boundary layer lays within or without the momentum boundary layer. In this case, subjected to a posteriori verification, it is reasonable to assume that $\delta_T \ll \delta$ when $Pe \gg Re$, i.e., when $Pe \gg 1$, while $\delta_T \gg \delta$ when $Pe \ll Re$, i.e., when $Pr \ll 1$.

- $Pr \ll 1$

Subjected to a posteriori verification, we assume that $\delta_T \gg \delta$, as sketched in Fig. 13.2. Consider the region at the edge of the thermal boundary layer, where $y \cong \delta_T$: since we are well outside the momentum boundary layer, here the longitudinal velocity equals its unperturbed value, V. Therefore, the continuity Eq. (13.2.3), with $x \approx L$, yields: $v_x \cong V(\delta_T/L)$, so that, in Eq. (13.2.4), the two convective terms have the same magnitude, while the first of the two diffusive terms is negligible compared to the second, i.e.,

$$v_x \frac{\partial T}{\partial x} \approx v_y \frac{\partial T}{\partial y} \approx \frac{V \Delta T}{L}, \quad \text{and} \quad \alpha \frac{\partial^2 T}{\partial y^2} \approx \frac{\alpha \Delta T}{\delta_T^2} \gg \alpha \frac{\partial^2 T}{\partial x^2} \approx \frac{\alpha \Delta T}{L^2}.$$

Now, by definition, at the edge of the boundary layer the convective terms must balance the diffusive terms. Consequently, equating the two expressions above, we obtain: $\delta_T^2 \approx \alpha L/V$; thus:

$$\delta_T/L \approx Pe^{-1/2} = Re^{-1/2} Pr^{-1/2}, \tag{13.2.9}$$

where $Pe = VL/\alpha$, $Re = VL/\nu$ and $Pr = \nu/\alpha$. Since $\delta \approx LRe^{-1/2}$, this final result reveals that $\delta_T/\delta \approx Pr^{-1/2} \gg 1$, in agreement with the assumption that we made at the start. Finally, from Eq. (13.2.7) we may conclude that, when $Pe \gg 1$ and $Sc \ll 1$ (so that $Re \gg 1$), we have:

$$Nu \approx L/\delta_T \approx Pe^{1/2} = Re^{1/2} Pr^{1/2}, \tag{13.2.10}$$

This result is in excellent agreement with the experimental data concerning liquid metals, where $Pr \ll 1$.

- $Pr \gg 1$

Subjected to a posteriori verification, we assume that now $\delta_T \ll \delta$, as sketched in Fig. 13.3. Thus, at the edge of the thermal boundary layer, the longitudinal velocity is much smaller than the unperturbed velocity, V. Being well inside the momentum boundary layer, from Chap. 8 we know that the pressure here is equal to its unperturbed value [cf. Eq. (8.1.6)], which is uniform in this case. Consequently, the velocity field consists of a Couette linear profile,[3] and $v_x \cong V(\delta_T/\delta)$. Now, repeating the previous analysis for $Pr \gg 1$, we see again that the two convective terms have the same magnitude, while the first of the two diffusive terms is negligible compared to the second, i.e.,

$$v_x \frac{\partial T}{\partial x} \approx v_y \frac{\partial T}{\partial y} \approx \frac{V \delta_T \Delta T}{\delta L}, \quad \text{and} \quad \alpha \frac{\partial^2 T}{\partial y^2} \approx \frac{\alpha \Delta T}{\delta_T^2} \gg \alpha \frac{\partial^2 T}{\partial x^2} \approx \frac{\alpha \Delta T}{L^2}.$$

---

[3]We can reach the same conclusion also observing that at the edge of the thermal boundary layer the convective terms in Eq. (13.2.1) are much smaller than their diffusive counterpart so that, pressure being constant, Eq. (13.2.1) reduces to $\partial^2 v_x/\partial y^2 = 0$, showing that v_x is linear with y.

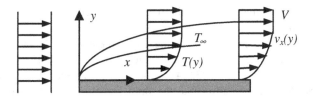

Fig. 13.3 Boundary layers when $Pr \gg 1$

Finally, balancing convective and diffusive terms we obtain: $\delta_T^3/\delta \approx \alpha L/V$, and so, since $\delta \approx LRe^{-1/2}$, we obtain:

$$\delta_T/L \approx Re^{-1/6}Pe^{-1/3} = Re^{-1/2}Pr^{-1/3}. \qquad (13.2.11)$$

Since $\delta \approx LRe^{-1/2}$, this final result reveals that $\delta_T/\delta \approx Pr^{-1/3} \ll 1$, in agreement with the assumption that we made at the start.

Finally, from Eq. (13.2.7) we may conclude that, when $Re \gg 1$ and $Sc \gg 1$ (so that $Pe \gg 1$), we have:

$$Nu \approx L/\delta_T \approx Re^{1/2}Pr^{1/3}. \qquad (13.2.12)$$

This relation is in excellent agreement with the experimental results summarized in Eqs. (10.2.7)–(10.2.9). Note that this relation continues to hold down to values of Pr as low as 0.5 (see Sect. 13.3.3 for a justification of this lucky circumstance).

All results concerning the case $Re \gg 1$ are summarized in Fig. 13.4, representing a log-log plot of $Nu/\sqrt{Re}$ as a function of Pr.

13.2.2 Small Reynolds Number

When $Re \ll 1$ there is no momentum boundary layer. Therefore, if the Peclet number is small, i.e., when $Pe \ll 1$, convection can be neglected and the Nusselt

Fig. 13.4 Log-log plot of $Nu/\sqrt{Re}$ at large Re as a function of Pr

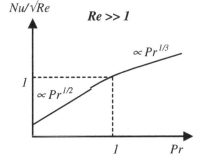

number is of $O(1)$ (to make it clear, that means that it could be 2 or ½, obviously, but not 100 or 0.01).

Different, and more interesting, is the case where $Pe \gg 1$. The dimensional analysis is similar to the previous ones, consisting of balancing convective and diffusive terms in the heat Eq. (13.2.4) at the edge of the thermal boundary layer, where $y \cong \delta_T$. Now, considering that the fluid velocity coincides with its unperturbed value at macroscopic distances from the wall, i.e., when $y \cong L$ (remind that when $Re \gg 1$ that happens at $y \cong \delta$), we can estimate the longitudinal velocity at $y \cong \delta_T$. In fact, assuming a linear velocity profile, we obtain: $v_x \cong V(\delta_T/L)$. Therefore, estimating v_y from the continuity equation, we see again that the two convective terms in Eq. (13.2.4) have the same magnitude, while the first of the two diffusive terms is negligible compared to the second, obtaining:

$$v_x \frac{\partial T}{\partial x} \approx v_y \frac{\partial T}{\partial y} \approx \frac{V \delta_T \Delta T}{L^2}, \quad \text{and} \quad \alpha \frac{\partial^2 T}{\partial y^2} \approx \frac{\alpha \Delta T}{\delta_T^2}.$$

Finally, balancing these two terms we obtain: $\delta_T^3 \approx \alpha L^2/V$, and so,

$$\delta_T/L \approx Pe^{1/3} = Re^{1/3}Pr^{1/3}. \tag{13.2.13}$$

Finally, applying Eq. (13.2.7), we conclude that, when $Re \gg 1$ and $Pe \gg 1$ (so that $Sc \gg Pe$), we have:

$$Nu \approx L/\delta_T \approx Pe^{1/3} \approx Re^{1/3}Pr^{1/3}. \tag{13.2.14}$$

All results concerning the case $Re \ll 1$ are summarized in Fig. 13.5, representing a log-log plot of Nu as a function of Pe.

Fig. 13.5 Log-log plot of Nu at small Re as a function of Pe

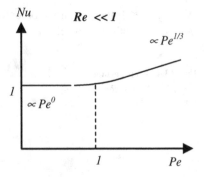

13.3 Colburn-Chilton Analogy

The Colburn-Chilton analogy establishes a correspondence between momentum, heat and mass transport, allowing the prediction of an unknown transfer coefficient when one of the other coefficients is known. As we will see, its range of application is far wider than what one would expect.

In this Section we will study the simplest version of the Colburn-Chilton analogy, also denoted as *film theory*, referring to the boundary layers that form at a wall wetted by a fluid flowing with large Reynolds and Peclet numbers.

First, we start from the definition of friction factor, f, that we have seen in Chap. 3,

$$f = \frac{(\text{diffusive momentum flux at the wall})}{1/2(\text{convective momentum flux far from the wall})} = \frac{\tau_w}{1/2 \rho V^2}, \qquad (13.3.1)$$

where V is the unperturbed velocity and ρ is the fluid density. We know that the fluid velocity at the wall is zero and therefore the momentum flux at the wall, that is the shear stress, is only diffusive and equal to $J_Q = \tau_w = \mu \, dv_x/dy$, where μ is the fluid viscosity, v_x is the longitudinal velocity (that is, parallel to the wall, along the x-direction), while y is the transverse coordinate. Since within the boundary layer, of thickness δ, the velocity v_x varies from 0 to V, we have: $\tau_w = \mu V/\delta$, where we have assumed a linear velocity profile within the boundary layer which, as we have seen in Chap. 7, amounts to neglecting the effects of the wall curvature, so that pressure is assumed to be uniform. Substituting these results into Eq. (13.3.1) we obtain: $\delta = 2\nu/fV$, where $\nu = \mu/\rho$ is the kinematic viscosity. Thus:

$$\frac{\delta}{L} = \frac{2}{f \, Re}, \quad \text{with } Re = \frac{VL}{\nu}, \qquad (13.3.2)$$

where L is a macroscopic typical length (e.g., the radius of the conduit or the size of the plate).

Now we may proceed, describing the heat flux at the wall in terms of the thickness of the thermal boundary layer, δ_T, obtaining: $J_U = kdT/dy = k\Delta T/\delta_T$, where $\Delta T = T_w - T_\infty$ is the temperature change across the thermal boundary layer (T_w is the wall temperature, while T_∞ is the fluid bulk temperature). Now, if the analogy between momentum and heat flux were complete, we should define a "thermal friction" coefficient as the ratio between the heat flux at the wall and the square of the temperature difference. Instead, while the convective momentum flux is proportional to the square of the unperturbed velocity, the corresponding dependence of the convective heat flux on the temperature difference is linear. Accordingly, as we saw, the heat flux at the wall is described through a heat transfer coefficient, h, defined as $J_U = h\Delta T$, so that $h = k/\delta_T$. Therefore, defining the Nusselt number as $Nu = J_U/J_{U,diff}$, where $J_{U,diff} = k\Delta T/L$ is the diffusive thermal flux, we obtain:

$$Nu = \frac{hL}{k} = \frac{L}{\delta_T}. \tag{13.3.3}$$

Finally, substituting Eq. (13.3.2) into (13.3.3), we find[4]:

$$Nu = \tfrac{1}{2}f\, Re(\delta/\delta_T). \tag{13.3.4}$$

At this point, we must evaluate the ratio (δ/δ_T). In the previous Section, we have seen that this ratio depends on the value of the Prandtl number, Pr. In particular, when $Pr > 1$, we have: $\delta_T/\delta \approx Pr^{-1/3}$, so that we obtain the following Colburn-Chilton equation,

$$Nu = \tfrac{1}{2}f\, RePr^{1/3} \Rightarrow h = \frac{k}{2R}f\, RePr^{1/3}. \tag{13.3.5}$$

At the end of this Section, we show that the relation $\delta_T/\delta \approx Pr^{-1/3}$ is valid under very general conditions, that is for turbulent flow as well, and also when Pr is not very large. Therefore, the Colburn-Chilton analogy can be applied when $Pr > 0.5$, provided that both Re and Pe are very large, so that both momentum and thermal boundary layers are present.

Let us see some application of the analogy.

13.3.1 Laminar Flow on a Flat Plate

Substituting into Eq. (13.3.5) the "exact" Blasius solution (8.2.14), $f = \tfrac{2}{3}Re_x^{-1/2}$, where $Re = Ux/v$, we obtain:

$$Nu_x = \frac{h_x x}{k} = \frac{1}{3}\, Re_x^{1/2} Pr^{1/3}, \tag{13.3.6}$$

where h_x is the local heat transfer coefficient. This same relation (apart from the $\tfrac{1}{3}$ coefficient) was obtained in Sect. 13.2, assuming $Pr \gg 1$.

Note that the local heat transfer coefficient,

$$h_x = \frac{1}{3}k(V/n)^{1/2}Pr^{1/3}x^{-1/2},$$

diverges at $x = 0$. The explanation is the same as for the divergence of the shear stress that we have seen in Sect. 7.2: at $x = 0$ a finite temperature difference is applied to a fluid layer of zero thickness. However, considering the average heat transfer coefficient,

---

[4]Note that Colburn assumed that $\delta = \delta_T$.

$$\bar{h} = \frac{1}{L} \int\limits_0^L h(x)dx,$$

we easily find:

$$\bar{h} = 2h_L \quad \text{and} \quad Nu = \frac{\bar{h}x}{k} = \frac{2}{3}Re_L^{1/2}Pr^{1/3}. \tag{13.3.6a}$$

This relation is in excellent agreement with Eq. (11.2.9), fitting the experimental data obtained for $Pr > 0.5$ and $Re < 5.10^5$. The same type of expressions, i.e., Eqs. (11.2.7) and (11.2.8), describe the heat flux at the surface of spheres and cylinders.

13.3.2 Turbulent Flow in a Pipe

From the Blasius correlation, $f = 0.08Re^{-1/4}$, where Re is evaluated in terms of the tube diameter, we obtain: $Nu = 0.04Re^{3/4}Pr^{1/3}$. This relation gives results that are very similar to those of the Dittus-Boelter correlation, and can be used in its stead.

13.3.3 The Relation between δ and δ_T

Let us start from Eqs. (13.2.1)–(13.2.5) within the boundary layer on a flat plate, neglecting the diffusive terms along the longitudinal direction (i.e., $\partial^2 v_x/\partial x^2$ and $\partial^2 T/\partial x^2$), and neglecting the pressure term as well since, as we saw, P is constant.[5] In terms of dimensionless variables $\tilde{v}_x = v_x/V$ and $\tilde{T} = (T - T_0)/(T_\infty - T_0)$, these equations are:

$$v_x \frac{\partial \tilde{v}_x}{\partial x} + v_y \frac{\partial \tilde{v}_x}{\partial y} = \alpha \frac{\partial^2 \tilde{v}_x}{\partial y^2}, \tag{13.3.7}$$

$$\frac{\partial v_x}{\partial x} + \frac{\partial v_y}{\partial y} = 0, \tag{13.3.8}$$

$$v_x \frac{\partial \tilde{T}}{\partial x} + v_y \frac{\partial \tilde{T}}{\partial y} = \alpha \frac{\partial^2 \tilde{T}}{\partial y^2}, \tag{13.3.9}$$

---

[5]These are all intuitive assumptions. Details can be found in the discussion of Sect. 7.2 about the Blasius solution.

The additional Navier-Stokes equation in the transversal, y-direction is totally equivalent to Eq. (13.3.7). These equations are subjected to the following boundary conditions

$$\tilde{v}_x = \tilde{T} = 0 \quad \text{per} \quad y = 0;$$
$$\tilde{v}_x = \tilde{T} = 1 \quad \text{per} \quad y \to \infty \quad \text{and} \quad \text{per } x = 0. \tag{13.3.10}$$

Obviously, when $v = \alpha$, the momentum equation becomes identical to the heat equation, and therefore the profile of $\tilde{T}$ is equal to that of $\tilde{v}_x$, with $\delta = \delta_T$. Otherwise, from the Blasius solution we know that $v_x = 0.332\eta + O(\eta^4)$, where $\eta = y/\sqrt{4vx/V}$, showing that the with excellent approximation we can assume that the velocity profile v_x, is linear with y. Accordingly, from the continuity equation we see that v_y can be assumed to be quadratic in y. So, in synthesis, we may write:

$$v_x(x,y) = a(x)y; \quad v_y(x,y) = b(x)y^2, \tag{13.3.11}$$

where $a(x)$ and $b(x)$ are two known functions. Now, defining the following stretched coordinate,

$$\zeta = yPr^{1/3}, \tag{13.3.12}$$

the heat equation can be written as:

$$a(x)\zeta\frac{\partial \tilde{T}}{\partial x} + b(x)\zeta^2 \frac{\partial \tilde{T}}{\partial \zeta} = v\frac{\partial^2 \tilde{T}}{\partial \zeta^2}, \tag{13.3.13a}$$

which is identical to that of v_x,

$$a(x)y\frac{\partial \tilde{v}_x}{\partial x} + b(x)y^2 \frac{\partial \tilde{v}_x}{\partial y} = v\frac{\partial^2 \tilde{v}_x}{\partial y^2}, \tag{13.3.13b}$$

Provided that the variable y replaces ζ. Therefore, we obtain:

$$\tilde{T}(x,\zeta) = \tilde{v}_x(x,y). \tag{13.3.14}$$

This means that the profiles of temperature and velocity (and therefore the respective boundary layers as well) when the transversal coordinate in the temperature profile is stirred by a factor $Pr^{1/3}$, obtaining:

$$\frac{\delta_T}{\delta} = Pr^{-1/3}. \tag{13.3.15}$$

This analysis is valid whenever the velocity profile within the momentum boundary layer can be assumed to be linear. As we have seen, this assumption is

Fig. 13.6 Plug flow across a fixed bed

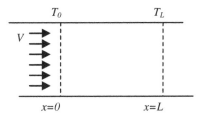

verified under very general conditions and, therefore, it is not surprising that Eq. (13.3.15) turns out to be applicable even when the Prandtl number is not very large.

Note that the same result (13.3.15) could be obtained using the approximate integral method of Sect. 8.4, as shown in Sect. 17.4.

13.4 Problems

13.1 Air with velocity $V = 6$ m/s and atmospheric pressure flows past a flat plate of length 1 m and width 2.5 m. Knowing that the plate temperature is $T_0 = 120$ °C, while that of the air is $T_\infty = 0$ °C, calculate the heat exchanged per unit time.

13.2 A fluid flows with uniform velocity V (i.e., in plug flow) across a fixed bed along the x-direction (see Fig. 13.6). The bed is maintained at temperature T_0 at $x < 0$, and at temperature T_L at $x > L$. (a) Calculate the temperature profile $T(x)$ and the heat flux J_U at $x = L$, in particular in the two limit cases $Pe \ll 1$ and $Pe \gg 1$. (b) Derive the heat flux in the two limit cases by using regular and singular expansion techniques.

13.3 Consider a shell and tube heat exchanger where the hot fluid, flowing inside the tubes of radius R, is liquid sodium, with Prandtl number $Pr \ll 1$. Assuming that the Reynolds number and the Peclet number are both very large, i.e., $Re \gg 1$ and $Pe \gg 1$, determine how the heat flux, i.e., the Nusselt number, depends on Re and Pr. Define clearly all the necessary hypotheses.

Chapter 14
Constitutive Equations for Transport of Chemical Species

Abstract In this chapter we begin to study the mass transfer of chemical species, defining its basic principles and defining the appropriate quantities of multicomponent flows. In Sect. 14.1 we show that, as the mean velocities of the individual species are different from each other, there are several ways to define the average velocity of a mixture. Accordingly, the separation between convective and diffusive fluxes appearing in the governing equations derived in Sect. 14.2 is not unique, as we stress in Sect. 14.3, where the constitutive relation of the diffusive material fluxes are investigated. The peculiarities of the transport of mass compared to that of heat and momentum are further illustrated in the discussion on boundary conditions in Sect. 14.4, and in the series of pointed questions that are listed in the final Sect. 14.5.

14.1 Fluxes and Velocities

As we have seen in Chaps. 1 and 6, the mass flux $\mathbf{j}$ of a single component fluid is simply the product of its density ρ and the mean fluid velocity $\mathbf{v}$. Microscopically, $\mathbf{j}$ is the sum of the momentum associated with the N identical molecules contained in a unit volume, $V = 1$:

$$\mathbf{j} = \rho\mathbf{v} = \frac{1}{V}\sum_{i=1}^{N} m\mathbf{v}_i, \qquad (14.1.1)$$

where m is the mass of each molecule. Considering that the density is defined as

$$\rho = \frac{1}{V}\sum_{i=1}^{N} m = \frac{Nm}{V}, \qquad (14.1.2)$$

© Springer International Publishing Switzerland 2015
R. Mauri, *Transport Phenomena in Multiphase Flows*,
Fluid Mechanics and Its Applications 112, DOI 10.1007/978-3-319-15793-1_14

this means defining the velocity $\mathbf{v}$ as the mass averaged velocity,

$$\mathbf{v} = \frac{\sum_{i=1}^{N} m\mathbf{v}_i}{\sum_{i=1}^{N} m} = \frac{1}{N}\sum_{i=1}^{N}\mathbf{v}_i. \qquad (14.1.3)$$

Obviously, as it appears from Eq. (14.1.3), mass averaging coincides here with arithmetic averaging (and therefore with molar averaging as well), since molecules in this case have the same mass; in the following, $\mathbf{v}$ will be indicated simply as the mean velocity.

Now, if the fluid is composed of two or more components, we can define in the same way many mass fluxes, one for each species. For example, when we have a binary mixture with two components, A and B, constituted of molecules of mass m_A and m_B, respectively, we define the following mass fluxes:

$$\mathbf{j}_A = \frac{1}{V}\sum_{i=1}^{N_A} m_A\mathbf{v}_{A,i}; \quad \mathbf{j}_B = \frac{1}{V}\sum_{i=1}^{N_B} m_B\mathbf{v}_{B,i}, \qquad (14.1.4)$$

where $\mathbf{v}_A$, is the velocity of the ith molecule of A-type, and the same applies to $\mathbf{v}_{B,i}$, while N_A and N_B, with $N = N_A + N_B$, are, respectively, the number of molecules of type A and B contained within the volume V. Naturally, the sum of the mass fluxes of all components equals the total mass flux,

$$\mathbf{j}_A + \mathbf{j}_B = \mathbf{j}. \qquad (14.1.5)$$

In addition, defining the mass concentration ρ_A and ρ_B as the mass of the A and B species contained within a unit volume,

$$\rho_A = \frac{1}{V}\sum_{i=1}^{N_A} m_A = \frac{N_A m_A}{V}; \quad \rho_B = \frac{1}{V}\sum_{i=1}^{N_B} m_B = \frac{N_B m_B}{V}, \qquad (14.1.6)$$

we obtain:

$$\mathbf{j}_A = \rho_A\mathbf{v}_A; \quad \mathbf{j}_B = \rho_B\mathbf{v}_B, \qquad (14.1.7)$$

where $\mathbf{v}_A$ and $\mathbf{v}_B$ are the mean velocities of the A and B species,

$$\mathbf{v}_A = \frac{\sum_{i=1}^{N_A} m_A\mathbf{v}_{A,i}}{\sum_{i=1}^{N_A} m_A} = \frac{1}{N_A}\sum_{i=1}^{N_A}\mathbf{v}_{A,i}; \quad \mathbf{v}_B = \frac{\sum_{i=1}^{N_B} m_B\mathbf{v}_{B,i}}{\sum_{i=1}^{N_B} m_B} = \frac{1}{N_B}\sum_{i=1}^{N_B}\mathbf{v}_{B,i}. \qquad (14.1.8)$$

Again, the mass average velocities of the A and B species coincide with the respective arithmetic, or molar, average velocities. Now we can define the mass average velocity, $\mathbf{v}$, of the mixture as the ratio between the total mass flux, $\mathbf{j} = \mathbf{j}_A + \mathbf{j}_B$, and the total mass density, $\rho = \rho_A + \rho_B$, in agreement with Eq. (14.1.3).

Therefore, we see that the mass average velocity of the mixture, $\mathbf{v}$, can be determined in terms of $\mathbf{v}_A$ and $\mathbf{v}_B$, as:

$$\mathbf{v} = \frac{\sum_{i=1}^{N} m\mathbf{v}_i}{\sum_{i=1}^{N} m} = \frac{\mathbf{j}}{\rho} = \frac{\rho_A \mathbf{v}_A + \rho_B \mathbf{v}_B}{\rho_A + \rho_B} = \omega_A \mathbf{v}_A + \omega_A \mathbf{v}_B. \tag{14.1.9}$$

Here, we have substituted Eqs. (14.1.4) and (14.1.6), denoting by ω_A and ω_B the mass fractions,

$$\omega_A = \rho_A/\rho; \quad \omega_B = \rho_B/\rho. \tag{14.1.10}$$

Similar relations can also be written in terms of the molar densities (or composition), c_A and c_B, defined as the number of moles of the A and B species contained within a unit volume, i.e.,

$$c_A = \rho_A/M_{W,A}; \quad c_B = \rho_B/M_{W,B}, \tag{14.1.11}$$

where $M_{W,A}$ and $M_{W,B}$ are the molecular weights of A and B.

We can also define the molar fluxes $\mathbf{J}_A$ and $\mathbf{J}_B$ as the number of moles of the A and B species that cross the unit section per unit time. Since $\mathbf{J}_A = \mathbf{j}_A/M_{W,A}$ and $\mathbf{J}_B = \mathbf{j}_B/M_{W,B}$, we see that, as expected, the mean velocities of the A and B species can be defined in terms of either mass or molar averaging, i.e.,

$$\mathbf{v}_A = \mathbf{j}_A/\rho_A = \mathbf{J}_A/c_A \quad \text{and} \quad \mathbf{v}_B = \mathbf{j}_B/\rho_B = \mathbf{J}_B/c_B.$$

Again, we can define the molar average velocity, $\mathbf{v}^*$, of the mixture as the ratio between the molar flux, $\mathbf{J} = \mathbf{J}_A + \mathbf{J}_B$, and the total molar density, $c = c_A + c_B$, in agreement with Eq. (14.1.3). Therefore, we see that the molar average velocity of the mixture, $\mathbf{v}^*$, can be determined in terms of $\mathbf{v}_A$ and $\mathbf{v}_B$ as:

$$\mathbf{v}^* = \frac{\mathbf{J}}{c} = \frac{c_A \mathbf{v}_A + c_A \mathbf{v}_B}{c_A + c_B} = x_A \mathbf{v}_A + x_A \mathbf{v}_B, \tag{14.1.12}$$

where we have defined the molar fractions,

$$x_A = c_A/c; \quad x_B = c_B/c. \tag{14.1.13}$$

In Sect. 14.5 we will see that mass quantities are better suited to describe liquid mixtures, while for gaseous mixtures molar quantities are more convenient. In the following, though, as the molar-based governing equations have the same form as their mass-based counterparts, we will refer to either of them as material quantities, such as material fluxes, or material balance equations.

14.2 Material Balance Equations

The principle of mass conservation in multi-component systems is formulated in the same way as in single component systems. Repeating the analysis of Sect. 6.1, consider a fixed volume element V and define for each point of the closed surface S enclosing it a unit vector $\mathbf{n}$ perpendicular to the surface and directed outward. Denoting by j_{An} the mass flux of the A-species leaving the control volume, the balance equation of the A species states that the mass of A accumulated in the volume V is equal to that entering through the surface S plus the mass of A generated in V, that is,

$$\frac{dM_A}{dt} = \int_V \frac{\partial \rho_A}{\partial t} dV = -\int_S j_{An} dS + \int_V r_A dV, \quad \text{where } M_A = \int_V \rho_A dV. \quad (14.2.1)$$

Here, r_A is the mass of A generated per unit volume and time, resulting, for example, from a chemical reaction. Note that M_A is the mass of A within the volume V and therefore we have to consider that V is fixed (i.e., it does not change with time) when we take its time derivative. Now, as stated by the Cauchy theorem shown in Sect. 6.1, j_{An} is the component along the direction $\mathbf{e}_n$ of a mass flux vector $\mathbf{j}_A$. As this latter can be expressed in terms of the mean velocity of the A species as $\mathbf{j}_A = \rho_A \mathbf{v}_A$, we obtain:

$$\int_S j_{An} dS = \int_S \mathbf{e}_n \cdot (\rho_A \mathbf{v}_A) dS = \int_V \nabla \cdot (\rho_A \mathbf{v}_A) dV, \quad (14.2.2)$$

where we have applied the divergence theorem. Finally we obtain:

$$\int_V \left[\frac{\partial \rho_A}{\partial t} + \nabla \cdot (\rho_A \mathbf{v}_A) - r_A \right] dV = 0. \quad (14.2.3)$$

Since this relation must apply for any volume V, the integrand of Eq. (14.2.3) must be identically zero, i.e.,

$$\frac{\partial \rho_A}{\partial t} + \nabla \cdot (\rho_A \mathbf{v}_A) = r_A. \quad (14.2.4)$$

An identical relation can be written for the B species, obtaining,

$$\frac{\partial \rho_B}{\partial t} + \nabla \cdot (\rho_B \mathbf{v}_B) = r_B. \quad (14.2.5)$$

Summing the last two equations, we obtain Eq. (6.2.2):

$$\frac{\partial \rho}{\partial t} + \nabla \cdot (\rho \mathbf{v}) = 0, \tag{14.2.6}$$

where Eqs. (14.1.6) and (14.1.9) have been applied, and we have considered that, since the total mass is conserved, $r_A + r_B = 0$.

This analysis can be repeated in terms of molar quantities, obtaining:

$$\frac{\partial c_A}{\partial t} + \nabla \cdot (c_A \mathbf{v}_A) = R_A, \tag{14.2.7}$$

$$\frac{\partial c_B}{\partial t} + \nabla \cdot (c_B \mathbf{v}_B) = R_B, \tag{14.2.8}$$

where R_A and R_B are the number of moles of the A and B species generated per unit volume and time. Summing these two equations we find:

$$\frac{\partial c}{\partial t} + \nabla \cdot (c \mathbf{v}^*) = R, \tag{14.2.9}$$

where Eqs. (14.1.11)–(14.1.12) have been applied, and we have defined $R = R_A + R_B$, considering that, in general, the total number of moles is not conserved (think, for example, of a chemical reaction $A \rightarrow 2B$, where $R_B = -2R_A$).

14.3 The Constitutive Equations of the Material Fluxes

We have seen that in a fluid mixture of N components we can write N equations of moles (or mass) conservation in terms of the molar (or mass) densities c_i of each species and their respective velocities $\mathbf{v}_i$. These latter, in principle, can be determined by solving N Navier-Stokes equations, i.e., one per each species. This, so-called many-fluid approach, is widely used in multiphase fluid mechanics; yet, it presents an insurmountable problem: the body forces in the Navier-Stokes equation for species A must include also all the interactions (attractive or repulsive) that the other species exert on the A-type molecules. Now, not only these forces are mostly unknown, but it is also very difficult, perhaps impossible, to measure them experimentally. Another approach, called mixture theory and mostly adopted in chemical engineering, expresses the molar fluxes (the same happens with the mass fluxes) as the sum of two parts: the former is defined in terms of the molar (or mass) average velocity of the mixture, $\mathbf{v}^*$, while the latter part expresses the relative flux of A with respect to $\mathbf{v}^*$,

$$\underbrace{\mathbf{J}_A = c_A \mathbf{v}_A}_{\substack{\text{Flux of A relative to}\\ \text{a fixed reference frame.}}} \quad = \quad \underbrace{c_A \mathbf{v}^*}_{\substack{\text{Flux of A with}\\ \text{velocity } \mathbf{v}^*.}} \quad + \quad \underbrace{\mathbf{J}_{A,d}}_{\substack{\text{Flux of A}\\ \text{relative to } \mathbf{v}^*.}} \qquad (14.3.1)$$

From this expression we have:

$$\mathbf{J}_{A,d} = c_A(\mathbf{v}_A - \mathbf{v}^*), \qquad (14.3.2)$$

indicating that the relative flux, $\mathbf{J}_{A,d}$, is a consequence of the fact that the A-type molecules, in general, can have a mean velocity, $\mathbf{v}_A$, that is different from the average velocity of the mixture. Obviously, the same relations (14.3.1) and (14.3.2) can be written for the other species as well; for example, for binary mixtures we have:

$$\mathbf{J}_B = c_B \mathbf{v}^* + \mathbf{J}_{B,d}, \qquad (14.3.3)$$

so that, summing (14.3.1) and (14.3.3), and considering that $\mathbf{J}_A + \mathbf{J}_B = \mathbf{J} = c\mathbf{v}^*$, [see Eqs. (14.1.12) and (14.1.13)], we obtain:

$$\mathbf{J}_{A,d} = -\mathbf{J}_{B,d}, \qquad (14.3.4)$$

indicating that, as it must be, the sum of all relative fluxed must be null.

The relative fluxes are due to two types of phenomena:

(a) **External forces**. These are forces that act differently upon the two species. For example, if species A is heavier than B, in a gravitational field A will *sediment*; or, if A and B are ions of different electric charges, in an electric field the two species will move with respect to one another, a phenomenon called *electrophoresis*. Analogous phenomenon occurs in the presence of temperature or pressure gradients.

(b) **Diffusion**. If the mass or molar density of the A-molecules is non-uniform, thermal fluctuations will induce a net flow of the A-molecules from regions with large densities to small, tending to restore a condition of chemical equilibrium, with uniform density. To understand this point, consider Fig. 14.1 and imagine to measure the net flux of the A-molecules, indicated as black dots,

Fig. 14.1 Typical diffusion process

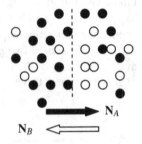

across the section indicated with the broken line: although the A-molecules can move left or right with the same probability, since there are more A-molecules on the left side, there will be a net flux of A-molecules from left to right. This is a typical diffusion process.

In the following, we will neglect the influence of the external forces, assuming that the relative flux, $\mathbf{J}_{A,d}$, is due exclusively to diffusion.

Now, we propose a constitutive equation to model the diffusive flux of the A-species as a function of the concentration field. This is the Fick equation (sometimes called Fick's first law), that we have derived at the microscopic level in Chap. 1, based on the kinetic theory of gases, i.e.,

$$\mathbf{J}_{A,d} = -cD_{AB}\nabla x_A, \tag{14.3.5}$$

where D_{AB} is the binary diffusion coefficient of A in B, expressed in m^2/s. Note the minus sign in Eq. (14.3.5), indicating that the mass flux is directed from regions of large density to low. It should be stressed that the Fick law is valid only for ideal mixtures, as we show in detail in Chap. 21 [see Eq. (21.5.6)].

Now, since [cf. Eq. (14.3.4)] $\mathbf{J}_{A,d} = -\mathbf{J}_{B,d}$, where $\mathbf{J}_{B,d} = -cD_{BA}\nabla x_B$, considering that $x_B = 1 - x_A$, we find:

$$D_{AB} = D_{BA}. \tag{14.3.6}$$

In the following, this coefficient of binary diffusion will be simply denoted as D. The equality (14.3.6) indicates that, since the diffusion of A into B can also be interpreted as the diffusion of B into A, it is natural that the two processes are described by the same transport coefficient.[1]

A more satisfactory constitutive relation is one involving chemical potentials instead of concentrations. In fact, a system at chemical equilibrium has uniform chemical potentials, just like a system at thermal equilibrium has a uniform temperature. Now, in heat transport, the Fourier constitutive relation indicates that a system responds to a temperature gradient by inducing a heat flux, trying to restore thermal equilibrium. In the same way, in the presence of a chemical potential gradient, a system will respond inducing a mass flux, proportional (and opposite) to the chemical potential gradient, trying to restore chemical equilibrium. Therefore, since chemical potentials are functions (at uniform temperature and pressure) of the composition of the mixture, we obtain again the Fick law (14.3.5), where D is, in general, a function of the composition. This point will be analyzed in detail in Chap. 20, where we will see that the diffusion coefficient D is a constant only for ideal mixtures (including the dilute limit case).

As a summary, substituting Eq. (14.3.5) into (14.3.1) and (14.3.3), we see that the molar fluxes in a binary A-B mixture can be expressed as the sum of a

---

[1]In Fig. 14.1 we see that when A diffuses from left to right, then B must be diffusing in the opposite direction.

convective term, depending on the molar average velocity of the mixture, and a diffusive term, as follows,

$$\mathbf{J_A} = c_A \mathbf{v}^* - cD\nabla x_A, \qquad (14.3.7)$$

$$\mathbf{J_B} = c_B \mathbf{v}^* - cD\nabla x_B. \qquad (14.3.8)$$

Often, it is convenient to express the convective terms in a different form, substituting Eqs. (14.1.12) and (14.1.13) into Eqs. (14.3.7) and (14.3.8), obtaining:

$$\mathbf{J_A} = x_A(\mathbf{J_A} + \mathbf{J_B}) - cD\nabla x_A \qquad (14.3.9)$$

$$\mathbf{J_B} = x_B(\mathbf{J_A} + \mathbf{J_B}) - cD\nabla x_B \qquad (14.3.10)$$

Finally, all the above constitutive relations can be expressed using mass-based quantities, obtaining:

$$\mathbf{j_A} = \rho_A \mathbf{v} - \rho D\nabla \omega_A = \omega_A(\mathbf{j_A} + \mathbf{j_B}) - \rho D\nabla \omega_A, \qquad (14.3.11)$$

$$\mathbf{j_B} = \rho_B \mathbf{v} - \rho D\nabla \omega_B = \omega_B(\mathbf{j_A} + \mathbf{j_B}) - \rho D\nabla \omega_B. \qquad (14.3.12)$$

Clearly, since molar and mass fluxes are linearly dependent on one another, we see that the coefficient of binary diffusion D is the same in the two cases.

An important consideration that immediately arises from the constitutive Eqs. (14.3.7)–(14.3.12) is that, while the diffusive term is linear in the composition x_A (remember however that the linear Fick law is valid only for ideal mixtures), the convective term depends on x_A in a complex way. Therefore, the great simplification that in heat transfer derives from the linearity between heat flux and temperature difference cannot be extended, in general, to mass transport, unless we are considering the dilute case, as we show in the following.

14.3.1 The Dilute Case

Consider a mixture consisting of a species B, say air or water, containing a contaminant, A, in very small proportion, so that $x_A \ll 1$ or $\omega_A \ll 1$ The mass and molar average velocities of the mixture (14.1.9) and (14.1.12) become:

$$\mathbf{v} = \omega_A \mathbf{v_A} + (1 - \omega_A)\mathbf{v_B} = \mathbf{v_B}[1 + O(\omega_A)].$$

and

$$\mathbf{v}^* = x_A \mathbf{v_A} + (1 - x_A)\mathbf{v_B} = \mathbf{v_B}[1 + O(\omega_A)],$$

This shows that the presence of a dilute solute does not affect appreciably the flow field of the solvent, which can be determined independently, i.e. neglecting the presence of the solute. Consequently, at leading order, the mean velocity of the mixture coincides with the velocity of the solvent. Now, substituting these expressions into Eqs. (14.3.7) and (14.3.11) for the mass and molar fluxes, we obtain, respectively:

$$\mathbf{j_A} = \rho[\omega_A \mathbf{v}_B - D\nabla \omega_A + O(\omega_A^2)]. \tag{14.3.13}$$

and

$$\mathbf{J_A} = c[x_A \mathbf{v}_B - D\nabla x_A + O(x_A^2)]. \tag{14.3.14}$$

Therefore, the constitutive equations for material transport in the dilute case are linear in the composition. That means that in the dilute limit there is a direct analogy between heat transport and mass transport.

14.3.2 Multi-component Mixtures

Generalizing Eq. (14.3.9), the constitutive relation for an n-component mixture is:

$$\mathbf{J}_i = x_i \sum_{J=1}^{n} \mathbf{J}_j - cD_{im}\nabla x_i, \tag{14.3.15}$$

where D_{im} is the multi-component diffusivity of the ith species in the mixture and is quite difficult to determine. Thus, instead of Eq. (14.3.15), it is preferred to use the following Stefan-Maxwell equations, that employs binary diffusivities:

$$\nabla x_i = \sum_{J=1}^{n} \frac{1}{cD_{ij}}(x_i \mathbf{J}_j - x_j \mathbf{J}_i). \tag{14.3.16}$$

If Eqs. (14.3.15) and (14.3.16) are combined, the multicomponent diffusion coefficient may be assessed in terms of binary diffusion coefficients. For gases, the values D_{ij} of this equation are approximately equal to the binary diffusivities for the ij pairs.

14.4 Boundary Conditions

Here we can repeat most of the comments that we made in Sect. 9.1 for heat transport. Indicating by c_A the material (i.e., mass or molar) concentration of species A, the boundary conditions at the interface between two phases, the material flux is

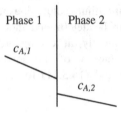

Fig. 14.2 The boundary conditions at the interface between two phases

continuous, while the compositions of the two phases may be different. Assuming local equilibrium at the interface, the ratio between the concentrations of the two phases, denoted as partition coefficient α, is a thermodynamic quantity that is uniquely determined in terms of pressure and temperature. Referring to Fig. 14.2, we have:

$$
c_{A,1}|_w = \alpha\, c_{A,2}|_w; \quad \mathbf{e}_n \cdot \mathbf{J}_{A,1}|_w = \mathbf{e}_n \cdot \mathbf{J}_{A,2}|_w \Rightarrow D_1 \frac{dc_{A,1}}{dz}\bigg|_w = D_2 \frac{dc_{A,2}}{dz}\bigg|_w,
$$

$$(14.4.1a, b)$$

where $\mathbf{e}_n$ denotes a unit vector perpendicular to the interface, while z is a coordinate along $\mathbf{e}_n$. Note that at the interface, due to the different diffusivities of the two phases, the concentration profiles have different slopes.

Sometimes, we know what happens on one side of the interface, so that the problem is greatly simplified. Consider the two following cases:

1. The concentration at the interface is fixed: $c_A|_w = c_0$. This is a particular case of the previous boundary conditions (14.4.1a, b) and applies, for example, when one of the two medium has a much larger inertia than the other. In this case, $c_{A,1} = \alpha\, c_0$ is uniform, so that $c_{A,2} = c_0$ at the interface.
2. The molar (or mass) flux at the interface is fixed, i.e., $\mathbf{e}_n \cdot \mathbf{J}_A|_w = J_0$. Since in general the velocity at the wall is zero, applying the Fick law, this means fixing the concentration gradient at the wall. The most common example consists of the no-flux boundary condition, with $J_0 = 0$, that can be applied at the interface between a fluid mixture and a solid wall.

Sometimes, above all in the case of convective mass transfer, we do not know exactly what happens at the interface. Therefore, at liquid-solid interfaces we can apply the equivalent of *Newton's law of cooling*, stating that the material flux at the wall is proportional to the difference between the concentration at the wall, c_{Aw}, and the bulk concentration, far from the wall, c_{A0} (see Fig. 14.3), i.e.,

$$
J_{Aw} = \mathbf{e}_n \cdot \mathbf{J}_A = k_M(c_{Aw} - c_{A0}), \tag{14.4.2}
$$

where $\mathbf{e}_n$ is the unit vector perpendicular to the surface and directed outward, i.e. towards the fluid, while k_M is denoted as the **molar or mass transfer coefficient**,

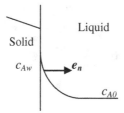

Fig. 14.3 A liquid-solid interface

expressed in m/s. Note that when $k_M = 0$ the Newton law reduces to $J_{Aw} = 0$, while when $k_M \rightarrow \infty$, since the flux must remain finite, it becomes $c_{Aw} = c_{A0}$.

As in the heat transport case, the Newton law is an empirical rule and k_M is a coefficient of ignorance. In fact, it predicts a concentration jump at the wall, while actually the concentration profile in the vicinity of a wall, as represented in Fig. 14.3, consists of a continuous line connecting the fluid bulk concentration with the wall concentration, where the boundary conditions (14.4.1a,b) can be applied. However, since the flow conditions are unknown, the detailed concentration profile near the wall cannot be determined and thus we are forced to introduce the concept of mass transport coefficient. There is nevertheless a fundamental difference between heat and mass transport: heat transport is always a linear process, while mass transport is linear only in the dilute limit. Therefore, since heat fluxes are proportional to temperature differences, the heat transfer coefficient is, approximately, independent of temperature and so it is a very convenient tool in modelling heat transport processes. On the other hand, as the dependence of the mass fluxes on composition in strongly non-linear, the mass transport coefficient is itself a function of x_A, and so, in general, it does not introduce any simplification in modelling processes of mass transport. In the dilute case, however, mass fluxes are linear with composition (see Sect. 14.3.1), and there is a direct analogy between heat and mass transport, showing that in dilute case it is convenient to use the mass transport coefficient.

14.5 Answers to Some Questions on Material Transport

Q. What is the difference between convection and diffusion in mass transport? A. Convection is a reversible process, while diffusion is not. Let us see why.

Based on the definition (14.3.1), one could think that, unlike what happens in momentum and energy transport, in mass transport the difference between convective flux and diffusive flux is not "objective", i.e., there are not two fundamentally different transport mechanisms, but instead the difference between diffusion and convection is only a question of definition. This is not true. Consider,

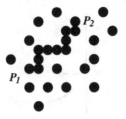

Fig. 14.4 Molecules in a liquid

for example, the ensemble of particles represented in Fig. 14.4: energy or momentum can be transferred from point P_1 to P_2 by diffusion, i.e., without any net particle movement, thanks to mutual particle interactions: one particle pushes another, who pushes a third one and so on, just like a train of billiard balls. Mass, however, can be transferred only when, physically, a particle move from P_1 to P_2, showing that mass transport is intrinsically a convective process.[2] In fact, for a single component fluid, mass flux has only a convective component, unlike momentum and heat fluxes, where the diffusive parts lead to the definition of viscosity and thermal conductivity.

Mass diffusion arises within multi-component systems, where, in fact, even in the absence of any macroscopic movement, we can have a dissipative process. Again, consider the black and white particles of Fig. 14.1, assuming that (a) they have the same mass and size; (b) their total concentration $c = c_1 + c_2$ is constant; (c) they oscillate chaotically by thermal fluctuations. If initially the two species are completely separated, as time progresses they will mix, with a consequent entropy increase. This will happen despite the system remaining macroscopically at rest, as its mean velocity is null, since the (diffusive) flux of the black particles from left to right is exactly equal to that of the white particles in the opposite direction. This process, intrinsically dissipative, is the simplest example of mass diffusion, and shows that its nature is not a question of definition, but instead reflects a very fundamental mechanism of mass transport.

Q. How much are the molecular diffusivities in different medium?

A. While the diffusivities of momentum, energy and mass are of the same magnitudein gases, in liquids and solids mass diffusivities are always smaller. Let us see why.

As we have seen in Sects. 1.5–1.7, in an ideal gas the diffusive transport of momentum, energy and mass occurs exclusively thanks to molecular thermal fluctuations: every molecule move randomly transporting, together with its mass, its energy and its momentum as well. Therefore, since the transport mechanism is the

---

[2]Actually, as we saw in Sect. 1.8, diffusion exists in single component systems as well, and can be studied by following the random movement of a molecule in time.

same in all three cases, it is not surprising that the three coefficients describing these transport phenomena, namely the diffusivities v, α and D, are equal to each other.

In real gases, multiple collisions must also be considered, where one particle interacts with many other particles contemporarily. As these collisions are not frequent,[3] the ideal gas model of Sects. 1.5–1.7 is still valid, explaining why the Prandtl number and the Schmidt number, $Pr = v/\alpha$ and $Sc = v/D$ for real gases are not very different from unity. For example, air has $v = 15.9$ mm²/s and $\alpha = 22.5$ mm²/s, while the molecular diffusivity of CO_2 in air is $D = 16$ mm²/s; in general, $D \sim 10$ mm²/s in gases.

In the case of liquids, instead, molecules are closely packed, so that momentum and energy transport occur primarily by multiple collisions among molecules. For example, referring to Fig. 14.4, energy and momentum can be transferred by diffusion from point P_1 to P_2 by utilizing the train of particles connecting these two points. This is a very efficient mechanism for the diffusive transport of momentum and energy; as a result, dynamic viscosity and thermal conductivity of liquids are much larger than in gases (see the table in Appendix 1). Then, kinematic viscosity and thermal diffusivity are obtained by dividing kinematic viscosity and thermal diffusivity by density, so that we see that for liquids, in most cases, v, $\alpha = 0.1$–1 mm²/s, that is approximately 10 times smaller than for gases. On the other hand, in quiescent liquids mass transport is very slow, as it is difficult for a molecule to move among the other closely packed molecules (and, of course, the particle train mechanism of momentum and energy transport does not apply to mass transport). Therefore, molecular diffusivity for liquids is much smaller than for gases, generally by a factor 10^4 or larger. For example, the molecular diffusivity of CO_2 in water is $D = 1.6 \times 10^{-3}$ mm²/s; when the solute consists of large molecules, it can easily be $D \cong 10^{-4}$ mm²/s. Therefore, we may conclude that in most cases the Schmidt number in liquids is very large, that is $Sc \cong 10^3$–10^4.

Consider the following numerical values:

The Prandtl number is $Pr = 0.015$ for mercury, $Pr = 0.7$–0.8 for most gases, including air, $Pr = 7$ for liquid water, $Pr = 100$ for olive oil, $Pr = 1300$ for glycerine, $Pr = 1000$–$40{,}000$ for motor oils.

The Schmidt number is $Sc = 1.1$ for CO_2 in air, $Sc = 10^3$ for ethyl acetate, or methane, in water, $Sc = 1.6 \times 10^4$ for albumin in water.

Q. When should we use molar and when mass quantities?
A. For liquid mixtures it is preferable to use mass quantities, while for gaseous mixtures molar quantities are more convenient. Let us see why.

In liquid mixtures, we may assume with good approximation that the species composing the mixture have the same density, which is therefore equal to the density of the mixture, ρ. Therefore, ρ is constant, (that is, independent of composition), i.e.,

---

[3]They are absent in ideal gases. For real gases, they correspond to the thermodynamic higher-order virial coefficients.

$$\rho = \text{constant} \qquad (14.5.1)$$

Consequently, mass fractions coincide with volumetric fractions. In fact,

$$\varphi_A = \frac{\text{volume of } A}{\text{total volume}} = \frac{\text{volume of } A}{\text{mole of } A} \frac{\text{mole of } A}{\text{mass of } A} \frac{\text{mass of } A}{\text{total volume}} = \frac{\tilde{V}_A}{M_{WA}} \rho_A = \frac{\rho_A}{\rho} = \omega_A,$$

where we have considered that the density of the A-component is equal to the ratio between molar mass and molar volume of A, i.e., $\rho = M_{WA}/\tilde{V}_A$. Considering that $\rho_A = \rho \omega_A$ and that the mass flux is $\mathbf{j}_A = \rho[\omega_A \mathbf{v} - D\nabla\omega_A]$, we see that when ρ is constant it is convenient to use mass quantities. In particular, the governing mass transport Eq. (14.2.4) reduces to the following expression:

$$\frac{\partial \omega_A}{\partial t} + \nabla \cdot (\omega_A \mathbf{v} - D\nabla\omega_A) = r_A{}' = \rho^{-1} r_A. \qquad (14.5.2)$$

This equation can be further simplified considering that, since ρ is constant, the incompressibility condition $\nabla \cdot \mathbf{v} = 0$ can be applied [see Eq. (14.2.6)], obtaining,

$$\frac{\partial \omega_A}{\partial t} + \mathbf{v} \cdot \nabla\omega_A = D \cdot \nabla^2 \omega_A + r'_A. \qquad (14.5.3)$$

With gases, instead, we can assume with good approximation that they consists of ideal mixtures of ideal gases. Therefore, the molar volumes of all the components are equal to each other, and are also equal to the molar volume of the mixture, $\tilde{V}$, where:

$$c = \frac{1}{\tilde{V}} = \frac{p}{RT} = \text{constant}, \qquad (14.5.4)$$

where we have assumed that temperature and pressure changes are negligible. Now, this case is totally analogous to the previous one, showing that mole fractions coincide with volumetric fractions. In fact,

$$\varphi_A = \frac{\text{volume of } A}{\text{total volume}} = \frac{\text{volume of } A}{\text{mole of } A} \frac{\text{mole of } A}{\text{total volume}} = \tilde{V}_A c_A = \frac{c_A}{c} = x_A.$$

Considering that $c_A = c x_A$ and $\mathbf{J}_A = c[x_A \mathbf{v}^* - D\nabla x_A]$, we see that when c is constant it is convenient to use molar quantities. In particular, the governing molar transport Eq. (14.2.4) reduces to the following expression:

$$\frac{\partial x_A}{\partial t} + \nabla \cdot (x_A \mathbf{v}^* - D\nabla x_A) = R_A{}' = c^{-1} R_A. \qquad (14.5.5)$$

In general, when c is constant, applying Eq. (14.2.9) we see that the mole average velocity is solenoidal, with $\nabla \cdot \mathbf{v}^* = 0$, only when $R = 0$, that is when the total number of moles is conserved.

Q. Why is it convenient to use the mixture theory?
A. Because it is simpler. Let us see why.

The approach to mass transport described in this chapter is called mixture theory: the mass (and molar) flux of a component is written as the sum of two contributions: one is the convection due to the mean velocity $\mathbf{v}$ of the mixture, while the other expresses the relative mass flux [see Eq. (14.3.1)]. This latter is modeled as a diffusive terms, that is, according to the Fick constitutive relation (14.3.5), it is the product between the diffusion coefficient, D, and the concentration gradients. Consequently, the evolution of a binary fluid mixture can be described by Eq. (14.5.3) of mass conservation,[4] in terms of the mole fraction ω_A of one of its components. This equation contains the mean velocity, $\mathbf{v}$, which in turn must satisfy the Navier-Stokes equation for a mixture with constant density and a composition-dependent viscosity.

As an alternative, in the so-called many fluid approach, the mass flux of any A-component is written in its "exact" form as $\mathbf{j}_A = \rho_A \mathbf{v}_A$, so that we obtain an Eq. (14.2.4) of mass conservation for each component A in terms of the its mass concentration ρ_A and its velocity $\mathbf{v}_A$. The latter, in turn, must be determined by solving a Navier-Stokes equation (one for each component). This approach, presents two flaws: the first is that for an N component system, we have to solve N Navier-Stokes equations, while in the mixture theory we only have one velocity, namely the mass (or molar) average velocity, and therefore we need to solve only one Navier-Stokes equation. The second flaw of the many fluid approach is that in the Navier-Stokes equation for the A component velocity we must include a body force $\mathbf{F}_A$ accounting for all the interactions (attractive or repulsive) that the other species exert on the A-type molecules. Clearly, not only these forces are mostly unknown, but they are also impossible to measure experimentally. Obviously, since the body forces $\mathbf{F}_A$ are internal forces, their sum is zero, so that they do not appear in the "average" Navier-Stokes equation for $\mathbf{v}$. Naturally, the great simplification introduced by the mixture theory has a price: the model includes two new parameters: the diffusion coefficient D appearing in the equation of mass conservation and the composition-dependent effective viscosity μ of the mixture appearing in the average Navier-Stokes equation. Although D and μ can be evaluated analytically only in a few simple cases, in general they can be measured experimentally rather easily. For example, the effective viscosity can be measured using a simple viscometer (see Sects. 7.1 and 7.2), while the diffusion coefficient can be determined by following the procedure described in Problems 15.5 and 15.6. Thus, on

---

[4]In general, for N-component mixtures, we can write $(N - 1)$ independent equations of mass conservation in terms of $(N - 1)$ mass fractions (consider that the sum of all diffusive mass fluxes is zero, and that the sum of all mass fractions is equal to unity).

one hand we have reduces the number of governing equations, eliminating the need to account for the inter-species body forces, $\mathbf{F}_A$; on the other hand, we have introduces new parameters, which are nevertheless rather easy to measure. Clearly, we believe that the advantages in using the mixture theory instead of the many fluid approach greatly outnumber the disadvantages.

Chapter 15
Stationary Material Transport

Abstract In this chapter we consider simple problems of stationary diffusive mass transport. In Sects. 15.1 and 15.2 we show that, since in general mass diffusion is non-linear, it is intrinsically more complicated than momentum diffusion (i.e., viscous flow) and heat diffusion (i.e., conduction). Then, in Sects. 15.3 and 15.4, we show that in the dilute case (e.g., a dilute contaminant in a water stream) mass transport problems are linear and hence they are identical to the respective heat transfer problems, as both diffusive and convective fluxes are linear functions of the concentration difference.

15.1 Diffusion Through a Stagnant Film

Consider an open tube with air (or pure nitrogen) at the open end and naphthalene at the other end (see Fig. 15.1). Naphthalene is a solid with a significant vapor pressure, so that it slowly evaporates. Calculate the concentration profile and the flux of naphthalene, knowing that air is not soluble in solid naphthalene.

Denoting by z_1 the position of the free naphthalene surface and by z_2 that of the open end of the tube, we know that in the region $z_1 < z < z_2$ a stagnant film is present, composed of a mixture of naphthalene (species A) and air (species B). Gaseous naphthalene diffuses, due to the concentration difference Δc_A between the A-concentration at the free surface, $c_{A,1} = c_A(z_1)$ and that at the open end of the tube, $c_{A,2} = c_A(z_2)$. Here, assuming local equilibrium at the interface $z = z_1, c_{A,1}$ is the saturation concentration c_A^{sat}, i.e., the maximum reachable concentration of naphthalene in air, at the given temperature and pressure; in addition, at $z = z_2$ there is pure air, so that $c_{A,2} = 0$. Finally, we will assume that the tube walls are impermeable, so that the molar fluxes are all directed along the z-direction.

Let us assume that the gaseous mixture is ideal, so that, at constant temperature, T, and pressure, p, we have:

© Springer International Publishing Switzerland 2015
R. Mauri, *Transport Phenomena in Multiphase Flows*,
Fluid Mechanics and Its Applications 112, DOI 10.1007/978-3-319-15793-1_15

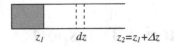

z_1 dz $z_2=z_1+\Delta z$

Fig. 15.1 Diffusion from an open tube

$$c = c_A + c_B = \frac{P}{RT} = \text{constant} \Rightarrow \frac{dc_A}{dz} = -\frac{dc_B}{dz}. \qquad (15.1.1)$$

We see that, since the gradient of c_A is non-zero, the same is true also for c_B, and therefore, in addition to the diffusion of A, there must be a diffusion of B as well, *despite B being stagnant*.

Let us consider a mass balance in the control volume $S dz$ (S is the area of the tube cross section) indicated in Fig. 15.1:

$$J_A(z)S = J_A(z+dz)S \Rightarrow J_A = \text{constant} = K_1, \qquad (15.1.2)$$

where K_1 is a constant to be determined. In the same way we find: $J_B = \text{constant}$. As air is insoluble in naphthalene, $J_B(z_1) = 0$, i.e., the flux of air at the interface must be zero. Consequently, since J_B is uniform, it must be zero everywhere, i.e.,

$$J_B = 0, \qquad (15.1.3)$$

and therefore, from $J_B = c_B v_B$, we obtain: $v_B = 0$, confirming that B is stagnant. So, we see that, since B is stagnant while A diffuses with a mean velocity $v_A = N_A/c_A$, applying Eq. (14.1.12) we find that the molar average velocity v^* (that determines the convective part of the mass transport) is a non-zero constant, i.e.,

$$v^* = \frac{1}{c}(J_A + J_B) = \frac{J_A}{c}. \qquad (15.1.4)$$

Therefore, applying the constitutive relation (14.3.7), we find the flux of A:

$$J_A = c_A v^* - cD\frac{dx_A}{dz} = x_A J_A - cD\frac{dx_A}{dz} \Rightarrow J_A = -\frac{cD}{1-x_A}\frac{dx_A}{dz} = K_1. \qquad (15.1.5)$$

The same result can be obtained applying Eq. (14.3.9). Now, integrating this expression between z_1 and z_2, and applying the following boundary conditions,

$$x_A(z_1) = x_{A1} = x_A^{sat}; \quad x_A(z_2) = x_{A2} = 0, \qquad (15.1.6)$$

we obtain:

$$K_1 = J_A = \frac{cD \ln[(1-x_{A2})/(1-x_{A1})]}{z_2 - z_1} = \frac{cD \ln (x_{B2}/x_{B1})}{z_2 - z_1} = \frac{cD(x_{A1} - x_{A2})}{(z_2 - z_1)(x_B)_{\ln}}. \qquad (15.1.7)$$

Here, $(x_B)_{\ln}$ is the logarithmic mean value of x_B,

$$(x_B)_{\ln} = \frac{x_{B2} - x_{B1}}{\ln(x_{B2}/x_{B1})}.$$ (15.1.8)

This result can also be expressed in terms of the mass transfer coefficient k_M in the the Newton law,

$$J_A = k_M(c_{A1} - c_{A2}),$$ (15.1.9)

with,

$$k_M = \frac{D}{\Delta z} \frac{1}{(x_B)_{\ln}},$$ (15.1.10)

where $\Delta z = z_2 - z_1$. Equation (15.1.10) manifests that k_M is a function of the composition, as one would expect since in general, as we saw in discussing Eqs. (14.3.7)–(14.3.12), mass transport is a non-linear process.

Another way to represent Eqs. (15.1.9)–(15.1.10) is to use the Sherwood number (i.e., the analog, in mass transport, of the Nusselt number in heat transfer),

$$Sh = \frac{k_M L}{D}.$$ (15.1.11)

Here, L is a characteristic linear dimension of the problem that, when it is possible, is chosen so that in the absence of convection Sh tends to unity. Therefore, the Sherwood number expresses the ratio between the molar, or mass, flux and the flux that we would obtain in the same conditions but in the absence of convection. In our case, defining $L = \Delta z$, we obtain:

$$Sh = \frac{1}{(x_B)_{\ln}}.$$ (15.1.12)

Finally, we can integrate Eq. (15.1.5) between z_1 and the generic distance z, obtaining:

$$\left(\frac{1 - x_A}{1 - x_{A1}}\right) = \left(\frac{1 - x_{A2}}{1 - x_{A1}}\right)^{\frac{z - z_1}{z_2 - z_1}}.$$ (15.1.13)

The concentration profile (15.1.13), represented in Fig. 15.2, is not linear, as one could predict. In fact, from Eq. (15.1.5) we see that, since J_A is constant and x_A decreases with z, then $|dx_A/dz|$ must increase.

In the limit case of a dilute mixture, with $x_A \ll x_B$ and $x_B \approx 1$, in Sect. 14.3.1 we have seen that, at leading order, the mean velocity coincides with the solvent velocity and therefore here $v^* = v_B = 0$, i.e., there is no convection. Thus, the

Fig. 15.2 Concentration
profiles

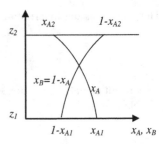

problem greatly simplifies, since the constitutive relation (14.3.7) contains only the
diffusive term,

$$J_A = -cD\frac{dx_A}{dz} = K_1 \quad \text{with } x_A(z_1) = x_{A1} \text{ and } x_A(z_2) = x_{A2}. \quad (15.1.14)$$

Therefore, dx_A/dz is constant and we see that in the dilute case we obtain the
following linear profile,

$$(x_A - x_{A1}) = \frac{z - z_1}{\Delta z}(x_{A2} - x_{A1}); \quad J_A = \frac{D}{\Delta z}(c_{A1} - c_{A2}); \quad k_M = \frac{D}{\Delta z}. \quad (15.1.15)$$

As expected, the coefficient of mass transport, k_M, is not a function of the concentration, as mass transport, in these conditions, is a linear process. In addition,
denoting $L = \Delta z$ in the definition (15.1.11) of the Sherwood number, we obtain
$Sh = 1$ (it shows that this is the most convenient choice for L). Naturally, these same
results can be obtained from Eqs. (15.1.13) and (15.1.7) in the limit $x_A \ll x_B$, with
$x_B \approx 1$.

Comparing Eq. (15.1.10) with (15.1.15), we see that the presence of convection
induces an increase in the mass flux that, by definition, equals the Sherwood
number,

$$\frac{J_A}{(J_A)_{diff}} = \frac{1}{(x_B)_{ln}} = Sh. \quad (15.1.16)$$

Again, in the dilute limit, when $x_B \approx 1$, we find: $Sh = 1$.

Note that at any position z there is a c_B gradient generating a diffusive flux of air
(i.e., the species B) from right to left, that exactly equilibrates the convective flux,
from left to right, so that the total flux is null,

$$J_B = c_B v^* - D\frac{dc_B}{dz} = 0. \quad (15.1.17)$$

Thus, convective and diffusive fluxes have the same sign in J_A, while they have
opposite signs in J_B.

Fig. 15.3 x_A profile when $x_{A1} \cong 1$

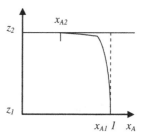

In the dilute case, the diffusive flux of B (which is equal and opposite to the flux of A) is of $O(cDx_A/L)$, and therefore it is of the same magnitude as the convective flux, as this latter is the product of c_B (with $c_B \cong c$) by v^*, which is of $O(Dx_A/L)$. Therefore, in the dilute case, the convective molar flux of B cannot be neglected, unlike the convective flux of A, $c_A v^*$, which is of $O(cDx_A^2/L)$, and therefore it is negligible compared to its $O(cDx_A/L)$ diffusive counterpart.

From Eq. (15.1.8) it appears that when $x_{A1} \cong 1$ then $(x_B)_{\text{ln}} \cong 0$ and so Eq. (15.1.7) gives $J_A \to \infty$. In the same way, when $x_{A2} \cong 1$ we find $J_A \to \infty$. Although neither of these boundary conditions is realistic, it is interesting to investigate the source of this divergence. From Fig. 15.3, representing the x_A profile when $x_{A1} \cong 1$, we see that the concentration profile remains flat, at $x_A \cong 1$, almost everywhere, with the exception of a thin layer in the vicinity of the upper boundary z_2, where it decreases abruptly to match the condition $x_A(z_2) = x_{A2}$. Now, from Eq. (15.1.5) we see that when either $x_A \cong 1$ (i.e., in the region not too close to the $z = z_2$ boundary) or when $dx_A/dz \to \infty$ (i.e., near the boundary $z = z_2$), $J_A \to \infty$, thus revealing the roots of the divergence. Actually, though, as J_A diverges logarithmically, even a tiny presence of B is enough to ensure that J_A remains finite. For example, keeping fixed $x_{A2} = 0.5$, when $x_{A1} = 0.8$ we obtain $J_A = 0.92 \, cD/\Delta z$, while when $x_{A1} = 0.99$, then $J_A = 3.9 \, cD/\Delta z$.

15.2 Diffusion with Heterogeneous Chemical Reaction

15.2.1 Plane Geometry

Consider the reaction of polymerization $nA \to A_n$ that takes place at the surface of a catalytic particle (see Fig. 15.4). Assume that both A and $B = A_n$ are gases and that the monomer, with molar fraction x_{A0}, flows past the surface forming a mass boundary layer of thickness δ. We want to determine the flux of A at the particle surface, assuming that the reaction speed is so high that A reacts (and disappears) as soon as it reaches the surface. In addition, δ is much smaller than the particle size, so that we may assume that the particle surface is plane, and located at $z = 0$.

As we know, convection within the boundary layer occurs predominantly in the longitudinal direction, so that mass transport in the z-direction, i.e., the molar flux

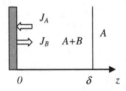

Fig. 15.4 Polymerization $nA \rightarrow A_n$ at the surface of a catalytic particle

J_A, is diffusive. Therefore, the problem is similar to that of Sect. 15.1. Here, too, from a simple mass balance we conclude that J_A = constant and J_B = constant. However, unlike the previous case, now J_B is not zero since it is not zero at the wall: in fact, whenever n moles of A reach the particle surface there is one mole of B that forms and goes back. Therefore,

$$J_B = -\frac{1}{n} J_A. \tag{15.2.1}$$

Consequently from Eq. (14.1.12) we obtain:

$$v^* = \frac{1}{c}(J_A + J_B) = \beta \frac{J_A}{c}, \quad \text{with } \beta = 1 - \frac{1}{n}. \tag{15.2.2}$$

Applying Eq. (14.3.7) we find the following molar flux of A,

$$J_A = c_A v^* - cD\frac{dx_A}{dz} = \beta x_A J_A - cD\frac{dx_A}{dz} \Rightarrow J_A = -\frac{cD}{1 - \beta x_A}\frac{dx_A}{dz} = K_1. \tag{15.2.3}$$

Now, integrating Eq. (15.2.3) between $z = 0$ and $z = \delta$, with the following boundary conditions,

$$x_A(z = 0) = 0 \quad \text{and} \quad x_A(z = \delta) = x_{A0}, \tag{15.2.4}$$

we obtain:

$$J_A = \frac{cD\ln(1 - \beta x_{A0})}{\beta\delta}. \tag{15.2.5}$$

Note that when $\delta \rightarrow \infty$ the molar flux tends to zero.
Defining the coefficient of mass transfer, k_M, as $J_A = -k_M c_{A0}$, we find:

$$Sh = \frac{k_M \delta}{D} = -\frac{\ln(1 - \beta x_{A0})}{\beta x_{A0}}, \tag{15.2.6}$$

where Sh is the Sherwood number, defined in Eq. (15.1.11) with $L = \delta$. Finally, integrating Eq. (15.2.3) between $z = 0$ and the generic distance z, we obtain:

$$(1 - \beta x_A) = (1 - \beta x_{A0})^{z/\delta}. \tag{15.2.7}$$

Note that the molar flux is negative, as it goes from right to left of Fig. 15.4. In addition, when $n \to \infty$, so that $\beta = 1$, this problem reduces to the one of Sect. 15.1.

As we have seen previously, mass transport is not a linear process, and therefore the mass transfer coefficient (and consequently the Sherwood number as well) is a function of the composition, unless the molar fraction of A is small, i.e., when $x_{A0} = \varepsilon \ll 1$. In fact, in the dilute limit, considering,

$$\log(1 - \varepsilon) = \varepsilon + O(\varepsilon^2) \quad \text{and} \quad (1 - \varepsilon)^\alpha = 1 - \alpha\varepsilon + O(\varepsilon^2),$$

we find, at leading order, a purely diffusive transport, with a linear concentration profile and a constant coefficient of mass transport:

$$x_A = x_{A0}(z/\delta); \quad J_A = -cDx_{A0}/\delta; \quad k_M = D/\delta \Rightarrow Sh = 1. \tag{15.2.8}$$

Countercurrent equimolar fluxes
When $n = 1$, so that $\beta = 0$, the flux of A and B are equimolar and countercurrent, that is they are opposed to one another. That means that for each mole of A reaching the particle surface there is one mole of B leaving it. In this case, the average molar velocity is null, with $v^* = 0$, i.e. there is no convection, and therefore the molar fluxes of A and B are only diffusive. Therefore, we find the same result (15.2.8) as in the dilute case[1] and we see that, as expected, the molar flux increment due to convection is the Sherwood number which, in general (i.e., in the non-dilute case) is a function of the composition,

$$\frac{J_A}{(J_A)_{diff}} = -\frac{\ln(1 - \beta x_{A0})}{\beta x_{A0}} = Sh. \tag{15.2.9}$$

15.2.2 Spherical Geometry

When the catalytic particle is a sphere of radius R (see Fig. 15.5), imposing that at steady state the total flux crossing any spherical crown of radius $r > R$ is constant we obtain:

---

[1]Note however that while in the countercurrent equimolar case this result is exact, in the dilute case it is only approximate, neglecting $O(\varepsilon^2)$-terms.

Fig. 15.5 Catalytic
spherical particle of radius R

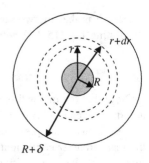

$$J_A\left(4\pi r^2\right) = \text{constant} \Rightarrow J_A = \frac{K_1}{r^2}. \qquad (15.2.10)$$

Here, K_1 is a constant to be determined, and we have assumed that the problem has a spherical symmetry and therefore depends only of the radial coordinate r. Now, applying Eq. (15.2.3) we find the following equation,

$$J_A = -\frac{cD}{1 - \beta x_A}\frac{dx_A}{dr} = \frac{K_1}{r^2}, \qquad (15.2.11)$$

to be integrated between R and $R + \delta$ with boundary conditions:

$$x_A(r = R) = 0 \quad \text{and} \quad x_A(r = R + \delta) = x_{A0}. \qquad (15.2.12)$$

So, we find:

$$K_1 = \frac{cD\ln(1 - \beta x_{A0})}{\beta\delta}R(R + \delta). \qquad (15.2.13)$$

Therefore, the molar flux of A at the particle surface $r = R$ is:

$$J_A(R) = \frac{cD\ln(1 - \beta x_{A0})}{\beta\delta}\left(1 + \frac{\delta}{R}\right). \qquad (15.2.14)$$

Note that, unlike the plane case, when $\delta \to \infty$ the molar flux does not tend to zero. Now, defining the coefficient of mass transport, $k_M = -J_A/(cx_{A0})$, we obtain the Sherwood number:

$$Sh = \frac{k_M}{D} = \frac{\ln(1 - \beta x_{A0})}{\beta x_{A0}}, \quad \text{where } L = \frac{R\delta}{R + \delta}. \qquad (15.2.15)$$

Clearly, the characteristic length L is defined so that, when $\beta = 0$, that is when the process is purely diffusive, the Sherwood number tends to unity. Therefore, we see that in spherical geometry the increment of the molar flux due to convection, that is the Sherwood number, has the same expression as in the plane case.

Countercurrent equimolar flux and dilute limit

In this case, $\beta = 0$ and therefore $v^* = 0$. Then, the constitutive relation for the molar flux becomes:

$$J_A = -cD\frac{dx_A}{dr} = \frac{K_1}{r^2}, \quad \text{obtaining:} \quad x_A(r) = A_1 + \frac{A_2}{r}, \tag{15.2.16}$$

where the constant A_1 and A_2 can be easily determined imposing the boundary conditions (15.2.12), obtaining:

$$x_A(r) = x_{A0}\frac{R+\delta}{\delta}\left(1 - \frac{R}{r}\right), \tag{15.2.17}$$

and so:

$$J_A(R) = \frac{cDx_{Ao}}{\delta}\left(1 + \frac{\delta}{R}\right) = \frac{cDx_{A0}}{L}, \quad \text{where } L = \frac{R\delta}{R+\delta} = \frac{R_i}{R_o}(R_o - R_i), \tag{15.2.18}$$

where $R_i = R$ and $R_o = R + \delta$. This result coincides with the analogous Eq. (9.2.13) for the heat flux case, when $B_i \to \infty$. Therefore we confirm that, defining L as in Eq. (15.2.15), the Sherwood number for a purely diffusive mass transport is $Sh = 1$.

Note that, when, $\delta \ll R$, Eqs. (15.2.17)–(15.2.18) reduce to the plane geometry case, while when $\delta \ll R$, considering that $L = R$, they become,

$$x_A(r) = x_{A0}\left(1 - \frac{R}{r}\right); \quad J_A(R) = -\frac{cDx_{A0}}{R}; \quad Sh = \frac{k_M R}{D} = 1. \tag{15.2.19}$$

Therefore we see that, having defined Sh in terms of the sphere radius R, in the absence of convection we find $Sh = 1$.[2]

15.3 Diffusion with Homogeneous, First-Order Chemical Reaction

Consider a spherical particle composed of a dilute binary mixture of A in B, with $x_A \ll 1$, where A transforms into B with an irreversible first-order reaction $A \to B$. The number of moles of A produced per unit time and per unit volume is $R_A = -kc_A$, where c_A is the concentration of A, while k denotes the reaction rate, with k^{-1}

---

[2]In most textbooks, Sh is defined in terms of the sphere diameter, so that, in the absence of convection, we have: $Sh = 2$.

Fig. 15.6 Simple material
balance on a spherical shell

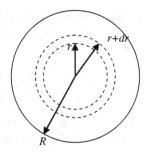

indicating a characteristic reaction time. We want to determine the flux of A entering
the particle, when at $r = R$ the concentration of A is kept constant and equal to c_{A0}.

From a simple material balance in the volume element represented in Fig. 15.6
we obtain

$$J_A(r)\left(4\pi r^2\right) - J_A(r + dr)\left(4\pi(r + dr)^2\right) - kc_A\left(4\pi r^2 dr\right) = 0,$$

i.e., dividing by dr,

$$\frac{1}{r^2}\frac{d}{dr}\left(J_A r^2\right) + kc_A = 0. \tag{15.3.1}$$

This equation must be solved with boundary condition, $c_A(R) = c_{A0}$. Since A is
dilute, $c_B \cong c$ and $v^* \cong 0$, so that the material flux is diffusive, with $J_A = -Ddc_A/dr$. Therefore, defining the dimensionless variables,

$$\tilde{c} = c/c_{A0}; \quad \xi = r/R; \quad Da = kR^2/D, \tag{15.3.2}$$

we obtain:

$$\frac{1}{\xi^2}\frac{d}{d\xi}\left(\xi^2\frac{d\tilde{c}}{d\xi}\right) - Da\tilde{c} = 0, \quad \text{con } \tilde{c}(1) = 1. \tag{15.3.3}$$

Here, Da is the Damköhler number that we have already encountered in Sect. 10.2,
representing the ratio between the velocity of the chemical reaction and that of
diffusion, that is the ratio between the diffusion time, R^2/D, and the reaction time, $1/k$.[3]
As for the molar flux,

---

[3]In catalysis, in place of the Damköhler number, it is often used the Thiele modulus, defined as
$Th = \sqrt{Da} = R\sqrt{k/D}$, where R is a characteristic length, that can be chosen as the pore size of
the catalyst or the size of the catalyst itself, when there are no pores. In this text, however, for
sake of simplicity we have preferred to denote the ratio between diffusion and reaction times as
the Damköhler number in all cases.

$$J_A(R) = -D\left(\frac{dc_A}{dr}\right)_{r=R},$$
(15.3.4)

it can also be expressed in non-dimensional form, considering that, in the absence of chemical reaction, i.e., in the purely diffusive case, the material flux is $J_{A,d}(R) \approx Dc_{A0}/R$, so that we may define the following Sherwood number,

$$Sh = \left|\frac{J_A(R)}{Dc_{A0}/R}\right| = \left|\frac{d\tilde{c}}{d\xi}\right|_{\xi=1}.$$
(15.3.5)

This problem can be solved defining an auxiliary dependent variable, $y = \xi\tilde{c}$, so that Eq. (15.3.3) and the Sherwood number become:

$$\frac{d^2y}{d\xi^2} - Day = 0, \quad \text{with } y(1) = 1 \text{ and } y(0) = 0,$$
(15.3.6)

and,

$$Sh = \left|\frac{dy}{d\xi} - y\right|_{\xi=1} = \left|\frac{dy}{d\xi}\right|_{\xi=1} - 1.$$
(15.3.7)

The second boundary condition in Eq. (15.3.6) derives from the fact that at $r = 0$, since $\tilde{c} \neq 0$, it must be: $y = 0$. Equation (15.3.6) coincides with Eq. (10.2.1), replacing $\Theta(\xi)$ with $y(1 - \xi)$ [note that the boundary conditions in the two problems are "inverted"], yielding the solution,

$$y(\xi) = \frac{\sinh\left(\sqrt{Da}\xi\right)}{\sinh\left(\sqrt{Da}\right)}, \quad \text{that is,} \quad \tilde{c}(\xi) = \frac{1}{\xi}\frac{\sinh\left(\sqrt{Da}\xi\right)}{\sinh\left(\sqrt{Da}\right)},$$
(15.3.8)

thus:

$$c_A(r) = c_{A0}\frac{R}{r}\frac{\sinh\left(\sqrt{Da}r/R\right)}{\sinh\left(\sqrt{Da}\right)}.$$
(15.3.9)

This solution is represented in Fig. 15.7, determining the Sherwood number,

$$Sh = \left|\sqrt{Da}Coth\left(\sqrt{Da}\right) - 1\right|.$$
(15.3.10)

As in the analogous heat transfer case examined in Sect. 10.2, at large Damköhler numbers, i.e., when $Da \gg 1$ a thin boundary layer forms near the outer surface and we find $Sh = \sqrt{Da}$. On the other hand, when $Da \ll 1$, the Sherwood number tends to zero, because, in the absence of chemical reaction, the concentration inside the

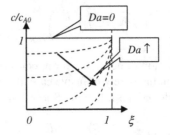

Fig. 15.7 Concentration profile as a function of *Da*

sphere is uniform, with $c_A = c_{A0}$, and the flux is null. In particular, expanding Eq. (15.3.10) in a power series of *Da*, we find, at leading order: $Sh = \frac{1}{3}Da$. Note that here the usual interpretation of the Sherwood number as the ratio between the material flux and its purely diffusive counterpart cannot be applied, as the latter quantity is zero.

In the following, the two limit cases will be examined separately, using the perturbation methods explained in Sect. 10.2.

15.3.1 Asymptotic Expansion for Small Da

Since the problem (15.3.6) is well-posed[4] even when $Da = 0$, we know that its solution can be expressed as a regular perturbation expansion in terms of, $\varepsilon = Da \ll 1$,

$$y(\xi) = y(\xi) + \varepsilon y_1(\xi) + \varepsilon^2 y_2(\xi) + \cdots \quad 0 \le \xi \le 1, \tag{15.3.11}$$

where we assume that $|y(\xi)| = O(1)$, i.e., each term in the expansion is smaller than the one preceding it. Now, substituting this expansion in Eq. (15.3.6), and imposing that terms of the same magnitude must balance each other, we find at the leading $O(1)$-order,

$$y_0''(\xi) = 0; \quad \text{with } y_0(0) = 0; \quad y_0(1) = 1$$

Then,

$$y_0(\xi) = \xi,$$

---

[4]That is, it consists of a second order differential equation with two boundary conditions.

and so,

$$Sh_0 = \frac{dy_0}{d\xi}(1) - 1 = 0.$$

This corresponds to the purely diffusive solution, $\tilde{c}(\xi) = 1$, that we would obtain by imposing $Da = 0$.

Now, collecting all at $O(\varepsilon)$-terms, we find:

$$y_1''(\xi) = y_0(\xi) = \xi; \quad \text{with } y_1(0) = y_1(1) = 0.$$

Integrating, we obtain:

$$y_1(\xi) = -\frac{1}{6}(\xi - \xi^3), \quad \text{and so: } Sh_1 = \frac{dy_1}{d\xi}(1) - 1 = \frac{1}{3}.$$

Finally, we can summarize our results as follows:

$$Sh = Sh_0 + Da\, Sh_1 + O(Da^2) = \frac{1}{3}Da + O(Da^2), \tag{15.3.12}$$

which coincides with the exact expression (15.3.10) when $Da \ll 1$. Naturally, we can continue the calculations, determining as many terms of the series solution (15.3.11) as we might wish.

15.3.2 Asymptotic Expansion for Large Da

In this case, as we approach the surface of the spherical particle, at $\xi = 1$, the normalized concentration of A increases abruptly from 0 to 1: this is a clear indication of the existence of a boundary layer. Since the thickness of the boundary layer is small, all curvature effects can be neglected and therefore the problem is identical to the plane case examined in Sect. 10.2.2, thus obtaining identical results. Let us see the details of this solution.

Denoting $\varepsilon = Da^{-1} \ll 1$, Eq. (15.3.6) becomes,

$$\varepsilon \frac{d^2 y}{d\xi^2} - y = 0; \quad y(0) = 0; \quad y(1) = 1. \tag{15.3.13}$$

When $\varepsilon = 0$, this problem reduces to a zeroth-order equation, $y = 0$, and therefore it becomes impossible to satisfy any of the two boundary conditions. Here, actually, the first boundary condition is satisfied, by chance, by the solution $y = 0$; however, the second boundary condition, at $\xi = 1$, is not satisfied, and this circumstance determines the formation of a boundary layer near $\xi = 1$. In fact, the source term in

Eq. (15.3.13) (i.e. the second term on the LHS) is generally much larger than the diffusive term (i.e. the first term on the LHS). Accordingly, the solution $y = 0$ is valid almost everywhere (see Fig. 15.7), with the exception of a thin boundary layer near the wall, at $\xi = 1$, where the concentration distribution jumps to satisfy the boundary condition $y(1) = 1$. Imposing that, within this region, the composition gradient is so large that the diffusive term in Eq. (15.3.13) balances the source term, we find the same boundary layer thickness, δ, as in Eq. (10.2.20), i.e.,

$$\delta = R/\sqrt{Da}. \tag{15.3.14}$$

Therefore, defining a stretched coordinate as in Sect. 10.2.2, $\zeta = (1 - \zeta)\varepsilon^{-1/2}$, we obtain[5]:

$$\frac{d^2y}{d\zeta^2} - y = 0; \quad y(0) = 1; \quad y\left(\frac{1}{\varepsilon^{1/2}} \to \infty\right) = 0.$$

with the solution,

$$y(\zeta) = e^{-\zeta} \Rightarrow y(\xi) = \exp\left(-(1 - \xi)/\sqrt{\varepsilon}\right) = \exp\left(-\sqrt{Da}(1 - \xi)\right),$$

and,

$$Sh = \sqrt{Da}. \tag{15.3.15}$$

Therefore, we find the same solution $Sh = \sqrt{Da}$ that can be derived from the exact result (15.3.10) and, as expected, it coincides with the Eq. (10.2.23) for the plane geometry.

15.4 Diffusion with Homogeneous, Second-Order Chemical Reaction

Consider a stagnant film, $0 < z < L$, composed of a ternary mixture of three species, A, B and C, with $x_A \ll 1$ and $x_B \ll 1$. At $z = 0$ the concentration of A is kept fixed, with $c_A(0) = c_{A0}$, while at the other interface, $z = L$, the concentration of B is kept fixed, with $c_B(L) = c_{AL}$. Besides diffusing in a mixture predominantly composed of C, A and B react with each other, with an irreversible reaction $A + B \to C$, so that $R_A = R_B = -kc_Ac_B$ is the number of moles of A and of B produced per unit time and per unit volume, with k denoting the reaction rate.

---

[5]The region that we want to "enlarge" lays near $\xi = 1$, and therefore the stretched coordinate is of the type $\zeta = (1 - \zeta)/\varepsilon^\alpha$. Here, α can be determined by imposing that in this region the two transport mechanisms (i.e., diffusion and generation) balance each other, obtaining: $\alpha = 1/2$.

Assuming that the mass transport is stationary and uni-directional along the z direction, from a materialbalance in a volume element Sdz, where S is the cross sectional area of the film, we obtain:

$$J_A(z)S - J_A(z + dz)S - kc_Ac_B(Sdz) = 0. \tag{15.4.1}$$

Considering that the material flux of A (and B) is only diffusive, since in the dilute limit (see Sect. 14.3) $v^* = v_c = 0$, the constitutive equation is: $J_A = -D_{AC}dc_A/dz$, where D_{AC} is the diffusion coefficient of A into C.[6] Now, dividing by dz each term of Eq. (15.4.1), we obtain:

$$D_{AC}\frac{d^2c_A}{dz^2} - kc_Ac_B = 0, \tag{15.4.2}$$

and a similar equation holds for B,

$$D_{BC}\frac{d^2c_B}{dz^2} - kc_Ac_B = 0. \tag{15.4.3}$$

These equations must be solved with boundary conditions,

$$c_A(0) = c_{A0}; \quad \frac{dc_B}{dz}(0) = 0; \quad c_B(L) = c_{BL}; \quad \frac{dc_A}{dz}(L) = 0, \tag{15.4.4}$$

where we have assumed that the fluxes of B at $z = 0$ and that of A at $z = L$ are null. The general solution of these two coupled differential equations cannot be found analytically and must be determined numerically. However, a much simpler solution can be determined when the reaction rate is much larger than the diffusion rate, that is for large Damköhler numbers, $Da = kc_{A0}L^2/D \gg 1$, where D is a typical value of the diffusion coefficient. In this case, as soon as a molecule A and a molecule B "meet", they both disappear, creating a molecule C. Therefore, species A and B cannot exist together, and consequently there must be a region, $0 < z < L_R$, where only the A-species is present, with $c_A \neq 0$ and $c_B = 0$, while in the region $L_c < z < L$ we have $c_B \neq 0$ and $c_A = 0$. The surface $z = L_R$ is called *reaction front*, where the chemical reaction takes place. In fact, in the region $0 < z < L_R$, we have $c_B = 0$, and therefore $R_A = 0$, and analogously for the region $L_c < z < L$. At the reaction front, $z = L_R$, both concentrations must be null, i.e., $c_A = c_B = 0$, otherwise the concentration profile would be discontinuous. Summarizing, Eqs. (15.4.3) and (15.4.4) become:

---

[6]Since each A molecule interacts predominantly with C molecules, diffusion of A into B can be neglected.

$$\frac{d^2 c_A}{dz^2} = 0; \quad c_A(0) = c_{A0}; \quad c_A(L_R) = 0, \tag{15.4.5}$$

yielding a linear profile,

$$c_A(z) = c_{A0}\left(1 - \frac{z}{L_R}\right). \tag{15.4.6}$$

In the same way we obtain:

$$c_B(z) = \frac{c_{BL}L_R}{L - L_R}\left(\frac{z}{L_R} - 1\right). \tag{15.4.7}$$

Now we have to determine the unknown position, L_R, of the reaction front. To do that, we need another condition, based on physical considerations. This is obtained by considering that for each mole of A reaching the reaction front from the left, there is one mole of B arriving from the right. That means that the fluxes of A and B are opposite to each other:

$$J_A(L_R) = -J_B(L_R). \tag{15.4.8}$$

Therefore,

$$L_R = \frac{L}{1 + K} \quad \text{where} \quad K = \frac{c_{BL}D_B}{c_{A0}D_A}, \tag{15.4.9}$$

and so we obtain the following molar fluxes:

$$J_A = -J_B\frac{c_{A0}D_A}{L}(1 + K). \tag{15.4.10}$$

It should be noted that, when the Damköhler number is very large, Eqs. (15.4.6) and (15.4.7) are obtained, irrespectively of the type of chemical reaction taking place at the reaction front. This information enters the solution of the problem through Eq. (15.4.8): for example, if the reaction were the following: $A + nB \rightarrow C$, then we would find:

$$J_A(L_R) = -nJ_B(L_R) \Rightarrow L_R = \frac{L}{1 + \frac{1}{n}K}. \tag{15.4.11}$$

Obviously, in reality the reaction front consists of a thin region, of thickness $\delta = L Da^{-1/2}$, in analogy with Eq. (15.3.14). Therefore, as seen in Sect. 15.3.2, the solution (15.4.10) constitutes the leading-order solution of the problem, where $O(Da^{-1})$-terms are neglected.

In all the examples of this Chapter we have seen that the material flux increases when a chemical reaction takes place in the system. This is important in designing gas absorbers, where the gas flux absorbed by a liquid film has to be maximized.

15.5 Problems

15.1 Calculate the mass and molar average velocities of the system naphthalene-nitrogen described in Sect. 15.1 (replaced air with nitrogen, for convenience), with a diffusion length $\Delta z = 10$ cm. Data : $p = 1$ atm, $T = 25\,°C$, $M_{WA} = 128$, $M_{WB} = 28$, $p_{vap} = 9.6\,Pa$, $D = 8 \times 10^{-2}\,cm^2/s$.

15.2 A little spherical water drop of radius R is sprayed in an atmosphere of dry air, where it evaporates. Supposing that around the drop there is a boundary layer of stagnant air of thickness δ, calculate the flux of the evaporating water. What happens in the dilute case, and when $\delta \ll R$ and $\delta \to \infty$?

15.3 Repeat the problem of Sect. 15.2 assuming cylindrical geometry.

15.4 Consider a cylindrical catalyst particle of radius R and length $L \gg R$. A dilute species A with molar fraction $x_A \ll 1$. diffuses within the particle with an effective diffusivity D and undergoes a homogeneous first-order reaction $A \to B$, with reaction constant k. Without solving the general case, determine:

  (a) The flux of A entering the particle assuming weak reaction rate and using a regular perturbation expansion.
  (b) The flux of A entering the particle assuming strong reaction rate and using the concept of boundary layer.

15.5 Consider the following experiment, intended to measure the diffusivity D of A into B, where A and B are gases that can be treated as ideal. In a tube of length L, the extremity $z = 0$ is kept in contact with a large reservoir containing pure A, while at the other tube end, at $z = L$, there is a catalyst, where the irreversible reaction $A \to B$ takes place. Determine the dependence of D from the flux J_A (that we can measure), pressure P and temperature T.

15.6 Repeat the previous problem assuming that on the catalyst the following equilibrium reaction takes place: $A \leftrightarrow 2B$, with known equilibrium constant $K = c_B^2/c_A$ (see Fig. 15.8).

15.7 The dominating life form on the planet γ-Valentinianus is the basilides, a telepathic worm of length *100 re* [rei *(re)* is the unit length on γ-Valentinianus]. The energy necessary to keep the basilides alive is produced through a chemical reaction of ammonia on the dorn, a sort of vertebral column of radius $R_i = 0.5\ re$ that runs along the basilides axis. Apart from the dorn, the rest of the basilides body is uniform. Ammonia is present within the atmosphere of γ-Valentinianus with a concentration $c_{A,atm} = 0.4\ mol/re^3$ and has a molar solubility $\alpha = 0.1$ with the basilides flesh. Knowing that the ammonia diffusivity in the basilides body is $D = 0.3re^2/p$ [panta *(p)* is the unit time on

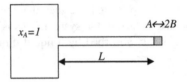

Fig. 15.8 Reaction-driven diffusion

γ-Valentinianus] and that the basilides needs a minimum amount $S = 1.35 \cdot 10^{-4} mol/p^{-1} re^{-3}$ of ammonia per unit time and per unit volume (larger worms consume more ammonia), calculate the maximum radius R_{max} that a basilides can reach on γ-Valentinianus.

15.8   Consider a container composed of two cubes of size L, separated by a partition. Initially, on the left side we have a dilute binary A-C mixture, with $x_A \ll 1$, while on the right side we have a dilute binary B-C mixture, with $x_B \ll 1$ When, at time $t = 0$, the partition is removed, A and B diffuse within C with the same diffusivity D and, in addition, react with each other, with an irreversible reaction $A + B \rightarrow C$, with reaction speed k. Show qualitatively the composition profiles, determining the characteristic times necessary to reach equilibrium as functions of the Damköhler number.

Chapter 16
Non-stationary Material Transport

Abstract In Sects. 16.1, 16.2 and 16.3 of this chapter we will solve three problems where we can apply the Quasi-Steady State (QSS) approximation, that we have already encountered in Sects. 5.5 and 9.4. This hypothesis is valid when the characteristic time needed to change the boundary conditions is much longer than the time needed to reach steady state. Then, in the last Sect. 16.4, we will see an examples where this assumption is not applicable, showing how much more difficult these problems become.

16.1 Transport Across a Membrane

Consider the two reservoirs (indicated with subscripts 1 and 2) of Fig. 16.1, having equal volume, V, and size, L, separated by a porous membrane of section $S = V/L$, porosity ε, and thickness δ. The two reservoirs contain a well-mixed A–B mixture; initially, the molar fractions of A in the two reservoirs are x_{A10} and $x_{A20} < x_{A10}$. Then, as time progresses, A will diffuse across the membrane from left to right, while B will diffuse in the opposite direction. We want to determine how the difference between the molar fractions of A in the two reservoirs, $x_{A1}-x_{A2}$, changes with time, so that the diffusion coefficient D of A into B within the membrane can be determine.

This problem is almost identical to the one that was solved in Sect. 9.4.

First of all, we observe that the total moles of A must be conserved. Therefore, as for long times the system must be at chemical equilibrium, with uniform concentration x_{Af}, assuming that the total concentration, c, is constant we obtain:

$$x_{A1} + x_{A2} = x_{A1,0} + x_{A2,0} = 2x_{Af} \Rightarrow x_{Af} = (x_{A1,0} + x_{A2,0})/2 \qquad (16.1.1)$$

where $x_{A1,0}$ and $x_{A2,0}$ are the initial compositions in the two reservoirs and we have neglected the number of moles within the membrane. Assuming that $x_A = O(1)$, this

© Springer International Publishing Switzerland 2015

R. Mauri, *Transport Phenomena in Multiphase Flows,*
Fluid Mechanics and Its Applications 112, DOI 10.1007/978-3-319-15793-1_16

Fig. 16.1 Material transport
across a membrane

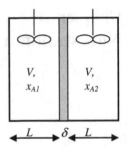

latter assumption means that the wetted volume V_w occupied by the fluid mixture
within the membrane is much smaller than the volume V of the reservoir. Therefore,
since the ratio between V_w and the membrane total volume, $S\delta$, is the porosity ε (see
Eq. 5.4.3), this assumption becomes:

$$\frac{\varepsilon\delta}{L} \ll 1. \tag{16.1.2}$$

Now, in order to find the transient solution of the problem, we must write a
separate molar balance of the A species for each reservoir separately, obtaining,

$$\frac{d}{dt}(cx_{A1}V) = -J_{A1}S\varepsilon, \tag{16.1.3a}$$

$$\frac{d}{dt}(cx_{A2}V) = -J_{A2}S\varepsilon, \tag{16.1.3b}$$

where J_{A1} and J_{A2} are the molar fluxes leaving the reservoirs 1 and 2, respectively.
Here, we have assumed that the porous membrane is isotropic, so that its porosity is
also equal to the ratio between the wetted area and the total cross sectional area (see
Eq. 4.4.6), and therefore the effective cross section area of the membrane is $S\varepsilon$.

Now, subjected to *a posteriori* verification, suppose that the hypothesis of Quasi
Steady State (QSS) can be applied. That means that the fluxes J_{A1} and J_{A2} correspond to the steady fluxes of A induced by a given, constant composition x_{A1} and
x_{A2} in the two reservoirs. In other words, if x_{A1} or x_{A2} change abruptly, the material
flux of A will adjust to the new condition, reaching a new steady state value within a
characteristic relaxation time τ_{ss}. Whenever the QSS approximation is valid, τ_{ss} is so
short, compared to the time τ characterizing the variation of x_A, that the relaxation
process can be considered to be instantaneous. Therefore, since at steady state there
cannot be any accumulation of A within the membrane, the flux of A leaving 1 is
equal to the flux of A entering 2, i.e.,

$$J_{A1} = -J_{A2}. \tag{16.1.4}$$

In addition, in the general constitutive relation (14.3.9),

$$J_A = x_A(J_A + J_B) - cD - \frac{dx_A}{dz},$$

the convective term is zero, because, as the total molar concentration $c = c_A + c_B$ is constant, for each mole of A moving from right to left there must be a mole of B moving in the opposite direction, so that the fluxes of A and B are equimolar and countercurrent,

$$J_A = -J_B. \tag{16.1.5}$$

Concluding, as the flux J_A (and J_B as well) is diffusive, in the QSS approximation the concentration profile is linear between x_{A1} and x_{A2}, that is:

$$J_A = \frac{cD}{\delta}(x_{A1} - x_{A2}). \tag{16.1.6}$$

Let us see when the QSS approximation can be applied. As we saw, the QSS hypothesis is applicable when the characteristic time, τ_{eq}, describing the temporal variation of x_{A1} and x_{A2} (in fact, x_{A1} and x_{A2} are the boundary values of the concentration profile x_A within the membrane) is much larger than the time τ_{ss} that takes the flux J_A to reach its stationary value (16.1.6). Thus, as x_{A1} and x_{A2} change, tending in a time τ_{eq}, to their common equilibrium value, $x_{Af} = x_{A1} = x_{A2}$, the flux J_A will adapt to the changing boundary conditions within a time, τ_{ss}, that is much smaller than τ_{eq}. From Eqs. (16.1.3a, b) we derive the following estimate,

$$\tau_{eq} \approx \frac{cV}{J_A S \varepsilon} \approx \frac{L\delta}{D\varepsilon}, \tag{16.1.7}$$

with $L = V/S$, where we have assumed that $x_A = O(1)$, and we have substituted Eq. (16.1.5) for the flux J_A. In addition, as in every diffusive process, the time necessary to reach steady state is $\tau_{ss} \approx \delta^2/D$. Therefore, the QSS hypothesis is valid when:

$$\tau_{ss} \ll \tau_{eq} \Rightarrow (\delta\varepsilon)/L \ll 1. \tag{16.1.8}$$

This condition is identical to Eq. (16.1.2), indicating that that QSS approximation means assuming that the wetted volume of the membrane is much smaller than the volumes of the reservoirs. It is satisfied when the membrane thickness, δ, is much smaller than the size of the reservoir, L, or/and when the membrane porosity, ε, is very small.

Now, let us find the time-dependent solution of the problem. Substituting Eq. (16.1.5) into (16.1.1) and (16.1.2) we obtain:

$$L\frac{dx_{A1}}{dt} = -\frac{\varepsilon D}{\delta}(x_{A1} - x_{A2}), \tag{16.1.9a}$$

$$L\frac{dx_{A2}}{dt} = \frac{\varepsilon D}{\delta}(x_{A1} - x_{A2}), \tag{16.1.9b}$$

to be solved with initial conditions $x_{A1}(0) = x_{A10}$ and $x_{A2}(0) = x_{A20}$. Summing these two equations we find again the condition (16.1.1). Instead, subtracting them from one another and integrating between $t = 0$ and t, we obtain:

$$\ln\frac{\Delta x_0}{\Delta x} = \frac{2D\varepsilon}{\delta L}t = 2\frac{t}{\tau_{eq}} \Leftrightarrow \Delta x = \Delta x_0 e^{-2t/\tau_{eq}}, \tag{16.1.10}$$

where $\Delta x = x_{A1} - x_{A2}$, while τ_{eq} is defined in (16.1.7). This solution confirms that τ_{eq} is indeed the characteristic time that the system needs to reach its equilibrium state, with uniform composition. Finally, substituting Eq. (16.1.10) into (16.1.9a) and integrating, we find:

$$x_{A1} = x_{A10} - \frac{\varepsilon D \tau_{eq} \Delta x_0}{2\delta L}\left(1 - e^{-2t/\tau_{eq}}\right). \tag{16.1.11}$$

Starting from Eq. (16.1.10), we see that the diffusion coefficient D can be determined by measuring the concentration at two times, $t = 0$ and t, i.e.,

$$D = \frac{\delta L}{2\varepsilon t}\ln\frac{\Delta x_0}{\Delta x}. \tag{16.1.12}$$

16.2 Evaporation of a Liquid from a Reservoir

This problem is the natural continuation of the steady diffusion across a stagnant film of Sect. 15.1, where we studied the sublimation of a solid species A, occupying the region $0 < z < z_1$ of a reservoir, with its vapor crossing a stagnant gas, B, laying in the region $z_1 < z < z_2$, with $\Delta z = z_2 - z_1$ (see Fig. 16.2). Obviously, the problem could also consist of the evaporation of a liquid from a container where it is initially confined. The molar fraction x_A of the vapor A at the interface $z = z_1$ is fixed at its saturation value, $x_{A1} = x_A^{sat}$, from the condition of local equilibrium, while at the upper boundary $z = z_2$ it is kept constant at x_{A2}. Unlike the case of Sect. 15.1, however, here the height z_1 is not constant, as A progressively sublimates (or evaporates), thus emptying the reservoir. We intend to determine the characteristic emptying time, assuming that initially $z_1 = z_{1,i}$.

Fig. 16.2 Evaporation of a
liquid from a reservoir

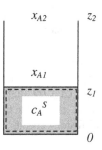

Consider the material balance within the volume indicated in Fig. 16.2,

$$\frac{d}{dt}\int_{V(t)} c_A dV = -\oint_{S(t)} J_{Ar} dS, \qquad (16.2.1)$$

where J_{Ar} is the flux of A exiting the solid volume V, relative to the velocity $w = dz_1/dt$ of the interface $z = z_1$. Therefore, since $J_A = c_A v_A$ is the flux of A referred to a fixed reference frame, we have:

$$J_{Ar} = C_A(v_A - w) = J_A - C_A \frac{dz_1}{dt}, \qquad (16.2.2)$$

so that:

$$C_A^S \frac{dV}{dt} = -J_{Ar}{}^S(z_1)S = -J_{Ar}{}^V(z_1)S, \qquad (16.2.3)$$

where S is the area of the interface. Here, $J_{Ar}^S(z_1)$ and $J_{Ar}^V(z_1)$ are, respectively, the flux of A on the solid side, $z = z_1^-$, and that on the vapor side, $z = z_1^+$, of the interface, and we have considered that, from the simple material balance illustrated in Fig. 16.3, by continuity the two fluxes are equal to each other. Therefore, considering that $V = Sz_1$, we finally obtain:

$$c_A^S \frac{dz_1}{dt} = -J_{A1} + c_{A1}\frac{dz_1}{dt}, \qquad (16.2.4)$$

Fig. 16.3 Continuity of the material flux at the interface

where $c_A{}^S$ is the concentration of A in the solid phase, while $c_{A1} = cx_{A1}$ and J_{A1} are the concentration and the flux of A in the vapor phase at the interface $z = z_1$, respectively.

Now, subjected to *a posteriori* verification, let us assume that the QSS approximation can be applied. That means that the characteristic relaxation time, $\tau_{ss} \approx (\Delta z)^2/D$, that is needed to reach steady state, is much shorter than the emptying time, τ, that is the time characterizing the interface motion. Consequently, the diffusion speed, $v_A \approx D/\Delta z$, is much larger than the speed of the interface, $w = dz_1/dt \approx \Delta z/\tau$. Therefore:

1. As $dz_1/dt \ll v_A$, from Eq. (16.2.2) we see that the relative flux can be approximated as the absolute flux, i.e. $J_{Ar} = J_A$, so that the last term in Eq. (16.2.4) can be neglected, showing that $c_{A1} \ll c_A{}^S$.
2. The flux J_{A1} can be expressed as the steady state expression (15.1.7), i.e.,

$$J_{A1} = \frac{cD(x_{A1} - x_{A2})}{(z_2 - z_1)(x_B)_{\ln}}. \tag{16.2.5}$$

Substituting Eq. (16.2.5) into (16.2.4) yields:

$$\frac{dz_1(t)}{dt} = -\frac{W}{z_2 - z_1(t)}, \quad \text{with} \quad W = \frac{cD(x_{A1} - x_{A2})}{c_A{}^S(x_B)_{\ln}}, \tag{16.2.6}$$

and integrating between $t = 0$ and t, with $z_1(t = 0) = z_{10}$, we obtain the final result:

$$(z_2 - z_1)^2 - (z_2 - z_{10})^2 = 2Wt. \tag{16.2.7}$$

In particular, we find the emptying time, τ, defined as $z_1(\tau) = 0$,

$$\tau = \frac{z_2{}^2 - (z_2 - z_{10})^2}{2W}. \tag{16.2.8}$$

Note that, as it always happens in diffusive processes, time is proportional to the square of a length.

Now, let us see when the QSS approximation can be applied:

$$\tau_{ss} \ll \tau \Rightarrow \frac{(\Delta z)^2}{D} \ll \frac{(\Delta z)^2}{W} \approx \frac{(\Delta z)^2 c_A^S}{Dc_{A1}} \Rightarrow c_{A1} \ll c_A^S, \tag{16.2.9}$$

where we have considered that $\tau \approx (\Delta z)^2/W$ and $W \approx c_{A1}/c_A^S$, with $(x_B)_{\ln} = O(1)$. We could obtain the same result by imposing that the diffusion speed, $v_A \approx D/\Delta z$, is much larger than the speed of the interface, $w = dz_1/dt \approx W/\Delta z$ (see Eq. 16.2.6).

The condition $c_{A1} \ll c_A^S$ is generally verified. For example, in the case of the evaporation of water, $c_A^S = 55$ mole/lt, while, assuming that the vapor behaves as an ideal gas, at ambient conditions $c_{A1} \cong 10^{-3}$ mole/lt, so that $c_{A1}/c_A^S \approx 10^{-4}$. Finally, note that in the dilute case, $x_A \ll 1$, we obtain the same result (16.2.7), with $x_B = 1$.

16.3 Slow Combustion of a Coal Particle

Consider a spherical particle of coal, having radius R and concentration c_C^S of carbon (species C), immersed in air, containing oxygen (species A) with molar fraction x_A^∞. Oxygen diffuses towards the particle surface, where it reacts with carbon to produce carbon dioxide (species B), with a first-order heterogeneous chemical reaction, with reaction rate k_M,[1]

$$C + O_2 \rightarrow CO_2 \quad J_A(r = R) = -k_M c_A(r = R). \tag{16.3.1}$$

Knowing that initially the particle radius is R_0, we want to calculate the time needed for the complete combustion.

A material balance of the moles of carbon within the particle volume yields:

$$\frac{d}{dt}\left(c_C^S \frac{4}{3}\pi R^3\right) = -J_{Br}\left(4\pi R^2\right) = J_{Ar}\left(4\pi R^2\right), \tag{16.3.2}$$

where J_{Br} and J_{Ar} are, respectively, the flux of carbon dioxide and of oxygen at the gas-solid interface, relative to the speed dR/dt of the interface,[2] and we have considered that for each mole of C consumed in the volume there is one mole of B leaving and one mole of A arriving at the interface.

Now we assume that the QSS hypothesis is satisfied. As we saw in the previous section, that amounts to assuming that the interface velocity, dR/dt, is negligible with respect to the diffusion speed, so that the relative material flux, J_{Ar}, is approximately equal to the absolute material flux, J_A, obtaining:

$$c_C^S \frac{dR}{dt} = J_A(r = R). \tag{16.3.3}$$

---

[1]More exactly, at the particle surface we have: $\mathbf{n} \cdot \mathbf{N}_A = -k_M c_A$, where $\mathbf{n}$ is the unit vector perpendicular to the surface and directed outward, therefore in a direction opposite to the oxygen flux.

[2]As in the problem of the previous Section, the flux of carbon on the inner side of the interface, at $r = R^-$, is equal to the one measured on the outer side, at $r = R^+$.

The material flux J_A satisfies the constitutive relation,

$$J_A = x_A(J_A + J_B + J_N) - cD\frac{dx_A}{dr} \tag{16.3.4}$$

where the subscript N indicates the nitrogen of the air which, being inert, is stagnant and therefore $J_N = 0$; in addition, as we saw, the flux of oxygen, J_A, and the flux of carbon dioxide, J_B, are countercurrent and equimolar, so that $J_A = -J_B$, and therefore the oxygen flux is purely diffusive,

$$J_A = -cD\frac{dx_A}{dr}. \tag{16.3.5}$$

Note that the diffusion coefficient D is an effective parameter, accounting for the fact that A diffuses in a multicomponent mixture. This approximation must be verified case by case, but in general it is always valid in the dilute case, for $x_A \ll 1$, when D indicates the diffusivity of oxygen in nitrogen.

Now, at steady state (congruent with the QSS approximation), imposing that $J_A r^2$ is constant, we obtain the usual spherically symmetric solution,

$$-\frac{dx_A}{dr}r^2 = B \Rightarrow x_A(r) = A + \frac{B}{r}, \tag{16.3.6}$$

where A and B are constant to be determined imposing that the following boundary conditions are satisfied,

$$x_A = x_A^\infty \quad \text{as } r \to \infty, \tag{16.3.7}$$

$$J_A = -cD\frac{dx_A}{dr} = -k_M c_A \quad \text{at } r = R. \tag{16.3.8}$$

Thus:

$$x_A(r) = x_A^\infty\left[1 - \left(\frac{k_M R^2}{D + k_M R}\right)\frac{1}{r}\right] \quad \text{and} \quad J_A(r = R) = -\frac{cDk_M x_A^\infty}{D + k_M R}. \tag{16.3.9}$$

Substituting this results into Eq. (16.3.3), we obtain:

$$\frac{dR}{dt} = \frac{J_A(r = R)}{c_C^S} = -\frac{Dk_M(c_A^\infty/c_C^S)}{D + k_M R}, \tag{16.3.10}$$

where $c_A^\infty = cx_A^\infty$. Finally, integrating with $R(t = 0) = R_0$ and $R(t = \tau) = 0$, we find:

$$\tau = \frac{R_0^2}{2D} \frac{c_C^S}{c_A^\infty} \left(1 + \frac{1}{Bi_M}\right) = \tau_{diff} + \tau_{react}, \quad \text{with } \tau_{diff} = \frac{c_C^S}{c_A^\infty} \frac{R_0^2}{2D} \quad \text{and } \tau_{react} = \frac{c_C^S}{c_A^\infty} \frac{R_0}{k_M},$$

$$(16.3.11)$$

where $Bi_M = k_M R_0/2D = \tau_{diff}/\tau_{react}$ is the material Biot number, expressing the ratio between the material flux at the wall, due to the heterogeneous chemical reaction, and the diffusive flux occurring at $r > R$. The material Biot number is analogous to its thermal counterpart, defined in Eq. (9.2.7). Note that the kinetics of the process is controlled by the slowest of the two processes, that is diffusion when $\tau_{diff} \gg \tau_{react}$, and reaction when $\tau_{diff} \gg \tau_{react}$.

The QSS hypothesis is verified assuming that $\tau \gg \tau_{ss} \approx R_0^2/D$, which means also assuming that the interface velocity, $dR/dt \approx R_0/\tau$, is much smaller than the diffusion speed, D/R_0, obtaining,

$$\left(1 + \frac{1}{Bi_{M0}}\right)\left(\frac{c_C^S}{c_A^\infty}\right) \gg 1. \qquad (16.3.12)$$

This condition is always satisfied, provided that $c_A^\infty \ll c_C^S$, which is normally true in normal conditions, as we saw in the previous section.

Now, it is instructive to consider the two limit cases, when $Bi_M \gg 1$ and $Bi_M \ll 1$.

(a) $Bi_M \gg 1$.

In this case, reaction is much faster than diffusion, meaning that the oxygen A-molecules react as soon as they reach the particle surface. Therefore, $k_M \to \infty$, and the boundary condition (16.3.8) reduces to $x_A (r = R) = 0$, so that the solution (16.3.9) becomes:

$$x_A(r) = x_A^\infty \left(1 - \frac{R}{r}\right); \quad J_A(r = R) = -\frac{Dc_A^\infty}{R}, \qquad (16.3.13)$$

and:

$$R_0^2 - R^2 = 2\frac{c_A^\infty}{c_C^S} Dt. \qquad (16.3.14)$$

Therefore, the characteristic time needed to consume a particle of initial radius R_0 is:

$$\tau = \tau_{diff} = \frac{R_0^2}{2D} \frac{c_C^S}{c_A^\infty}, \qquad (16.3.15)$$

and the QSS hypothesis is satisfied provided that $c_A^\infty \ll c_C^S$. As expected, diffusion here is the controlling process, since it is much slower than reaction, with $\tau_{diff} \gg t_{react}$.

(b) $Bi_M \ll 1.$

In this case, diffusion is much faster than reaction, meaning that the oxygen A-molecules diffuse very rapidly towards the particle surface, where they slowly react. Therefore, $k_M \rightarrow 0$, and the boundary condition (16.3.8) reduces to $J_A(r = R) = 0$, so that the solution (16.3.9) becomes:

$$x_A(r) = x_A^\infty. \tag{16.3.16}$$

Substituting this result into Eqs. (16.3.1) and (16.3.3), we obtain: $dR/dt = -k_M c_A^\infty / c_C^S$, and thus:

$$R_0 - R = \frac{c_A^\infty}{c_C^S} k_M t. \tag{16.3.17}$$

Therefore, the characteristic time needed to consume a particle of initial radius R_0 is:

$$\tau = \tau_{react} = \frac{R_0}{k_M} \frac{c_C^S}{c_A^\infty}, \tag{16.3.18}$$

and the QSS hypothesis is satisfied provided that $Bi_M c_A^\infty \ll c_C^S$. This condition is satisfied even more easily than in the previous case. As expected, the controlling process here is reaction, since it is much slower than diffusion, with $\tau_{diff} \ll \tau_{react}$.

16.4 Unsteady Evaporation

Consider a liquid A that evaporates into a vapor B in a tube of infinite length, assuming that the liquid level is kept at position $z = 0$ and that B is insoluble in A. In addition, the system is kept at constant pressure and temperature and the vapors A and B form an ideal gas mixture, so that the concentration c is constant in the gas phase. Therefore, as A starts to evaporate, it pushes B forward, along the tube z-axis.

The continuity equations for A and B in the semi-infinite region $z \geq 0$ are:

$$\frac{\partial c_A}{\partial t} + \frac{\partial J_A}{\partial z} = 0, \tag{16.4.1}$$

$$\frac{\partial c_B}{\partial t} + \frac{\partial J_B}{\partial z} = 0, \tag{16.4.2}$$

where J_A and J_B are the material fluxes, satisfying the following boundary and initial conditions for x_A (and similar ones for x_B),

$$x_A(z, t = 0) = 0; \quad x_A(z = 0, t) = x_{A0}; \quad x_A(z \to \infty, t) = 0. \tag{16.4.3}$$

Here, we have assumed that at time $t = 0$, B fills the whole region, while at the interface $z = 0$ the A vapor is at local equilibrium with its liquid, so that $x_{A0} = x_A^{sat}$.

Now, adding Eqs. (16.4.1) and (16.4.2), and considering that $c = c_A + c_B$ is constant, we conclude that $J_A + J_B$ is independent of z and therefore it is a function of time alone. Therefore, since $J_B (z = 0, t) = 0$, as B is insoluble into liquid A, we see that,

$$J_A(z, t) + J_B(z, t) = J_A(0, t) \equiv J_{A0}(t). \tag{16.4.4}$$

Obviously, J_{A0} decreases with time and is larger at shorter times (it is infinite at $t = 0$). In addition, since J_A is zero as $z \to \infty$, then $J_{A0}(t) = J_B(z \to \infty, t)$, indicating that, as expected, A pushes B along the tube axis, z.

The material flux J_A is given by the constitutive relation,

$$J_A(z, t) = x_A(J_A + J_B) - cD\frac{\partial x_A}{\partial z} = x_A(z, t)J_{A0}(t) - cD\frac{\partial x_A(z, t)}{\partial z}. \tag{16.4.5}$$

At $z = 0$, we obtain:

$$J_{A0}(t) = -\frac{cD}{1 - x_{A0}}\left(\frac{\partial x_A}{\partial z}\right)_{z=o}. \tag{16.4.6}$$

At this point, substituting Eq. (16.4.6) into (16.4.5) and then applying Eq. (16.4.1), we obtain a partial differential equation (of second order in z and first order in t) to be solved with two boundary conditions and one initial condition [see Eqs. (16.4.3)].

In the dilute case, with $x_{A0} \ll 1$, we can neglect the convective contribution of the material transport [i.e., the first term on the RHS of Eq. (16.4.5)] and therefore, as we have seen in Sect. 7.4, we obtain the typical self-similar solution,

$$\frac{x_A(z, t)}{x_{A0}} = f(\eta) = erfc(\eta) = [1 - erf(\eta)], \quad \text{with } \eta = \frac{z}{\sqrt{4Dt}}, \tag{16.4.7}$$

where erf indicates the error function (7.4.9), represented in Fig. 7.6. Substituting this result into Eq. (16.4.6) we find the flux at the interface, $J_{A0}(t) = cx_{A0}\sqrt{D/\pi t}$ [see Eq. (7.4.10)], that is,

$$(J_{A0})_{diff} = cD\frac{x_{A0} - x_{A\infty}}{\partial_{diff}}, \quad \text{where } \partial_{diff} = \sqrt{\pi Dt}. \tag{16.4.8}$$

Here, $x_{A\infty} = 0$, while δ_{diff} is the diffusion length, that is a typical distance covered by an A-molecule within a time t.

In the general, non-dilute case, we again look for a self-similar solution $f(\eta)$, expressed in terms of the dimensionless variable η defined in Eq. (16.4.7), Then, the following equation is obtained:

$$\frac{d^2 f}{d\eta^2} + 2(\eta - \phi)\frac{df}{d\eta} = 0, \tag{16.4.9}$$

where ϕ is a constant, still to be determined, due to the contribution of convection,

$$\phi = -\frac{1}{2}\frac{x_{A0}}{1 - x_{A0}}\frac{df}{d\eta}(0). \tag{16.4.10}$$

Naturally, when $\phi = 0$, we find the error function solution of the purely diffusive, or dilute, case. As for the conditions to be imposed, following the discussion of Sect. 7.4, here we note again that the most stringent condition in finding self-similar solutions is that Eq. (16.4.9) can only satisfy 2 conditions, while our initial problem imposes the 3 conditions (16.4.3). In our case, however, these three relations collapse into the following two conditions,

$$f(0) = 1 \quad \text{and} \quad f(\infty) = 0, \tag{16.4.11}$$

and so, from Eq. (16.4.9) we obtain:

$$f(\eta) = \frac{1 - erf(\eta - \phi)}{1 + erf(\phi)}. \tag{16.4.12}$$

In addition, substituting Eq. (16.4.12) into (16.4.10), we can determine ϕ implicitly as:

$$x_{A0} = \frac{2\phi}{2\phi - f'(0)} = \left\{1 - \frac{f'(0)}{2\phi}\right\}^{-1} = \left\{1 + [\sqrt{\pi}(1 + erf(\phi))\phi\exp(\phi^2)]^{-1}\right\}^{-1}, \tag{16.4.13}$$

where we have considered that $f'(0) = -2\exp(-\phi^2)/\sqrt{\pi}(1 + erf\,\phi)$. Physically, ϕ is a function of x_{A0} and represents a dimensionless molar velocity.

Now, let us calculate the material flux (16.4.6) of A at the interface,

$$J_{A0} = -\frac{cD x_{A0}}{1 - x_{A0}}\left(\frac{\partial f}{\partial z}\right)_{z=0} = -\frac{cD x_{A0}}{1 - x_{A0}}\frac{1}{\sqrt{4Dt}}\frac{df}{d\eta}(0) = c\phi\sqrt{\frac{D}{t}}. \tag{16.4.14}$$

Finally, we can determine the Sherwood number, that is the ratio between the flux J_{A0} and its purely diffusive part, i.e.,

$$Sh = \frac{J_{A0}}{(J_{A0})_{diff}} = \frac{\sqrt{\pi\varphi}}{x_{A0}}. \qquad (16.4.15)$$

x_{A0}	ϕ	Sh
0	0	1
0.25	0.156	1.11
0.50	0.358	1.27
0.75	0.662	1.56
1	∞	∞

As we see from the Table above, the convective contribution starts to be relevant only when the molar fraction of A at the interface is almost unity. Finally, the flux diverges at $x_{A0} = 1$, as we have already seen at the end of Sect. 15.1.

In analogy with Eq. (16.4.8), the result (16.4.14) can be written as:

$$J_{A0} = cD\frac{x_{A0} - x_{A\infty}}{\delta_{diff}}, \quad \text{where } \delta_{diff} = \frac{\sqrt{\pi Dt}}{Sh}. \qquad (16.4.16)$$

Here we see that in the presence of convection, with $Sh > 1$, the diffusion length δ_{diff} decreases, as one would expect, since the concentration profile at the interface becomes steeper.

16.5 Problems

16.1 Derive Eq. (16.2.1) by applying the integral Reynolds transport theorem (6.2.7).

16.2 A little drop of water (species A) of radius R_i is suspended at time $t = 0$ in an atmosphere of dry air (species B), that is, with molar fraction of water vapor, $x_\infty = 0$. As the drop evaporates, at the liquid-vapor interface the local equilibrium condition determines the molar fraction of water vapor, x_0. Using the QSS approximation, determine the evaporation time.

16.3 Model the combustion of methane (A) immersed in an atmosphere of nitrogen and oxygen (B), according to the reaction $2O_2 + CH_4 \rightarrow CO_2 + 2H_2$. Assume that the reaction is very rapid, A and B are dilute and the problem can be schematized as being plane, with the following conditions: (a) at $z < 0$ there is only A; (b) initially, at $z > 0$, there is only B. Determine the expression of the methane flux.

16.4 A spherical catalyst of radius R is immersed in a gas mixture containing a dilute species A, with concentration c_A, with which it reacts instantaneously. Determine the flux of A at the surface of the catalyst.

Chapter 17
Convective Material Transport

Abstract In this Chapter we study the material transport (in terms of mass or moles) of a dilute solute dissolved in a solvent. As we have seen in Sect. 14.3, at leading order the average velocity of a mixture in the dilute limit equals the solvent velocity, so that the material transport reduces to a linear problem, as both diffusive and convective fluxes are linear functions of the concentration difference. Accordingly, the governing equations of mass transport in the dilute limit are identical as those of heat transport, so that most of the considerations that we made in Chap. 13 can be extended here. In particular, in Sects. 17.1 and 17.2 we study the mass boundary layer, with its dependence on the fluid velocity and the geometry of the problem. Then, in Sect. 17.3, we focus on the case where there is a mass boundary layer, with no momentum boundary layer, occurring when the material Peclet number is large and concomitantly the Reynolds number is small. The material boundary layer is further examined in Sect. 17.4, using the integral approximation described in Sect. 7.7. Finally, in Sect. 17.5, we apply the quasi steady state approximation to solve important problems of mass transfer, related to particle growth or consumption.

17.1 Mass Transport Through a Fixed Bed

A fluid is pumped across a fixed bed located at $0 < x < L$, with a uniform velocity field V (plug flow). Downstream, a dilute solute is dissolved into the fluid, so that the concentration is $c = c_L$ at $x \geq L$, while upstream the mixture is kept solute-free, so that $c = 0$ at $x \leq 0$ (see Fig. 17.1). (a) Calculate the concentration profile $c(x)$ and the material flux J_L at $x = L$, analyzing the limit cases $Pe \ll 1$ and $Pe \gg 1$, where $Pe = VL/D$ is the Peclet number[1] and D denotes the diffusivity of the solute in the fluid. (b) Derive the results of the two limit cases using perturbation methods.

---

[1]Here, Pe denotes the material Peclet number, that is the ratio between convective and diffusive fluxes in the material (i.e., mass- or mole-based) transport of a chemical species. As we saw in Sects. 3.2 and 13.1, Pe may also indicate the thermal Peclet number; therefore, in this Chapter, the material Peclet number is sometimes denoted as Pe_M.

© Springer International Publishing Switzerland 2015
R. Mauri, *Transport Phenomena in Multiphase Flows*,
Fluid Mechanics and Its Applications 112, DOI 10.1007/978-3-319-15793-1_17

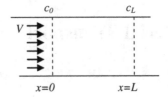

Fig. 17.1 Concentration profile $c(x)$ in a plug flow across a porous medium

The solute concentration is a function of x only and in $0 < x < L$ it satisfies the following equation and boundary conditions,

$$\frac{d}{dx} J = 0; \quad J(x) = Vc - D\frac{dc}{dx}; \quad \text{with } c(0) = 0; \ c(L) = c_L, \tag{17.1.1}$$

where, due to the condition of diluteness, the mean velocity appearing in the constitutive equation of the material flux has been replaced by the solvent velocity. So, the material flux J_L at $x = L$ is:

$$J_L = J(L) = Vc_L - D\frac{dc}{dx}(L). \tag{17.1.2}$$

In dimensionless terms, defining $\tilde{c} = c/c_L$, $\xi = x/L$ and $Pe = VL/D$, we obtain:

$$Pe\frac{d\tilde{c}}{d\xi} = \frac{d^2\tilde{c}}{d\xi^2}; \quad \tilde{c}(0) = 0; \ \tilde{c}(1) = 1. \tag{17.1.3}$$

It is convenient, as usual, to express the material flux in term of the Sherwood number, defined as $Sh = J_L/J_{L,diff}$, where $J_{L,diff}$ is the material flux due exclusively to diffusion. In our case, $J_{L,diff} = -Dc_L/L$ and therefore we obtain:

$$Sh = \frac{J_U(L)}{J_{U,diff}} = \frac{d\tilde{c}}{d\xi}(1) - Pe. \tag{17.1.4}$$

Note that here, in addition to the usual diffusion part [i.e., the first term on the RHS of Eq. (17.1.4)], the molar flux, J_L, and the Sherwood number, Sh, are composed of a convective component as well. Note that the convective component is identically zero at a wall, where the fluid velocity is null, or when the boundary conditions are inverted in (17.1.1), i.e., imposing $\tilde{c}(0) = 1$ and $\tilde{c}(1) = 0$ (see Problem 17.1).

Equation (17.1.3) has the general solution, $\tilde{c}(\xi) = C_1 e^{pe\xi} + C_2$; then, imposing that the boundary conditions are satisfied, we obtain the solution:

$$\tilde{c}(\xi) = \frac{e^{pe\xi} - 1}{e^{pe} - 1}, \tag{17.1.5}$$

and thus,

$$Sh = \frac{Pe}{e^{Pe} - 1}.$$

(17.1.6)

The concentration profile is represented in Fig. 17.2. As expected, it is linear when $Pe = 0$, that is when the flux is purely diffusive, and so $Sh = 1$, while at $Pe > 0$ solute diffusion is opposed by fluid convection so that Sh progressively decreases. Finally, at $Pe \gg 1$, the concentration profile appears to form a boundary layer near $\xi = 1$ but, unexpectedly at first, we find $Sh = 0$. As we will see in the following, this is due to the fact that the diffusive flux arising within the boundary layer exactly balances its convective counterpart.

(a) $Pe = \varepsilon \ll 1$

Since $e^{\varepsilon} = 1 + \varepsilon + \frac{1}{2}\varepsilon^2 \cdots$ and $(1 + e)^{-1} = 1 - \varepsilon + \cdots$, we obtain:

$$\tilde{c}(\xi) = \frac{\xi(1 + \frac{1}{2}Pe\xi + \ldots)}{1 + \frac{1}{2}Pe} \cong \xi\left(1 + \frac{1}{2}Pe\xi + O\left(Pe^2\right)\right)\left(1 - \frac{1}{2}Pe + O\left(Pe^2\right)\right),$$

that is,

$$\tilde{c}(\xi) = \xi - \frac{1}{2}Pe\xi(1 - \xi) + O\left(Pe^2\right),$$

(17.1.7)

Fig. 17.2 Concentration profile and Sherwood number as functions of Pe

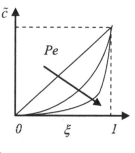

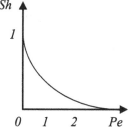

and

$$Sh = 1 - \frac{1}{2}Pe + O(Pe^2). \tag{17.1.8}$$

As expected, when $Pe = 0$ the concentration profile is linear and $Sh = 1$. In fact, when $Pe = 0$, the problem (17.1.3) remains well-posed (i.e., a second-order differential equation with two boundary conditions), yielding $\tilde{c}(\xi) = \xi$, corresponding to $Sh = 1$. Therefore, we expect that this limit case can be solved by using a regular perturbation expansion,

$$\tilde{c}(\xi) = \tilde{c}_0(\xi) + \varepsilon \tilde{c}_1(\xi) + O(\varepsilon^2) \tag{17.1.9}$$

Here, we remind, $\varepsilon = Pe$, and we have assumed that $\tilde{c}_0(\xi) \gg \varepsilon \tilde{c}_1(\xi)$ everywhere within the region $0 < \xi < 1$; if in a certain region this equality were not satisfied, we would need to apply a singular perturbation analysis (see Sects. 10.2.2 and 15.3.2).

Substituting the expansion (17.1.9) into (17.1.3) we obtain:

$$\begin{aligned}
&\varepsilon \tilde{c}_0' + \varepsilon^2 \tilde{c}_1' - \tilde{c}_0'' - \varepsilon \tilde{c}_1'' - \varepsilon^2 \tilde{c}_2'' + \ldots = 0 \\
&\tilde{c}_0(0) + \varepsilon \tilde{c}_1(0) + \ldots = 0; \qquad \tilde{c}_0(1) + \varepsilon \tilde{c}_1(1) + \ldots = 1
\end{aligned} \tag{17.1.10}$$

and

$$Sh = \tilde{c}_0'(1) + \varepsilon \left[\tilde{c}_1'(1) - 1\right] + O(\varepsilon^2). \tag{17.1.11}$$

Collecting all terms of the same magnitude, at $O(1)$ we have:

$$\tilde{c}_0'' = 0; \quad \tilde{c}_0(0) = 0; \quad \tilde{c}_0(1) = 1; \quad \text{with } Sh^{(0)} = \tilde{c}_0'(1), \tag{17.1.12}$$

obtaining:

$$\tilde{c}_0(\xi) = \xi; \quad \text{and} \quad Sh^{(0)} = 1. \tag{17.1.13}$$

This is the usual solution, that is valid when $Pe = 0$.
Now let us collect all $O(\varepsilon)$-terms, finding the $O(Pe)$ correction as follows.

$$\tilde{c}_1'' = \tilde{c}_0' = 1; \quad \tilde{c}_1(0) = 0; \quad \tilde{c}_1(1) = 0; \quad \text{with } Sh^{(1)} = \tilde{c}_1'(1) - 1, \tag{17.1.14}$$

obtaining:

$$\tilde{c}_1(\xi) = \frac{1}{2}\xi^2 - \frac{1}{2}\xi; \quad \text{and} \quad Sh^{(1)} = -\frac{1}{2}. \tag{17.1.15}$$

Therefore we have:

$$\tilde{c}(\xi) = \xi - \tfrac{1}{2}Pe\xi(1 - \xi) + O(Pe^2) \quad \text{and} \quad Sh = 1 - \tfrac{1}{2}Pe + O(Pe^2), \quad (17.1.16)$$

which coincides with Eqs. (17.1.7) and (17.1.8).

(b) $Pe = 1/\varepsilon \gg 1$

In this case, the exact solution reduces to:

$$\tilde{c}(\xi) \cong e^{-Pe(1-\xi)} \quad \text{and} \quad Sh = Pe - Pe = 0. \qquad (17.1.17)$$

To analyze this case, we must use a singular perturbation expansion. In fact, imposing that in Eq. (17.1.3) $Pe^{-1} = 0$, and so neglecting diffusion altogether, we would obtain: $\tilde{c}'(\xi) = 0$, with $\tilde{c}(0) = 0$ and $\tilde{c}(1) = 1$, which is an ill-posed problem (i.e., a first-order differential equation with two boundary conditions). Therefore, if we are not too close to the boundary $\xi = 1$, and therefore ignoring the boundary condition $\tilde{c}(1) = 1$, the problem is well-posed and would give the solution $\tilde{c}(\xi) = 0$. Instead, in the vicinity of $\xi = 1$, the concentration profile jumps suddenly from 0 to 1, creating a material boundary layer. Since, by definition, at the edge of this region the diffusive flux must balance the convective flux, considering that this latter is proportional to Pe, then the boundary layer thickness is proportional to Pe^{-1}. Repeating this analysis more rigorously, we can determine the boundary layer thickness defining the stretched coordinate, $\zeta = (1 - \xi)/\varepsilon^\alpha$ in the region very close to $\xi = 1$, where the exponent α is determined by balancing convective and diffusive terms. In fact, rewriting Eq. (17.1.3) with the new coordinate we obtain:

$$-\frac{d\tilde{c}(\zeta)}{d\zeta}\varepsilon^{-\alpha} = \varepsilon\frac{d^2\tilde{c}(\zeta)}{d\zeta^2}\varepsilon^{-2\alpha}. \qquad (17.1.18)$$

Therefore, imposing that the two terms have the same magnitude, we find $\alpha = 1$, and therefore considering that $\varepsilon = Pe^{-1}$, we obtain: $\zeta = Pe(1 - \xi)$. So, within the boundary layer (i.e., using the stretched coordinate), we have to solve the following problem:

$$\frac{d\tilde{c}(\zeta)}{d\zeta} + \frac{d^2\tilde{c}(\zeta)}{d\zeta^2} = 0; \quad \tilde{c}(\zeta = 0) = 1; \; \tilde{c}(\zeta \to \infty) = 0, \qquad (17.1.19)$$

where we have considered that, when $\zeta \to \infty$, the solution that is valid within the boundary layer must match the solution that is valid outside, that is $\tilde{c} = 0$ in this case. So at the end we find:

$$\tilde{c}(\zeta) = e^{-\zeta} \Rightarrow \quad \tilde{c}(\zeta) = e^{-Pe(1-\zeta)} \Rightarrow Sh = Pe - Pe = 0, \qquad (17.1.20)$$

that coincides with the exact result (17.1.17).

Note that, using dimensional analysis, we could correctly predict, when $Pe \gg 1$, the existence of a boundary layer whose thickness is of $O(Pe^{-1})$, thus inducing a diffusive flux of $O(Pe)$. However, we could not predict that this flux identically cancels the $O(Pe)$ convective flux, directed in the opposite direction. This example indicates that, after performing a dimensional analysis, it is recommended, when it is possible, to do the detailed calculation as well. In fact, besides getting the exact numerical result [like the ½ factor in the $O(Pe)$ correction of Eq. (17.1.16)], the complete perturbation analysis is indispensable to check whether unpredictable events [like the cancellation of terms having the same magnitude, occurring in Eq. (17.1.20)] would not change dramatically the final result.

17.2 Laminar Material Boundary Layer

Consider an fluid mixture composed of a solvent, with density ρ and kinematic viscosity v, and a dilute solute, diffusing in the fluid with diffusion coefficient D. The mixture has concentration c_∞, and flows with uniform unperturbed velocity V and pressure P_∞ past a submerged still object of size L (see Fig. 17.3). Knowing that on the surface of the object the solute concentration is c_0 (for example, assume that $c_0 < c_\infty$), we want to determine the material flux of the solute being adsorbed at the object surface, assuming large Peclet numbers, i.e., when $Pe_M = VL/D \gg 1$.

As always in the dilute case, the mean velocity of the mixture coincides at leading order with the solvent velocity. Accordingly, the material balance equation becomes linear with respect to the concentration difference $\Delta c = (c_\infty - c_0)$, and therefore the mass transport coefficient and the Sherwood number will not depend on Δc (see discussion in Sect. 14.3.1).

The unknowns of this problem are the velocity and pressure of the solvent, and the concentration of the solute; in principle, they can be determined by solving the Navier-Stokes and continuity equations for the solvent flow, together with the convection-diffusion equation for the solute concentration,

$$\frac{D\mathbf{v}}{Dt} = \mathbf{v} \cdot \nabla \mathbf{v} = -\frac{1}{\rho} \nabla P + v \nabla^2 \mathbf{v}, \qquad (17.2.1)$$

$$\nabla \cdot \mathbf{v} = 0, \qquad (17.2.2)$$

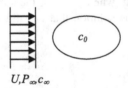

$$U, P_\infty, c_\infty$$

Fig. 17.3 Fluid flowing past a body having a different composition concentration

$$\frac{Dc}{Dt} = \mathbf{v} \cdot \nabla c = D\nabla^2 c, \tag{17.2.3}$$

with the following boundary conditions:

$\mathbf{v} = 0;$ $c = c_0$ at the surface; $\mathbf{v} = \mathbf{V}; \ P = P_\infty; \ c = c_\infty$ far from the surface.

$$\tag{17.2.4}$$

As usual, at the end we are interested in finding the material flux at the body surface,

$$J_M = -D\mathbf{e}_n \cdot \nabla c \text{ at the surface,} \tag{17.2.5}$$

where $\mathbf{e}_n$ is the unit vector perpendicular to the surface of the object and directed outward. Note that the material flux at the surface is only diffusive, since the fluid velocity is zero.

This problem is identical to the heat transport studied in Sect. 13.1. In fact, defining the dimensionless quantities,

$$\widetilde{\mathbf{r}} = \frac{\mathbf{r}}{L}; \quad \widetilde{t} = \frac{t}{(L/V)}; \quad \widetilde{P} = \frac{P - P_\infty}{(\mu V/L)}; \quad \widetilde{\mathbf{v}} = \frac{\mathbf{v}}{V}; \quad \widetilde{c} = \frac{c - c_0}{c_\infty - c_0}, \tag{17.2.6}$$

Equations (17.2.1)–(17.2.4) become:

$$Re\frac{D\widetilde{\mathbf{v}}}{D\widetilde{t}} = Re\, \widetilde{\mathbf{v}} \cdot \widetilde{\nabla}\widetilde{\mathbf{v}} = -\widetilde{\nabla}\widetilde{P} + \widetilde{\nabla}^2\widetilde{\mathbf{v}}, \tag{17.2.7}$$

$$\widetilde{\nabla} \cdot \widetilde{\mathbf{v}} = 0, \tag{17.2.8}$$

$$Pe_M\frac{D\widetilde{c}}{D\widetilde{t}} = Pe_M\, \widetilde{\mathbf{v}} \cdot \widetilde{\nabla}\widetilde{c} = \widetilde{\nabla}^2\widetilde{c}, \tag{17.2.9}$$

$\widetilde{\mathbf{v}} = \widetilde{c} = 0$ at the surface; $\widetilde{\mathbf{v}} = \widetilde{c} = 1$ far from the surface. $\tag{17.2.10}$

From this analysis, two dimensionless numbers arise naturally. They are the Reynolds number, $Re = VL/\nu$, and the material Peclet number, $Pe_M = VL/D$, equal to the ratio between convective and diffusive fluxes in the transport of fluid momentum and solute molecules, respectively. The Peclet number can also be expressed as the product between the Reynolds number and the Schmidt number, $Sc = \nu/D$. This latter represents the ratio between the efficiency of the diffusive momentum transport and that of the diffusive material (mass or molar) transport. In synthesis,

$$Re = \frac{VL}{v}; \quad Pe_M = \frac{VL}{D}; \quad Sc = \frac{v}{D}. \tag{17.2.11}$$

The material flux, too, can be expressed in a dimensionless form, defining the Sherwood number as the ratio between the flux at the wall and its purely diffusive part, that is,

$$Sh = \frac{|J_M|}{D\Delta c/L} = \left| \mathbf{e}_n \cdot \widetilde{\nabla} \tilde{c} \right| \text{ at the surface,} \tag{17.2.12}$$

where $\Delta c = |c_\infty - c_0|$.

Now, repeating the analysis of Sect. 13.1, we can define a material boundary layer of thickness δ_M, such that $Sh \approx L/\delta_M$, finding the results that are summarized Figs. 17.4 and 17.5, where:

$$Re \gg 1: \quad \begin{array}{l} Sh \propto Pe_M^{1/2} = Re^{1/2} Sc^{1/2} \quad \text{when} \quad Sh \ll 1 \\ Sh \propto Re^{1/2} Sc^{1/3} \quad \text{when} \quad Sh \gg 1 \end{array} \tag{17.2.13}$$

and

$$Re \ll 1: \quad \begin{array}{l} Sh \approx 1 \quad \text{when } Pe_M \ll 1 \\ Sh \propto Pe_M^{1/3} = Re^{1/3} Sc^{1/3} \quad \text{when } Pe_M \gg 1 \end{array} \tag{17.2.14}$$

Fig. 17.4 Log-log plot of Sh at large Re as a function of Sc

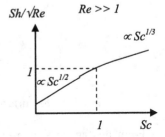

Fig. 17.5 Log-log plot of Sh at small Re as a function of Pe_M

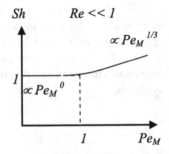

These results could also be expressed in terms of the coefficient of mass transport, k_M, defined in Eq. (14.4.2),

$$k_M = \frac{J_M}{\Delta c} = \frac{D}{L} Sh. \qquad (17.2.15)$$

Now, the case $Re \gg 1$ can be analyzed using the Colburn-Chilton analogy, described in Sect. 13.3. Instead, in the next section, we will concentrate on the case $Re \ll 1$.

17.3 Mass Boundary Layer for Small Reynolds Number

In the previous section we have seen that when $Re \ll 1$ we must distinguish the case with $Pe_M \gg 1$ from that with $Pe_M \ll 1$.

(a) $Re \ll 1$; $Pe_M \ll 1$

When convection can be neglected altogether, Eq. (17.2.9) becomes: $\tilde{\nabla}^2 \tilde{c} = 0$, with boundary conditions (17.2.10). In the case of a spherical object, we would obtain: $\tilde{c} = 1 - 1/\tilde{r}$ and therefore the Sherwood number (17.2.12), defined with $L = R$, gives, as expected, $Sh = 1$.

When Pe_M is small but not zero, we obtain a correction to this simple result, that can be obtained using a singular perturbation expansion method, as shown by Acrivos and Taylor in 1962. For a spherical object, they obtained:

$$Sh = 1 + \frac{1}{2} Pe + O\left(Pe^2 \ln Pe\right), \qquad (17.3.1)$$

where $Pe_M = VR/D$. Proof of this result goes beyond the scope of this textbook; a clear treatment can be found in Leal (pp. 456–477).

(b) $Re \ll 1$; $Pe_M \gg 1$

First of all, we must realize that this case is very common in the transport of mass in liquid mixtures, considering that the Schmidt numbers is often of $O(10^4)$ or larger. For example, in the transport of nutrients or waste products within a lymph vessel of a human body, we have typically $Re \approx 10^{-2}$ and $Pe_M \approx 10^3$.

Let us model this system as a duct transporting a liquid Newtonian solvent (i.e., the lymph), in which a dilute solute (i.e., a large macromolecule) is dissolved, being gradually absorbed at the walls. Since mass transport is very slow, it takes place within a thin layer near the wall of the vessel, which is therefore the only region of interest to us. Accordingly, we can neglect curvature and consider the wall as flat, assuming that the velocity profile in that region is linear, i.e., $v_x = \gamma y$, with γ

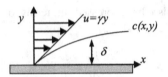

Fig. 17.6 Material boundary layer in shear flow at small Re

constant, and $v_y = 0$ (see Fig. 17.6). This corresponds to the so-called Leveque approximation, which therefore amounts to the following problem[2]:

$$\gamma y \frac{\partial c}{\partial x} = D \frac{\partial^2 c}{\partial y^2}, \tag{17.3.2}$$

where D is the diffusion coefficient, to be solved with boundary conditions,

$$c(x > 0, y = 0) = c_0; \quad c(x < 0, y) = c(x, y \to \infty) = c_\infty. \tag{17.3.3}$$

Let us look for a self-similar solution, $c = c(\eta)$, where $\eta = y/g(x)$, and $g(x)$ is a function to be determined. Considering that $\partial c/\partial x = -\eta g' c'/g$ and $\partial c/\partial y^2 = c''/g^2$, Eq. (17.3.2) becomes,

$$\frac{d^2 \tilde{c}}{d\eta^2} + \left(\frac{1}{a^2} g^2 \frac{dg}{dx} \right) \eta^2 \frac{d\tilde{c}}{d\eta} = 0, \quad \text{with } \tilde{c} = \frac{c - c_0}{c_\infty - c_0}, \tag{17.3.4}$$

where $a = \sqrt{D/\gamma}$ is a characteristic distance. Since the term in parenthesis of Eq. (17.3.4) must be an arbitrary constant (see the discussion in Sect. 8.2 about this point), we obtain:

$$\left(\frac{1}{3a^2} \frac{dg^3}{dx} \right) = 3 \Rightarrow g^3 = 9a^2 x + C \Rightarrow g = \sqrt[3]{9a^2 x}, \tag{17.3.5}$$

where we have imposed that the integration constant C is zero, so that the three original boundary conditions (17.3.3) reduce to the following two[3]:

$$\tilde{c}(0) = 0; \quad \tilde{c}(\infty) = 1. \tag{17.3.6}$$

---

[2] The Leveque approximation describes the mass or heat exchange of a fluid in laminar flow in the entrance region of a tube, when the solute concentrations (or the temperatures) at the entrance and at the wall are specified (see Problem 17.1). The case of a fluid flowing past a submerged body is more complicated, because γ depends on x and, consequently, by continuity, the transversal velocity component is not null. Nevertheless, the results that are obtained in the general case are qualitatively similar to the ones that are obtained using the Leveque approximation.

[3] As we discussed in Sect. 8.2, this reasoning is based on the assumption that the problem admits a unique solution. Therefore, if at the end of this procedure we find a solution that satisfies all the assumptions that we have made on the way, we do not have to go on looking for another solution.

From the resulting equation,

$$\frac{d^2\tilde{c}}{d\eta^2} + 3\eta^2 \frac{d\tilde{c}}{d\eta} = 0,$$

we have: $\tilde{c}(n) = C_1 \int_0^\eta e^{-\zeta^3} d\xi + C_2$; finally, imposing that the boundary conditions are satisfied, we finally find the self-similar Leveque solution,

$$\tilde{c}(\eta) = \frac{1}{\Gamma(4/3)} \int_0^\eta e^{-\zeta^3} d\xi, \qquad (17.3.7)$$

where $\Gamma(4/3)$ is the gamma function,

$$\Gamma(4/3) = \int_0^\infty e^{-\zeta^3} d\xi = 0.893. \qquad (17.3.8)$$

Now, calculate the material flux at the surface,

$$J_{M0} = -D\left(\frac{\partial c}{\partial y}\right)_{y=0} = \frac{D\Delta c}{g(x)}\left(\frac{d\tilde{c}}{d\eta}\right)_{\eta=0} = \frac{\Delta c}{\Gamma(4/3)}\left(\frac{\gamma D^2}{9x}\right)^{1/3}$$
$$= 0.538\Delta c (\gamma D^2)^{1/3} x^{-1/3}. \qquad (17.3.9)$$

The total flux, that is the mass of the solute A adsorbed per unit time from $x = 0$ to $x = L$ is:

$$\dot{m}_A = W \int_0^L J_{M0} dx = \frac{3^{1/3}\Delta c}{2\Gamma(4/3)} (\gamma D^2)^{1/3} WL^{2/3} = 0.808\Delta c (\gamma D^2)^{1/3} WL^{2/3},$$

$$(17.3.10)$$

where W is the width of the plane surface. In dimensionless terms, defining the Sherwood number as $Sh = \dot{m}_A/\dot{m}_{A,diff}$, where $\dot{m}_{A,diff} = (D\Delta c/L)(LW)$, we obtain:

$$Sh = \frac{\dot{m}}{DW\Delta c} = \frac{3^{1/3}}{2\Gamma(4/3)}\left(\frac{\gamma L^2}{D}\right)^{1/3} = 0.808 Pe_M^{1/3}, \qquad (17.3.11)$$

where $Pe_M = VL/D = \gamma L^2/D$ is the Peclet number, with $V = \gamma L$ denoting a characteristic velocity. This confirms the scaling $Sh \propto Pe_M^{1/3}$ that we have seen in Eq. (17.2.14).[4]

17.4 Integral Methods

In Sect. 7.6. we have seen that sometimes it is convenient to use approximate, integral methods, when we lack all the necessary information to find the exact result. In the case of the material, or thermal, boundary layer forming on a flat plate, using the same notation as in Sect. 17.2, consider the material balance equation of a species A,

$$\frac{\partial}{\partial x}(v_x c) + \frac{\partial}{\partial y}(v_y c) = D\frac{\partial^2 c}{\partial y^2}, \tag{17.4.1}$$

where we have considered that the velocity field is solenoidal and we have neglected the diffusive term $\partial^2 c/\partial x^2$. Integrating this equation over the y-direction we obtain:

$$D\int_0^\infty \frac{\partial^2 c}{\partial y^2}dy = -D\frac{\partial c}{\partial y}(0) = J_w, \tag{17.4.2}$$

$$\frac{\partial}{\partial y}\int_0^\infty v_y c\, dy = [v_y c]_0^\infty = v_\infty c_\infty = c_\infty \int_0^\infty \frac{\partial v_y}{\partial y}dy = -c_\infty \int_0^\infty \frac{\partial v_x}{\partial x}dy$$

$$= -c_\infty \frac{d}{dx}\int_0^\infty v_x dy, \tag{17.4.3}$$

where we have applied the continuity equation and we have defined v_∞ as the transversal velocity far from the wall, i.e., $v_\infty = v_y(y \to \infty)$, as seen in Sect. 8.4. Therefore we find:

$$J_w = -D\frac{\partial c}{\partial y}(0) = \frac{d}{dx}\int_0^\infty v_x(c - c_\infty)dy. \tag{17.4.4}$$

---

[4]In the case of a flow past a sphere, we obtain: $Sh = 1.249\ Pe^{1/3}$, where Sh and Pe are defined in terms of the sphere radius R [see Leal (pp. 513–525)]. This confirms that the results obtained with the Leveque approximation can be applied also to more complex geometries.

This integral equation reflects the fact that the net mass loss of the species A is equal to the flux of A adsorbed at the wall.

Now we assume that, as for the velocity profile of Eq. (7.7.4), the concentration profile has also a given specific form, depending on the thickness of the mass boundary layers, δ_M, i.e.,

$$\frac{c - c_0}{c_\infty - c_0} = \begin{cases} f_c(\eta_c) & \text{at } 0 < \eta_c < 1 \\ 0 & \text{at } \eta_c > 1 \end{cases} \quad \text{with } \eta_c = \frac{y}{\delta_M(x)}. \tag{17.4.5}$$

Here, $f_c(\eta_c)$ is any reasonable function, which we choose to be any appropriate polynomial that satisfies the boundary conditions and offers a smooth transition from the region inside to that outside of the boundary layer.

Now, substituting Eqs. (7.7.4) and (17.4.5) into Eq. (17.4.4) we obtain:

$$-D f_c'(0) = V \delta_M \frac{d\delta_M}{dx} \int_0^1 f(n)[f_c(\eta_c) - 1] d\eta_c. \tag{17.4.6}$$

Assuming that $\delta_M/\delta = \Delta$, where Δ is a constant, find:

$$-\beta_c D = \frac{1}{2} \alpha_{2c} V \Delta^3 \frac{d\delta^2}{dx}, \tag{17.4.7}$$

where

$$\beta_c = f_c'(0) \quad \text{and} \quad \alpha_{2c} = \Delta^{-1} \int_0^1 f(n)[f_c(\eta_c) - 1] d\eta_c. \tag{17.4.8}$$

Choosing for both f and f_c the same Eq. (7.7.10), i.e., $f(x) = f_c(x) = \frac{3}{2} x - \frac{1}{2} x^3$, we find that $\beta_c = 3/2$ while, from (17.4.6) to (17.4.7) we have: $\delta^2 = (280/13)(vx/V)$. In addition, assuming that $\delta_M \ll \delta$ (i.e., when $\Delta \ll 1$), in the integral of Eq. (17.4.8) we can replace f with the linear function, $f(\eta) \cong \frac{3}{2}\eta = \frac{3}{2}\Delta\eta_c$, so that we find: $\alpha_{2c} = -3/20$, obtaining again:

$$\Delta^3 = \frac{13}{14} Sc^{-1} \Rightarrow \Delta = \frac{\delta_M}{\delta} \cong Sc^{-1/3}, \tag{17.4.9}$$

where $Sc = v/D$ is the Schmidt number.

17.5 Quasi Steady State (QSS) Approximation

A uniform air flow streams past a spherical coal particle located on a grill. Oxygen reacts very rapidly at the particle surface with the chemical reaction $C + O_2 \rightarrow CO_2$. Denoting by V the fluid velocity, v its kinematic viscosity and D the (effective) diffusivity of oxygen in air, we want to determine the typical combustion time of the coal particle, assuming that its initial radius is R_0 and considering that the molar fraction of oxygen is rather small.

In this problem, we can safely assume that $Re = VR/v \gg 1$. In addition, since diffusion takes place in air, $Sc = v/D \cong 1$. Therefore, since $Pe_M = VR/D \cong Re \gg 1$, we have two boundary layers, one of momentum and one of mass, having the same thickness, $\delta = \delta_M$. In these conditions, then, $Sh \approx Pe_M^{1/2}$.

Proceeding as in Sect. 16.3, from a material balance on the coal particle, we obtain:

$$c_C^S d\left(\tfrac{4}{3}\pi R^3\right)/dt = -J_{Br}(4\pi R^2),$$

where c_C^S is the carbon concentration within the solid particle, while J_{Br} is the flux of carbon dioxide relative to the speed dR/dt of the interface. Now, consider (a) for each mole of carbon dioxide (B) leaving the coal surface there is one mole of oxygen (A) arriving, so that $J_{Br} = -J_{Ar}$; (b) assuming that the QSS hypothesis is satisfied (subjected to a posteriori verification), the interface velocity is negligible compared to the diffusion speed, and therefore relative fluxes can be replaced by absolute fluxes. So we obtain Eq. (16.3.3) as follows:

$$c_C^S \frac{dR}{dt} = J_A(R) = -k_M c_A^\infty. \tag{17.5.1}$$

Here, we have considered that the oxygen flux towards the particle (note that it is negative) is due to its diffusion across the mass boundary layer, induced by a concentration difference. In fact, the oxygen concentration is null at the particle surface, while it is c_A^∞ just outside the boundary layer. Accordingly, since in the dilute limit the problem is linear, $J_A(R) = -k_M c_A^\infty$, where the mass transport coefficient k_M is proportional to the Sherwood number, $k_M = (D/R)Sh$. So we obtain:

$$k_M = \frac{D}{R} Sh = a\frac{D}{R}\left(\frac{VR}{D}\right)^{1/2} = a\frac{D^{1/2}V^{1/2}}{R^{1/2}}, \tag{17.5.2}$$

where a is an $O(1)$ dimensionless constant. Substituting Eq. (17.4.3) into (17.4.2) we have:

$$R^{1/2}dR = a(DV)^{1/2}\frac{c_A^\infty}{c_C^S}dt,$$

and integrating between $t = 0$, when $R = R_0$, and $t = \tau$, when $R = 0$, we obtain:

$$\tau = \frac{2}{3a}\frac{R_0^{3/2}}{(DV)^{1/2}}\frac{c_C^S}{c_A^\infty}. \tag{17.5.3}$$

Now, let us determine when the QSS condition is satisfied, imposing that the combustion time, τ, is much larger than the characteristic relaxation time within the boundary layer (in fact, outside of it, the concentration is uniform and constant). Instead of using Eq. (17.5.3), τ can be evaluated through Eq. (17.5.1), i.e., $\tau \approx (R_0/k_M)(c_C^S/c_A^\infty)$, while the diffusion time is $\tau_{ss} \approx \delta_M^2/D \approx \delta_M/k_M$, where we have used the equality $\delta_M = R/Sh$ and $Sh = k_M R/D$. So we may conclude that the QSS approximation is valid when the following condition is satisfied:

$$\tau \approx \frac{R_0}{k_M}\frac{c_C^S}{c_A^\infty} \gg \tau_{ss} \approx \frac{\delta_M}{k_M} \Rightarrow \frac{1}{Sh}\frac{c_A^\infty}{c_C^S} \ll 1. \tag{17.5.4}$$

Clearly, this condition is generally satisfied, as $Sh \gg 1$ and $c_C^S \gg c_A^\infty$.

17.6 Problems

17.1 Repeat the problem of Sect. 17.1, assuming that $c_0 > 0$ and $c_L = 0$.

17.2 Consider a fluid flowing in laminar regime in a circular pipe of radius R, containing a solute with very small molar fraction $x_{A\infty}$. In the region $z \geq 0$, where z is the axial coordinate, the tube is coated with a catalyst that adsorbs the solute irreversibly, so that $x_A = 0$ at the wall. Calculate the total flux (*mole/s*) of solute adsorbed by the catalyst in the region $0 \leq z \leq L$, as a function of the mass flow rate and for small values of L (how small is small?).

17.3 A water drop of radius R falls under gravity in an atmosphere of dry air, with terminal velocity $V \propto \sqrt{gR}$, where g is the acceleration of gravity. Modeling the problem with all the necessary hypotheses, evaluate the time that is needed for a water drop of initial radius R_0 to evaporate completely.

17.4 A long and thin cylindrical candy bar made of sugar (A), of initial radius R_0 and length $L \gg R_0$ is slowly dissolving in a mildly agitated water basin. Assuming that $Re < 1$ and $Sc \gg 1$, so that $Pe \gg 1$, evaluate the time that is needed for the candy to melt completely.

17.5   A spherical nucleus of radius R composed of a B species grows progressively in a binary $A + B$ alloy. The mixture is stagnant and diluted in A, so that A and B diffuse into one another purely by diffusion. At $r \gg R$ the concentration of A is constant and equal to c_A^∞, while at the nucleus surface $r = R$ the concentration depends on the radius R, in agreement with Kelvin's equation, $c_A^R = c_A^\infty + c_B^S \delta / R$, where c_B^S is the concentration of B inside the nucleus, while δ is a typical intermolecular distance. Applying the QSS approximation and considering that $c_A + c_B = \rho$ constant, show that the radius of the nucleus grows in time like $t^{1/3}$. Hint: Assume that initially $R(t = 0) = R_0 \cong 0$.

Chapter 18
Transport Phenomena in Turbulent Flow

Abstract In this chapter we intend to provide a brief introduction to the study of transport phenomena in the presence of turbulent flows. This is a very large and quite specialized topic, and here we will confine our treatment to some basic features, above all in the context of pipe flow, referring the interested readers to more specialized texts (See, for example, *Turbulent Flows* by S.B. Pope, Cambridge University Press.). Most of the discussion on convective transport phenomena in the preceding chapters assumes that the flow is laminar, while the case of turbulent flows has been treated so far using empirical correlations, that are often subjected to stringent geometric constraints. The reason for concentrating on laminar flows is quite obvious: transport phenomena involving laminar flows can be treated using appropriate simplifying hypothesis, such as the fact that the process is stationary or uni-directional, so that the problem can be treated systematically, often even finding an analytical solution. No such simplifying features can be assumed to apply to turbulent flows, which are always irregular, transient and three-dimensional, and therefore very complex to treat. These difficulties notwithstanding, turbulent flows pop up everywhere in engineering and the natural world, regardless of whether we are optimizing a heat exchanger or watching a cloud moving in the sky. The importance of this subject has stimulated the interest of many investigators, leading to an impressive amount of experimental and theoretical studies. Unfortunately, while the experiments have produced many useful empirical relationships (albeit often with narrow applicability conditions), a complete theory of turbulence is still lacking, although its fundamental characteristics are well understood. After a brief introduction (Sect. 18.1), in Sect. 18.2 we describe the Kolmogorov scaling arguments, thus determining the characteristic time- and length-scales of turbulence. Then, in Sect. 18.3, we present the Reynolds decomposition and the associated governing equations, involving the turbulent fluxes of mass, momentum and energy. Next, Sect. 18.4, a closure model for the diffusive turbulent fluxes is described, based on the Reynolds analogy and Prandtl's mixing length model. This approach is applied in Sect. 18.5 to determine the logarithmic velocity profile near a wall, and then to describe turbulent pipe flow. Finally, in Sect. 18.6, more complex models are briefly sketched.

© Springer International Publishing Switzerland 2015   299
R. Mauri, *Transport Phenomena in Multiphase Flows*,
Fluid Mechanics and Its Applications 112, DOI 10.1007/978-3-319-15793-1_18

18.1 Fundamental Characteristics of Turbulence

As we have seen in Sect. 3.5, turbulent flows are always time-dependent and three-dimensional. Notwithstanding, the local velocity at any instant can be expressed as the sum of an average velocity, $\langle \mathbf{v} \rangle$, and a fluctuation, $\tilde{\mathbf{v}}$, i.e.,

$$\mathbf{v}(\mathbf{r}, t) = \langle \mathbf{v} \rangle(\mathbf{r}, t) + \tilde{\mathbf{v}}(\mathbf{r}, t). \tag{18.1.1}$$

The mean value of velocity (and of temperature and concentration as well) at a location $\mathbf{r}$ and time t may have two different meanings. First, it may be the result of an ensemble average: if we repeat the same identical experiment many times using the same nominal conditions (and there is no dependence between different repetitions), the ensemble average velocity is the arithmetic mean of the fluid velocities $\mathbf{v}_i(\mathbf{r}, t)$ measured on the i-th experimental realization at the same position $\mathbf{r}$ and time t, i.e., $\langle \mathbf{v} \rangle(\mathbf{r}, t) = N^{-1} \sum_1^N \mathbf{v}_i(\mathbf{r}, t)$, when $N \to \infty$.

The other meaning of average velocity is that of a time, or space, average; in the following, we will consider only the case of time averaging, although space averaging is totally analogous. Time averages are meaningful only when the characteristic time of the velocity fluctuations is much smaller than the time over which the macroscopic velocity changes significantly. In other words, we must assume that there is a clear separation of scales between the time of variation of the mean velocity, τ_{macr}, and that of the fluctuations, $\tau_\delta \ll \tau_{macr}$. Thus, on an intermediate timescale, τ, with $\tau_\delta \ll \tau \ll \tau_{macr}$, the instantaneous velocity will oscillate innumerable times around a mean velocity whose changes occur on a much larger timescale, so that we can define,

$$\langle \mathbf{v} \rangle(\mathbf{r}, t) = \frac{1}{2\tau} \int_{t-\tau}^{t+\tau} \mathbf{v}(\mathbf{r}, t')dt', \tag{18.1.2}$$

and,

$$\int_{t-\tau}^{t+\tau} \tilde{\mathbf{v}}(\mathbf{r}, t')dt' = \mathbf{0}. \tag{18.1.3}$$

Naturally, when a temporal mean velocity can be defined through Eq. (18.1.2), its value coincides with the ensemble average and the corresponding random process is called *ergodic*.

In particular, when the velocity field is statistically stationary and uni-directional, as it is the case with pipe flow along the z-direction, the mean velocity is:

$$\langle \mathbf{v} \rangle(\mathbf{r}, t) = \langle v \rangle(\mathbf{r}) \mathbf{1}_z, \tag{18.1.4}$$

while the fluctuating velocity $\tilde{v}$ is transient and three-dimensional.

Turbulence intensity is defined in terms of the covariance matrix $\langle \tilde{\mathbf{v}}\tilde{\mathbf{v}} \rangle$. In general, turbulence is assumed to be locally isotropic,[1] and therefore the velocity covariance matrix reduces to a diagonal tensor, $\frac{1}{3}u^2\mathbf{I}$, where u is the root mean square of the velocity fluctuations, $u = \sqrt{\langle \tilde{\mathbf{v}} \cdot \tilde{\mathbf{v}} \rangle}$. As a reference value of the velocity fluctuations, consider that if we are not too close to the walls[2] we have:

$$u/V \approx 0.1, \qquad (18.1.5)$$

where V is a typical value of the mean velocity, that is the ratio between volumetric flow rate and tube cross section. In Sect. 18.4, we will see that in Eq. (18.1.5) the constant 0.1 should be replaced with $\sqrt{f}$, where f is the friction factor.

The most important effect of the velocity fluctuations is a strong increase in cross-stream mixing which, in the laminar regime, is due exclusively to molecular diffusion. The enhanced transversal mixing is due to the velocity fluctuations, resulting in a zero-mean cross-stream convective transport of momentum, heat and mass. Note that, in the turbulent motion of a fluid flowing in a pipe, enhanced transports of heat and mass from the fluid to the walls are generally desirable, while an analogous enhanced transfer of momentum determines an increase in the pressure drop, which is not desirable.

An obvious consequence of the large increase in cross-stream transport occurring in turbulent flows is that the entrance length[3] is greatly reduced. In fact, a typical entrance length for turbulent flow at $Re = 10^5$ are $L/R \approx 50$, compared to $L/R \approx 0.1Re$ in the laminar regime. Also, since pressure drops are larger in turbulent flows, shear stresses and velocity gradients at the walls are also larger than in laminar flow. Therefore, for a given mean velocity, it follows that the velocity profile far from the wall must be flatter, and it will approach an ideal plug flow as Re is increased.

Let us consider a few numerical examples, to determine the magnitude of the turbulent exchanges of momentum and heat. Concerning the momentum transport, we know that it coincides with the shear stress, and therefore, in the case of flow in a circular conduit, the momentum transfer at the wall can be written in terms of the pressure drop as $J_{Qw} = \tau_w = (R/2)(\Delta P/L)$, where R and L are the tube radius and length, respectively [see Eq. (4.4.2)]. Therefore, defining the Fanning friction factor (4.5.4), we have:

$$f = \frac{\tau_w}{\rho V^2/2} = \frac{\Delta p}{\rho V^2}\frac{R}{L}, \qquad (18.1.6)$$

---

[1]This, so called, principle of local isotropy is the basic assumption of the most commonly used turbulence model.

[2]Very near the walls the velocity field is laminar and therefore there is no turbulence.

[3]The entrance length is the distance from the tube entrance that is required for the velocity or temperature profile to reach their respective fully developed profile.

where ρ is the fluid density, and V its mean velocity. In the case of laminar flow, the friction factor can be determined exactly, as $f = 16/Re$, where Re is the Reynolds number, while in turbulent flow many semi-empirical correlations are used. Among them, a convenient expression is the Blasius correlation, $f = 0.0791Re^{-1/4}$, that is valid when $3 \times 10^3 < Re < 10^5$. So we see that in a turbulent flow regime with $Re = 10^5$, the Blasius correlation predicts $f = 4.45 \times 10^{-3}$, while if laminar flow could be maintained at such a large Re, we would obtain $f = 1.6 \times 10^{-4}$, that is a value 28 times smaller. So, we may conclude that turbulent flow requires a pressure drop much larger than that for laminar flow at the same conditions, and therefore, compared to the laminar case, there is an additional power that has to be dissipated.

The same analysis can be applied to heat (or mass) transport. This is described in terms of the heat transfer coefficient h, defined as the ratio between the heat flux at the wall and the temperature difference. Again, using the expressions for h described in Sect. 11.2, we see that, at the same conditions, the heat flux in the turbulent regime is much larger than that in laminar flow.

18.2 Time- and Length-Scales in Turbulence

As mentioned in Chap. 3, macroscopically, we see that in the laminar to turbulent transition the flow becomes unstable and eddies start to form, with a large range of sizes and lifetimes. As we have explained qualitative in Sect. 3.5, it looks as if the fluid finds itself with an energy surplus that cannot be dissipated in laminar regime, and therefore it starts to turn around randomly (as it must be a zero average motion) at the same macroscopic scale that characterizes the laminar flow. So, the additional energy input initially supplies the kinetic energy of the larger, macro-scale eddies where, however, mechanical energy is conserved. This is due to the fact that the Reynolds number based on the size and velocity of these eddies is very large, so that viscous effects are negligible and therefore energy dissipation (i.e., the conversion of mechanical energy into heat) is null. As a result, energy is transferred from the larger to smaller eddies, due to the non-linearity of the momentum convective flux. This can be seen assuming that the velocity field is spatially periodic, i.e., $v(x,t) = V_0(t) \sin(k_0 x)$, with $k = 2\pi/\lambda_0$, where λ_0 is the size of the eddy, that initially coincides with a macroscopic length, such as the tube radius. Then, considering that the viscous term in the Navier-Stokes equation can be neglected, we find:

$$\partial v/\partial t = -v(dv/dx) = -\frac{1}{2}V_0^2(t) \sin(2k_0 x),$$

revealing that new eddies will be formed, with size $\lambda_1 = \lambda_0/2$. In turn, these second-generation eddies will induce the formation of even smaller eddies with size $\lambda_2 = \lambda_1/2$, and so on, in a kind of energy cascade. At the end, this mechanism generates eddies that are sufficiently small that the corresponding Reynolds number is roughly unity: at that micro-scale of motion, the mechanical energy is finally dissipated.

In the energy cascade that has been described above, the same energy that is pumped into the kinetic energy of the macroscopic eddies is finally dissipated at the micro-scale. This surplus power (energy per unit time) per unit mass is customary denoted as ε, expressed in $m^2 s^{-3}$.

The basic assumption concerning the macro-scale is that the large eddies are long-lived (this has been also verified experimentally), so that the energy surplus ε is transferred to the smaller eddies within a time $\tau_L \approx L/u$, where L and u denote the macro-scale length and the associated turbulent velocity fluctuation [see Eq. (18.1.5)], respectively. Therefore, ε can be estimated as,

$$\varepsilon \approx \frac{u^2}{\tau_L} = \frac{u^3}{L}. \tag{18.2.1}$$

As for eddy sizes much smaller than L, the key assumption, known as the first Kolmogorov's 1941 assumption, is that turbulent statistics depend only on the mean dissipated power, ε (which is the same as for the large eddies), and the kinematic viscosity, ν, while it does not depend on L, as the dissipation process takes place on very small distances, where the macro-scale is not directly perceived. Using dimensional analysis, i.e., imposing that the typical size, δ, of the small eddies scale as $\delta = \nu^\alpha \varepsilon^\beta$, we find $\alpha = 3/4$ and $\beta = -1/4$.[4] Applying the same technique, the characteristic micro-scale velocity, u_δ, and time, τ_δ, can be determined, finding at the end:

$$\delta = \left(\frac{\nu^3}{\varepsilon}\right)^{1/4}; \quad u_\delta = (\nu\varepsilon)^{1/4}; \quad \tau_\delta = \left(\frac{\nu}{\varepsilon}\right)^{1/2}. \tag{18.2.2}$$

These are the turbulent micro-scales, that are often called the *Kolmogorov scale*, that characterize the smallest eddies. Note that, as expected, $\tau_\delta \approx \delta/u_\delta$, and, in addition, the micro-scale Reynolds number is unity,

$$Re_\delta = u_\delta \delta/\nu = 1. \tag{18.2.3}$$

The micro-scale and macro-scale quantities are related as:

$$\frac{\delta}{L} \approx Re_L^{-3/4}; \quad \frac{u_\delta}{u} \approx Re_L^{-1/4}; \quad \frac{\tau_\delta}{\tau_L} \approx Re_L^{-1/2}, \tag{18.2.4}$$

---

[4]This result can be obtained by substituting the units of ν (m^2/s) and ε (m^2/s^3), and solving the algebraic equation in the unknowns α and β that result from equating the exponents of m (meter) and of s (second). Note that we would obtain the same result had we assumed that δ depends separately on the dynamic viscosity, μ, and on the density, ρ.

where

$$Re_L = uL/v. \tag{18.2.5}$$

is the Reynolds number based on the turbulence macro-scale.

These relations are very important, as they reveal that, as Re_L increases, micro- and macro-scales diverge more and more, so that turbulence develops a progressively finer microstructure that is difficult both to determine experimentally and to simulate numerically.

Consider, for example, a pipe flow with $Re = 10^5$, where Re is the "usual" Reynolds number, evaluated in terms of the mean velocity, V and the tube diameter, D. Since the larger eddies have a size $L \approx D$ and the velocity fluctuation is $u \approx 0.1\ V$ [see (18.1.5)], we see from Eq. (18.2.5) that $Re_L = 0.1Re = 10^4$. Therefore, applying Eq. (18.2.4), we obtain: $\delta \approx 10^{-3}D$ and then, when $D = 10$ cm, we see that $\delta \approx 100\,\mu$m. By the same token, applying Eq. (18.2.4) we obtain the turbulence temporal micro-scale, $\tau_\delta \approx 10^{-1}D/V$. Therefore, when the fluid is water, $V \approx 1$ m/s and $\tau_L = 1s$, so that $\tau_\delta \approx 10$ ms, while when the fluid is air, at the same Re, then $V \approx 10$ m/s and $\tau_L = 0.1$ s, so that $\tau_\delta \approx 1$ ms. This is the duration of the most rapid fluctuations, which must be smaller than the characteristic time associated with the variation of the mean flow, so that the definition (18.1.2) of the mean value as a temporal average is valid.[5]

Clearly, 100 μm is much larger than the molecular mean free path which, in air, is about 0.1 μm, and the same considerations hold also for the timescales. Accordingly, turbulent flow at the scale of the small eddies can be described using all the tools of continuum mechanics, such as solving the Navier-Stokes equations and applying the no slip boundary conditions. This task, however, is far from being easy. For example, if we want to simulate the turbulent flow at the conditions described above, i.e. water flowing in a pipe at $Re = 10^5$, solving the unsteady Navier-Stokes equation (this is the so-called DNS, i.e., *Direct Numerical Simulation*) would need a spatial grid with at least 10^9 points and at least 10^3 time steps for each second of simulation, which at the moment is quite impossible to do.

18.3 Reynolds-Averaged Equations

18.3.1 Mean Quantities

As we have seen in Sect. 18.1, in turbulent flows all variables can be considered as the sum of an average quantity and a fluctuating one, where by mean value we mean a time average evaluated over a time interval τ that is much larger than the

---

[5]This condition is always satisfied when turbulence is statistically stationary, which is the case that is considered here. However, in the presence of shock waves or other discontinuities, this condition is not satisfied any more, and therefore the definition (18.1.2) ceases to be valid.

micro-scale time τ_δ given in Eq. (18.2.4), and yet much shorter than the macro-scopic time describing the variation of the mean flow, which in our case is infinitely long, as the process is assumed to be statistically stationary. Therefore, for velocity, pressure, temperature and concentration, we have:

$$\mathbf{V} = \langle \mathbf{V} \rangle + \tilde{\mathbf{V}}; \quad p = \langle p \rangle + \tilde{p}; \quad T = \langle T \rangle + \tilde{T}; \quad c = \langle c \rangle + \tilde{c}, \tag{18.3.1}$$

where the brackets indicate the mean values, defined in Eq. (18.3.2) as

$$\langle f \rangle (\mathbf{r},t) = \frac{1}{2\tau} \int\limits_{t-\tau}^{t+\tau} f(\mathbf{r},t')dt', \tag{18.3.2}$$

so that

$$\left\langle \tilde{\mathbf{V}} \right\rangle = 0; \quad \langle \tilde{p} \rangle = 0; \quad \langle \tilde{T} \rangle = 0; \quad \langle \tilde{c} \rangle = 0. \tag{18.3.3}$$

For future reference, consider the following properties:

(a) Double averages equal simple averages, i.e.,

$$\langle \langle f \rangle \rangle = \langle f \rangle. \tag{18.3.4}$$

In fact, $\langle f \rangle = \langle \langle f \rangle + \tilde{f} \rangle = \langle \langle f \rangle \rangle + \langle \tilde{f} \rangle = \langle \langle f \rangle \rangle$.

(b) The means of spatial and temporal derivatives equal the derivatives of the mean, i.e.,

$$\langle \nabla f \rangle = \nabla \langle f \rangle \quad \text{and} \quad \left\langle \frac{df}{dt} \right\rangle = \frac{d}{dt} \langle f \rangle. \tag{18.3.5}$$

In fact, $\langle \nabla f \rangle (\mathbf{r}, t) = \frac{1}{2\tau} \int_{t-\tau}^{t+\tau} \nabla f(\mathbf{r}, t')dt' = \frac{1}{2\tau} \nabla \int_{t-\tau}^{t+\tau} f(\mathbf{r}, t')dt' = \nabla \langle f \rangle$, and the same applies for time derivatives as well.[6]

18.3.2 Conservation of Mass

Starting from the continuity equation for incompressible flow, Eq. (6.2.13),

$$\nabla \cdot \mathbf{v} = 0, \tag{18.3.6}$$

---

[6]We assume that the characteristic time of the variation of $\langle f \rangle$ is much longer than τ_δ.

we take a time average, and apply Eqs. (18.3.1) and (18.3.5), so that:

$$\nabla \cdot \langle \mathbf{v} \rangle = 0. \tag{18.3.7}$$

Subtracting Eq. (18.3.7) from (18.3.6), we see that the fluctuating velocity is solenoidal, as well,

$$\nabla \cdot \tilde{\mathbf{v}} = 0. \tag{18.3.8}$$

18.3.3 Conservation of Momentum

Consider the Cauchy Eq. (6.3.6) for incompressible flow,

$$\rho \frac{\partial \mathbf{v}}{\partial t} + \nabla \cdot \mathbf{J}_Q = \mathbf{F}, \tag{18.3.9}$$

where ρ is the fluid density, $\mathbf{F}$ a body force and $\mathbf{J}_Q$ is the momentum flux,

$$\mathbf{J}_Q = P\mathbf{I} + \rho \mathbf{vv} - \mu(\nabla \mathbf{v} + \nabla \mathbf{v}^+), \tag{18.3.10}$$

where μ is the viscosity and P is the modified pressure (6.7.4), $P = p + \rho gz$. Now we take an average of Eqs. (18.3.9) and (18.3.10), considering that $\langle \mathbf{vv} \rangle = \langle \mathbf{v} \rangle \langle \mathbf{v} \rangle + \langle \tilde{\mathbf{v}}\tilde{\mathbf{v}} \rangle$ and that $\langle \mathbf{v} \rangle$ is solenoidal, obtaining:

$$\rho \frac{\partial \langle \mathbf{v} \rangle}{\partial t} + \nabla \cdot \left(\langle \mathbf{J}_Q \rangle + \mathbf{J}_Q^* \right) = \langle \mathbf{F} \rangle. \tag{18.3.11}$$

Here,

$$\langle \mathbf{J}_Q \rangle = \langle P \rangle \mathbf{I} + \rho \langle \mathbf{v} \rangle \langle \mathbf{v} \rangle - \mu\left(\nabla \langle \mathbf{v} \rangle + \nabla \langle \mathbf{v} \rangle^+ \right) \tag{18.3.12}$$

is the mean momentum flux, while

$$\mathbf{J}_Q^* = \rho \langle \tilde{\mathbf{v}}\tilde{\mathbf{v}} \rangle \tag{18.3.13}$$

is the *turbulent momentum flux*, that is the opposite of the *turbulent* stress, or *Reynolds* stress. Equation (18.3.11) shows that the net effect of the turbulence consists in increasing the shear stresses, by adding to the mean stress an additional term, depending on the velocity fluctuations. It should be stressed that both the viscous and the Reynolds stresses describe the cross-stream transport of momentum as the effect of chaotic fluctuations of the fluid particle velocities. However, the physical natures of the two shear stresses are very different from each other: the mean stress describes a viscous momentum transport, that is *diffusive* in nature, while the turbulent stress is fundamentally *convective*. In fact, in diffusive transport, fluctuations are due to the thermal agitation, i.e. the Brownian motion that we have studied in Sect. 1.8, with typical molecular scales of $O(10^{-3}\,\mu m)$; on the other hand, turbulent fluctuations take place over the Kolmogorov scale, which is $O(100\,\mu m)$, and larger.

Substituting Eqs. (18.3.12) and (18.3.13) into (18.3.11) and also considering Eqs. (18.3.7), (18.3.8) and (18.3.10), we conclude that the Reynolds-averaged Navier-Stokes equation becomes,

$$\rho\left[\frac{\partial\langle\mathbf{v}\rangle}{\partial t} + \langle\mathbf{v}\rangle\cdot\nabla\langle\mathbf{v}\rangle\right] = -\langle P\rangle\mathbf{I} + \mu\nabla^2\langle\mathbf{v}\rangle - \rho\nabla\cdot\langle\tilde{\mathbf{v}}\tilde{\mathbf{v}}\rangle. \tag{18.3.14}$$

Consider also that $\nabla\cdot\langle\tilde{\mathbf{v}}\tilde{\mathbf{v}}\rangle = \langle\tilde{\mathbf{v}}\cdot\nabla\tilde{\mathbf{v}}\rangle$.

18.3.4 Conservation of Energy and of Chemical Species

Consider the energy balance equation, describing the transport of heat in an incompressible flow,

$$\rho c_p \frac{\partial T}{\partial t} + \nabla\cdot\mathbf{J}_U = \dot{q}. \tag{18.3.15}$$

Here, $\dot{q}$ is the source term and c_p is the specific heat, while,

$$\mathbf{J}_U = \rho c_p \mathbf{v} T - k\nabla T \tag{18.3.16}$$

is the heat flux, with k denoting the thermal conductivity. Averaging Eqs. (18.3.15) and (18.3.16), considering that $\langle\mathbf{v}T\rangle = \langle\mathbf{v}\rangle\langle T\rangle + \langle\tilde{\mathbf{v}}\tilde{T}\rangle$ and that $\langle\mathbf{v}\rangle$ is divergence-free, we obtain:

$$\rho c_p \frac{\partial\langle T\rangle}{\partial t} + \nabla\cdot\left(\langle\mathbf{J}_U\rangle + \mathbf{J}_U^*\right) = \langle\dot{q}\rangle. \tag{18.3.17}$$

Here,

$$\langle\mathbf{J}_U\rangle = \rho c_p\langle\mathbf{v}\rangle\langle T\rangle - k\nabla\langle T\rangle \tag{18.3.18}$$

is the mean heat flux, while

$$\mathbf{J}_U^* = \rho c_p\langle\tilde{\mathbf{v}}\tilde{T}\rangle \tag{18.3.19}$$

is the *turbulent heat flux*. Proceeding as in the case of the turbulent momentum flux. Equations (18.3.17)–(18.3.19) can be rewritten as,

$$\rho c_p \left[\frac{\partial \langle T \rangle}{\partial t} + \langle \mathbf{v} \rangle \cdot \nabla \langle T \rangle \right] = k\nabla^2 \langle T \rangle - \rho c_p \nabla \cdot \langle \tilde{\mathbf{v}} \tilde{T} \rangle + \langle \dot{q} \rangle, \tag{18.3.20}$$

where $\nabla \cdot \langle \tilde{\mathbf{v}} \tilde{T} \rangle = \langle \tilde{\mathbf{v}} \cdot \nabla \tilde{T} \rangle$.

The transport of a dilute chemical species A can be described in the same way, denoting by c_A the material (mole- or mass-based) concentration and by R_A the source term. Finally, the following mean equation is obtained:

$$\frac{\partial \langle c_A \rangle}{\partial t} + \nabla \cdot (\langle \mathbf{J}_A \rangle + \mathbf{J}_A^*) = \langle R_A \rangle. \tag{18.3.21}$$

Here,

$$\langle \mathbf{J}_A \rangle = \langle \mathbf{v} \rangle \langle c_A \rangle - D\nabla \langle c_A \rangle \tag{18.3.22}$$

is the mean material flux of A, while

$$\mathbf{J}_A^* = \langle \tilde{\mathbf{v}} \tilde{c}_A \rangle \tag{18.3.23}$$

is the *turbulent material flux*. This equation can be rewritten as,

$$\frac{\partial \langle c_A \rangle}{\partial t} + \langle \mathbf{v} \rangle \cdot \nabla \langle c_A \rangle = D\nabla^2 \langle c_A \rangle - \nabla \cdot \langle \tilde{\mathbf{v}} \tilde{c}_A \rangle + \langle R_A \rangle. \tag{18.3.24}$$

18.3.5 Turbulent Fluxes

The quantities $\mathbf{J}_Q^*$, $\mathbf{J}_U^*$ and $\mathbf{J}_A^*$ in Eqs. (18.3.13), (18.3.19) and (18.3.23) are called *turbulent fluxes*. They depend on the cross correlation between velocity, temperature and concentration fluctuations, and indicate the contribution to the fluxes of momentum, heat and A-species concentration, due to the turbulence. For example, when the velocity field is statistically stationary and uni-directional along a longitudinal z-direction, while the mean velocity is a function of a transverse, y-direction, as is the case with pipe flow, the turbulent fluxes are:

$$J_{Q,yz}^* = \rho \langle \tilde{v}_y \tilde{v}_z \rangle, \tag{18.3.25}$$

$$J_{U,y}^* = \rho c_p \langle \tilde{v}_y \tilde{T} \rangle, \tag{18.3.26}$$

$$J_{A,y}^* = \langle \tilde{v}_y \tilde{c}_A \rangle. \tag{18.3.27}$$

Since the turbulent fluxes are the average values of the products between two zero-mean fluctuating field variables, they will be non-zero only when these fluctuating quantities are correlated with each other. For example, the turbulent heat flux (18.3.26) will be non-zero when the fluctuating transversal velocity field, $\tilde{v}_y$, is correlated with the temperature fluctuation, $\tilde{T}$, despite the fact that the average of both fluctuating quantities is zero. In this case, the ability of the fluctuations to yield a net flux becomes evident if we assume that both $\tilde{v}_y$ and $\tilde{T}$ oscillate with the same frequency, with $\tilde{v}_y = v_0 \sin(\omega t)$ and $\tilde{T} = T_0 \sin(\omega t + \alpha)$, finding $\langle \tilde{v}_y \tilde{T} \rangle = \frac{1}{2} v_0 T_0 \cos \alpha$. So, now the question that arises spontaneously is the following: why the temperature and velocity fluctuations should be correlated with each other? The answer, as we have repeatedly stressed, is that turbulence arises because of an instability in the velocity field, due to a strong non-linearity in the convective momentum flux. In fact, the convection-diffusion equations governing the evolutions of the temperature and concentration (in the dilute limit) fields are linear; so, temperature and concentration alone would never exhibit a turbulent behavior. Accordingly, temperature (and concentration) fluctuations are not independent quantities, as they are driven by velocity fluctuations; therefore, it is not surprising that $\tilde{v}_y$ and $\tilde{T}$ are strongly correlated.

As we have stressed in our comments on Eq. (18.3.13), turbulent fluxes are determined by the chaotic fluctuations of the fluid velocity around its mean value. In this respect, turbulent transport is similar to diffusion, the difference being that, while the latter takes place on a molecular scale, over $O(10^{-3} \ \mu m)$ distances, turbulent transport occurs over the Kolmogorov micro-scale, and larger. An important consequence of this observation is that turbulent fluxes vanish at solid surfaces, due to a no-slip boundary condition for the velocity that is usually imposed there. In addition, very near to any solid wall, the fluid flow is laminar, and therefore the usual constitutive equations can be applied, with no fluctuations. However, far from the surface, turbulent fluxes are typically several orders of magnitude larger than their diffusive counterparts, and therefore they have to be taken into account in order to determine the mean values, $\langle \mathbf{v} \rangle$, $\langle T \rangle$ and $\langle c_A \rangle$, using the Reynolds-averaged Eqs. (18.3.11), (18.3.17) and (18.3.21). Clearly, since the smoothed fields cannot be determined without some statistical knowledge of the velocity fluctuations, it is evident that we have more unknowns than equations. This shortage of governing equations is called the *closure problem*. Thus, the cost of averaging is the need to find additional relations, expressing the turbulent fluxes as a function of the average field variables, through appropriate turbulent closure relations. Among these approaches, the best-known is the *eddy diffusivity model*, that is described in the next section.

18.4 Turbulent Diffusion

18.4.1 Eddy Diffusivities

We have seen that turbulence increases the cross-stream transport, due to a chaotic motion of the fluid particles. In that respect, as previously noted, turbulent transport can be modeled as being similar to diffusion, apart from the very different scales over which the two mechanisms take place (i.e., molecular vs. Kolmogorov length scales). Therefore, it seems natural to model turbulent transport in much the same way as a diffusive process, following the same approach, based on kinetic theory, that we have seen in Sects. 1.5–1.7. In fact, as turbulence speeds up equilibration by enhancing cross-stream transport, we see that turbulent eddies tend to transport heat (as well as momentum and chemical species) from regions of high to low temperature (as well as velocity and concentration). Accordingly, we obtain the following turbulent constitutive equations:

$$\left(\mathbf{J}_Q^*\right)^d = \rho\langle\tilde{\mathbf{v}}\tilde{\mathbf{v}}\rangle^d = -\mu_t\left(\nabla\langle\mathbf{v}\rangle + \nabla\langle\mathbf{v}\rangle^+\right); \quad \mu_t = \rho\nu_t; \tag{18.4.1}$$

$$\mathbf{J}_U^* = \rho c_p\langle\tilde{\mathbf{v}}\tilde{T}\rangle = -k_t\nabla\langle T\rangle; \quad k_t = \rho c_p\alpha_t; \tag{18.4.2}$$

$$\mathbf{J}_A^* = \langle\tilde{\mathbf{v}}\tilde{c}_A\rangle = -D_t\nabla\langle c_A\rangle. \tag{18.4.3}$$

Here, μ_t, k_t, and D_t are the turbulent, or eddy, viscosity, thermal conductivity and diffusivity, respectively, while ν_t, and α_t are the eddy diffusivities for momentum and heat, having the same m²/s units as D_t. The d-superscript in Eq. (18.4.1) denotes the deviatoric part of a tensor, i.e., $\mathbf{T}^d = \mathbf{T} - \frac{1}{3}\mathbf{I}(\mathbf{I}:\mathbf{T})$. For example, when the velocity field is statistically stationary and uni-directional, the turbulent fluxes (18.3.25)–(18.3.27) become:

$$J_{Qyz}^* = \rho\langle\tilde{v}_y\tilde{v}_z\rangle = -\rho\nu_t\frac{\partial\bar{v}_z}{\partial y}, \tag{18.4.1a}$$

$$J_{Uy}^* = \rho c_p\langle\tilde{v}_y\tilde{T}\rangle = -\rho c_p\alpha_t\frac{\partial\bar{T}}{\partial y}, \tag{18.4.2a}$$

$$J_{iy}^* = \langle\tilde{v}_y\tilde{c}_i\rangle = -D_t\frac{\partial\bar{c}_i}{\partial y}, \tag{18.4.3a}$$

where the mean value has been denoted with an overbar (the overbar notation will be used interchangeably with bracket notation, unless otherwise noted).

Since the turbulence is due to the chaotic fluctuations of the fluid velocity, the three diffusivities have the same origin and therefore it is reasonable to assume that they are equal to each other, i.e.,

$$\nu_t = \alpha_t = D_t. \tag{18.4.4}$$

This equality is called the *Reynolds analogy*; it is very similar to the equality between molecular diffusivity, thermal diffusivity and kinematic viscosity for an ideal gas. In that case, too, the transport of momentum, energy and mass have the same nature, i.e., thermal fluctuations at the molecular scale, and so it is not surprising that the three corresponding diffusivities are equal to each other. Unlike molecular diffusion, though, turbulent diffusivity is not a physical property of the fluid, but it depends on the flow field. We can, however, use the analogy between the two mechanisms to estimate the eddy diffusivity as $u\ell$, where u is a characteristic eddy velocity and ℓ is an eddy length scale. These two quantities, and in particular the latter, must be modeled with care, considering that the eddy diffusivity strongly depends on position, as it must vanish at solid surfaces, while far from them it is much larger than its constant, molecular counterpart (i.e., the molecular mean free path). Despite the many attempts at finding a general expression for the eddy diffusivity, the results so far have not been completely satisfactory. This is reflected in the fact that commercial codes that simulate turbulent flows still leave much to be desired.

18.4.2 Dimensionless Wall Variables

In most applications, we are particularly interested in the flow behavior near a wall, because in this region momentum, heat and mass are exchanged, thus determining both the drag and heat fluxes. Based on dimensional considerations, we see that the typical velocity, V_w, and distance, δ_w, in the wall region depend on the shear stress at the wall, τ_w, on the density, ρ, and on the kinematic viscosity, ν, of the fluid, obtaining,

$$V_w = \sqrt{\tau_w/\rho} \quad \text{and} \quad \delta_w = \nu/V_w. \tag{18.4.5}$$

So, the Reynolds number based on V_w, and δ_w is equal to unity, that is $V_w \delta_w/\nu = 1$.

These velocity and length scales are not so obvious, and it is helpful to see what they mean in a familiar case, namely the fluid flow in a circular tube of radius R. Then, we know from Eq. (4.5.2) that the shear stress at the wall is proportional to the pressure drop, as $\tau_w = (R/2)(\Delta p/L)$. Thus, substituting into Eq. (18.4.5) the definition (4.5.4) of the friction factor, $\tau_w = \frac{1}{2} f \rho V^2$, where V is the mean velocity,[7] we obtain:

$$V_w/V = \sqrt{f/2}, \quad \text{and} \quad \frac{\delta_w}{R} = \frac{2}{Re\sqrt{f/2}}, \tag{18.4.6}$$

with $Re = 2VR/\nu$ denoting the usual Reynolds number, based on the tube diameter. Thus, turbulence intensity is proportional to $\sqrt{f}$, in agreement with the comments following Eq. (18.1.5). For example, when $Re = 10^5$, we find $f = 0.00450$, so that $V_w/V = 4.7 \times 10^{-2}$ and $\delta_w/R = 4.2 \times 10^{-4}$. Considering that $f \propto Re^{-1/4}$, we also see that $V_w/V \propto Re^{-1/8}$ and $\delta_w/R \propto Re^{-7/8}$.

---

[7]Here V is the mean volume flow rate per a unit cross sectional area, as distinct from the mean velocity at the centerline.

Based on the characteristic wall velocity and distance above, we finally define the dimensionless, so-called, *wall variables*,

$$v^+ = \frac{v}{V_w} = \frac{v}{\sqrt{\tau_w/\rho}}; \quad y^+ = \frac{y}{\delta_w} = \frac{\sqrt{\tau_w/\rho}}{v} y, \qquad (18.4.7)$$

where y is the distance from the wall. In the example that we have considered above, the full range of the wall coordinate is $0 < y^+ < y_R^+$, where $y_R^+ = R/\delta_w \cong 2,380$ corresponds to the tube centerline.

18.4.3 Mixing Length Model

Among all the models for evaluating the eddy diffusivity, the most influential is the mixing length model, introduced in 1925 by Prandtl, who proposed the following closure equation:

$$v_t = \ell^2 \left| \frac{\partial \bar{v}_z}{\partial y} \right|, \qquad (18.4.8)$$

where ℓ is the so-called *mixing length*. This model is based on the presumption that the eddy viscosity is expected to grow with the square of the distance, as $v_t \approx vy \approx (dv/dy)y^2$. Rewriting Eq. (18.4.8) in terms of the wall variables, we obtain:

$$\frac{v_t}{v} = \ell^{+2} \left| \frac{\partial \bar{v}_z^+}{\partial y^+} \right|, \qquad (18.4.9)$$

where $\ell^+ = \ell/\delta_t$. Since ℓ must vanish at the wall and grow with y, Prandtl assumed a simple linear relation, that is:

$$\ell = \kappa y \quad \Rightarrow \quad \ell^+ = \kappa y^+. \qquad (18.4.10)$$

Here, κ is a dimensionless constant, to be determined experimentally; comparisons of the predictions based on this model with experimental measurements yields a best-fit value of $\kappa = 0.40$.

However, although Prandtl's mixing length linear model works remarkably well near a wall, it breaks down in the so-called *viscous sublayer*, corresponding to the region when $y^+ < 5$. In fact, in that region, the flow is laminar and therefore $v_t = 0$, that is $\ell = 0$. Instead of using a piecewise continuous expression for the mixing length, though, it is preferable to assume that ℓ is a continuous function of y, decreasing more rapidly than linearly very near the wall. A useful expression of this type is that of van Driest,

$$\ell = \kappa y (1 - e^{-y/A\delta_t}) \quad \Rightarrow \quad \ell^+ = \kappa y^+ (1 - e^{-y^+/A}), \tag{18.4.11}$$

where A is a dimensionless constant to be determined experimentally.[8] Obviously, this, as any other correlation, does not apply very far from the wall. Experimentally, we see that in the case of a pipe flow, Prandtl's linear relation ceases to be valid when $y/R \geq 0.1$. However, it should be stressed that this corresponds to a large value of the wall variable: in fact, we saw that when $Re = 10^5$, at $y = 0.1R$ we have $y^+ \cong 240$. In addition, we should also consider that at that distance from the wall the mean velocity profile is flat, so that the flow behavior is quite predictable.

18.5 Logarithmic Velocity Profile

Consider the turbulent flow of a Newtonian fluid in a circular pipe. Proceeding as in Sect. 5.1, we can write the simple force balance (5.1.2), where the shear stress is the sum of a viscous component and its turbulent counterpart, obtaining:

$$\tau = \tau_w \frac{r}{R} = \mu \left(1 + \frac{v_t}{v} \right) \frac{d\bar{v}_z}{dr}. \tag{18.5.1}$$

Defining the wall variable (18.4.7), with $y = R - r$ indicating the distance from the wall, and considering the near wall region, with $y/R \ll 1$, we obtain:

$$\frac{dv_z^+}{dy^+} = \frac{1}{1 + (v_t/v)} = \frac{1}{1 + \ell^{+2}(dv_z^+/dy^+)}. \tag{18.5.2}$$

Now this equation is solved asymptotically in the two limit cases, when $y^+ \to 0$ and $y^+ \to \infty$. In the first case, $v_t = 0$ and therefore we easily find $dv^+/dy^+ = 1$, to be solved with the no slip boundary condition $v^+(y^+ = 0) = 0$, obtaining:

$$v_z^+ = y^+, \quad \text{as} \quad y^+ \to 0. \tag{18.5.3}$$

In the second case, when $y^+ \to \infty$, (a) we can apply Prandtl's linear relation, and (b) $v_t \gg v$, as the turbulent stresses are much larger than the viscous stresses. Therefore we have:

$$\frac{dv_z^+}{dy^+} = \frac{1}{v_t/v} = \frac{1}{\ell^{+2}(dv_z^+/dy^+)} \quad \Rightarrow \quad \frac{dv_z^+}{dy^+} = \frac{1}{\ell^+} = \frac{1}{\kappa y^+} \quad \Rightarrow \quad v_z^+ = \frac{1}{\kappa} \ln y^+ + c,$$

---

[8]For a turbulent flow past a flat plate it is found that $A = 25$.

Fig. 18.1 Velocity
distribution for turbulent
flow near a wall

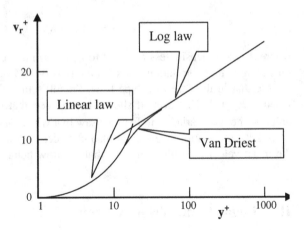

where c is a constant to be determined experimentally. This is the *logarithmic velocity profile*, also denoted as the *law of the wall*, stating that the average velocity of a turbulent flow at a certain point is proportional to the logarithm of the distance of that point from the wall. This law was first published by Theodore von Kármán, in 1930. The values of the integration constant κ and c were determined by Nikuradse, who performed a series of celebrated experiments that led to Moody's famous chart, finding $\kappa = 0.4$ and $c = 5.5$. Today, it is preferred to use the values $\kappa = 0.41$ and $c = 5.2$, obtaining:

$$v_z^+ = 2.44 \ln y^+ + 5.2 \quad \text{as} \quad y^+ \to \infty. \tag{18.5.4}$$

Using the van Driest (18.4.11) correlation for the mixing length and solving numerically Eq. (18.5.2), we find (see Fig. 18.1) that the linear solution (18.5.3) is valid when $y^+ < 5$: this is the so-called *viscous sublayer*, where the flow is laminar. On the other hand, the logarithmic profile (18.5.4) is applicable when $y^+ > 30$: this is the so-called *log-law region*. Finally, in the so-called *buffer layer*, for intermediate values, $5 < y^+ < 30$, we obtain a curve smoothly connecting the previous two.

The most important feature of the velocity dependence represented in Fig. 18.1 is its universality, that is the fact that the velocity profile, expressed in dimensionless wall variables, does not depend on the Reynolds number, nor on any other dimensionless parameter of the flow.[9]

Note that the previous analysis could not be extended to calculate the velocity profile in the entire tube cross section, because Prandtl's linear model $\ell = \kappa y$ is not valid far from the walls, that is when $y/R > 0.1$. However, it turns out that Eq. (18.5.4) is in excellent agreement with the complete velocity profile that has

---

[9]Naturally, once we go back to the dimensional variables v and y, the resulting velocity profile strongly depends on Re.

Fig. 18.2 Turbulent velocity
profile in a pipe

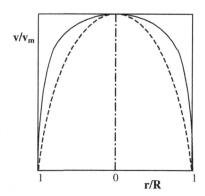

been measured experimentally. This is due, in part, to a lucky series of error cancellations, but above all to the fact that far from the walls the fluid flow is statistically homogeneous (see below), so that any model gives good results in that region. In fact, the logarithmic profile (18.5.4) is well approximated by the following power-law expression,

$$v^+ = 8.56 y^{+1/7}. \tag{18.5.5}$$

Now, considering that the fluid velocity reaches its maximum value, V_{max}, at the center of the conduit, when $y = R$, we obtain:

$$V_{max} = 8.56 V_w (R/\delta_w)^{1/7}. \tag{18.5.6}$$

Now, denoting by $V = \frac{49}{60} V_{max}$ the mean fluid velocity, Eq. (18.5.5) yields:

$$v = V_{max} \left(\frac{y}{R}\right)^{1/7} = 1.22 \, V \left(1 - \frac{r}{R}\right)^{1/7}. \tag{18.5.7}$$

This profile is represented in Fig. 18.2, where it appears that the turbulent velocity profile is blunt and flattened compared to the laminar Poiseuille profile, represented by the broken line in the figure.[10] Near the wall, the velocity goes to zero very rapidly; in fact, in Eq. (18.5.7), and in the logarithmic profile as well, we find a diverging derivative, i.e., $dv/dr \to \infty$ when $r \to R$. Actually, at the wall, the velocity profile is linear, with $v^+ = y^+$ [see Eq. (18.5.3)], so that we find, trivially: $dv/dy = \tau_w/\mu$. These relations hold in the viscous sublayer, when $y^+ \leq 5$: when $Re = 10^5$, it means a distance from the wall $y/R < 10^{-3}$, which is so small that it cannot be visible in Fig. 18.2.

---

[10]In fact, the maximum velocity is only 1.22 times larger than the mean velocity, instead of twice larger, as in laminar flow.

From the velocity profile (18.5.7) it is easy to determine the pressure drops in terms of the mass flow rate or, equivalently, the friction factor as a function of the Reynolds number. In fact, from Eq. (18.5.6) we obtain:

$$\frac{V}{V_w} = \frac{8.56}{1.224 \times 2^{1/7}} \left(\frac{VD}{v}\right)^{1/7} \left(\frac{v}{V_w \delta_w}\right)^{1/7} \left(\frac{V_w}{V}\right)^{1/7}$$

$$= 6.33 Re^{1/7} \left(\frac{V_w}{V}\right)^{1/7} \quad \Rightarrow \quad \frac{V}{V_w} = 5.03 Re^{1/8},$$

where $D = 2R$ is the tube diameter, $Re = VD/v$ is the Reynolds number, and we have considered that $V_w \delta_w / v = 1$. From the definition (18.4.6) of the friction factor, $f = 2(V_w/V)^2$, we find the Blasius expression (4.5.7),

$$f = 0.0791 Re^{-1/4}. \tag{18.5.8}$$

Using the logarithmic velocity profile (18.5.4), instead of the power law (18.5.5), we find the von Karman-Nikuradse equation,

$$1/\sqrt{f} = 4 \log\left(\sqrt{f} Re\right) - 0.4, \tag{18.5.8a}$$

that is very accurate for smooth pipes. Naturally, once we know the expression for the friction factor f, we can apply the Colburn-Chilton analogy,

$$Nu = \frac{1}{2} f \, Re \, Pr, \tag{18.5.9}$$

and find the heat (and mass) flux at the walls.

So, it appears that the mixing length approach can model the turbulent flow in a tube quite successfully. However, for more complicated geometries, Prandtl's simple linear relation (18.4.10) does not provide satisfactory results and therefore more complicated models must be applied.

18.6 More Complex Models

A key quantity of all the more complicated turbulence models is the turbulent kinetic energy per unit mass, defined as,

$$K = \frac{1}{2} \langle \tilde{v} \cdot \tilde{v} \rangle. \tag{18.6.1}$$

Dotting the Navier-Stokes equation for the fluctuating velocity into $\tilde{\mathbf{v}}$ and averaging the result, after some algebra we obtain the following balance equation for K[11]:

$$\frac{DK}{Dt} = -\nabla \cdot \mathbf{J}_K + P - \langle \tilde{\mathbf{v}}\tilde{\mathbf{v}} \rangle : \nabla \bar{\mathbf{v}} - \varepsilon, \qquad (18.6.2)$$

where $P = -\langle \tilde{\mathbf{v}}\tilde{\mathbf{v}} \rangle : \nabla \bar{\mathbf{v}}$ is the production rate of turbulent kinetic energy. Also, D/Dt is the material derivative referred to the mean velocity $\bar{\mathbf{v}}$, $\mathbf{J}_K$ is the diffusive flux of K (i.e., relative to $\langle \mathbf{v} \rangle$), while ε is the turbulent (pseudo) dissipation rate,

$$\varepsilon = v \langle (\nabla \tilde{\mathbf{v}}) : (\nabla \tilde{\mathbf{v}}) \rangle; \quad \text{in 2D,} \quad \varepsilon = v \left(\frac{\partial \tilde{v}_x}{\partial y} \right)^2. \qquad (18.6.3)$$

This conservation equation must be coupled with a constitutive relation for $\mathbf{J}_K$, the most common being,

$$\mathbf{J}_K = -\left(v + v_t^K \right) \nabla K, \qquad (18.6.4)$$

where v_t^K is the eddy diffusivity for turbulent kinetic energy. Extending the Reynolds analogy (18.4.4), we can assume that v_t^K is equal to the other eddy diffusivities, i.e., $v_t^K = v_t$, thus assuming that kinetic energy diffuses like any other conserved quantity. Finally, assuming that the derivatives along the direction orthogonal to the wall, y, are much larger than those in the x- and z-directions, we obtain, after substituting the turbulent viscosity assumption from Eq. (18.4.1) for the Reynolds stresses,

$$\frac{DK}{Dt} = \frac{\partial}{\partial y} \left[(v + v_t) \frac{\partial K}{\partial y} \right] + v_t \left(\frac{\partial \bar{v}_x}{\partial y} \right) - \varepsilon. \qquad (18.6.5)$$

This equation can be used to create a *one-equation model* for turbulent boundary layers, postulating an a priori relation between ε, K and v_t. Based on dimensional analysis, we obtain:

$$v_t = aK^{1/2}\ell_t; \quad \varepsilon = bK^{3/2}/\ell_t, \qquad (18.6.6)$$

where a and b are tunable constant to be determined by numerical experimentation, while ℓ_t is a turbulent length scale, that depends on position. Now, the hypothesis on ℓ_t that have been made are not less arbitrary than those concerning the mixing length, ℓ. Therefore, we see that Eq. (18.6.5) presents the same problems that have been encountered using the mixing length model.

---

[11]See W.M. Deen, *Analysis of Transport Phenomena*, Oxford University Press, pp. 541–544.

The difficulties encountered using the one-equation model have stimulated the research to find a more satisfactory approach. This is why the K-ε *model* has been developed. It is a *two-equation model*, where one equation consists of (18.6.5), while the other is a transport equation for the turbulent dissipation rate, ε. The latter can be obtained after long manipulation, obtaining[12]:

$$\frac{D\varepsilon}{Dt} = -\nabla \cdot \mathbf{J}_\varepsilon + G - S. \tag{18.6.7}$$

Here, $\mathbf{J}_\varepsilon$ is the diffusive flux of ε, with the "usual" constitutive relation,

$$\mathbf{J}_\varepsilon = -(v + v_t^\varepsilon)\nabla\varepsilon, \tag{18.6.8}$$

while G and S are, respectively, the generation and the sink terms, that are correlated as follows,

$$G = C_1 \frac{\varepsilon}{K} P; \quad S = -C_2 \frac{\varepsilon^2}{K}, \tag{18.6.9}$$

where P is the production rate of turbulent kinetic energy defined above, while C_1 and C_2 indicate two constants, to be determined by numerical experimentation. Imposing that Eqs. (18.6.5)–(18.6.7) are mutually compatible, we have:

$$v_t = C_0 \frac{K^2}{\varepsilon}, \tag{18.6.10}$$

where C_0 is a another constant, to be determined experimentally. This result is important, as it shows that now the eddy diffusivity can change freely, following the variations of turbulent kinetic energy and dissipation, without any need for debatable models to evaluate it. At this point, we must solve the continuity and momentum balance equations for the mean velocity, in addition to the turbulent viscosity assumption Eq. (18.4.1), where the eddy diffusivity is given by Eq. (18.6.10), K and ε being functions of position and time, and solve Eqs. (18.6.5) and (18.6.7). So, the strong point of the K-ε model is that we do not have to specify upfront any mixing length, nor any turbulent length scale, while the constants C_0, C_1 and C_2 can be determined by numerical experimentation, obtaining,

$$C_0 = 0.09; \quad C_1 = 1.44; \quad C_2 = 1.92. \tag{18.6.11}$$

Sometimes, an extension of the Reynolds analogy is not postulated. In these cases, the ratios between the turbulent diffusivities must also be indicated.

---

[12]An accurate treatment of the K-ε model can be found in C.G. Speziale, *Annu. Rev. Fluid Mech.* **23**, 107–157 (1991).

Although the results that have been obtained applying the K-ε model go beyond the most optimistic expectations, much work is still required before we can trust it to model turbulent flows with complex geometries. In fact, it should be borne in mind that the turbulent viscosity model is inherently flawed in that the anisotropy tensor (i.e., the traceless component of the Reynolds stress tensor) is normally not aligned with the mean rate of strain tensor, clearly indicating that turbulence is not locally isotropic.

Chapter 19
Free Convection

Abstract In the previous chapters we have confined ourselves to the case of forced convention, where the fluid movement is caused by an external cause, such as the movement of one of the fluid boundaries, or the pressure difference induced by a pump or by gravity. We have not considered, so far, the important case of *free convection*, named also *natural convection*, or *buoyancy-driven flows*, where the fluid movement is caused by the density differences due to changes of temperature or concentration of the fluid. Free convection is encountered very often in nature. For example, air heats up, as it flows past the slopes of a mountain which has been heated by the sun, so that its density decreases and consequently moves upward. This mechanism generates those updrafts that are well known to all birds and, generally, to all those who delight of gliding. Also, as the water at the surface of warm seas evaporates, its salt concentration increases, so that the water becomes heavier, sediments and is then replaced by the lighter water coming from colder seas, thus generating marine currents. The crucial aspect of free convection is that velocity and temperature (and concentration as well) are tightly coupled to one another. That means that, while in forced convection the velocity field is imposed and can be determined (almost) independently of temperature (or concentration), in free convection fluid motion is due directly to a temperature difference, so that velocity and temperature must be determined in parallel, thus making the problem much more difficult to solve. This chapter intends to be only a short introduction to the free convection problem, determining in Sect. 19.1 the governing equation and their scaling, and in Sect. 19.2 solving them in one of the very few cases where an exact solution does exist. The boundary layer theory in free convection is then explained in Sects. 19.3 and 19.4, followed by Sect. 19.5, where a few experimental correlations are presented. Finally, in Sect. 19.6, a few examples of flow with phase transition are described, where the density difference between the phases plays obviously a very important role.

© Springer International Publishing Switzerland 2015   321
R. Mauri, *Transport Phenomena in Multiphase Flows*,
Fluid Mechanics and Its Applications 112, DOI 10.1007/978-3-319-15793-1_19

19.1 The Boussinesq Approximation

Consider a fluid in which a dilute solute is dissolved, so that its density depends on the solute concentration, c, as well as on its temperature, T. Let us rewrite the Navier-Stokes equation in a convenient way, starting from,

$$\rho \frac{Dv}{Dt} = -\nabla p + \rho \mathbf{g} + \nabla \cdot \mathbf{T}', \tag{19.1.1}$$

where $\mathbf{T}' = \mathbf{T} - p\mathbf{I}$ is the viscous part of the stress tensor. Now define a sort of mean dynamic pressure, P, as:

$$\nabla P = \nabla p - \rho_0 \mathbf{g}, \tag{19.1.2}$$

where ρ_0 is the fluid density, measured at a convenient reference state, at temperature, T_0, and solute concentration, c_0. Substituting Eq. (19.1.2) into (19.1.1) we obtain:

$$\rho \frac{Dv}{Dt} = -\nabla P + (\rho - \rho_0)\mathbf{g} + \nabla \cdot \mathbf{T}'. \tag{19.1.3}$$

Here it appears evident that the driving force is the buoyancy body force, $(\rho - \rho_0)\mathbf{g}$.

In general, the density ρ is a function of T and c, which in turn can be determined by solving the respective balance equations of heat and chemical species.[1]

At this point, we introduce the Boussinesq[2] approximation, which is based on the assumption that the maximum density variation, $\Delta\rho_{max}$, is much smaller than the mean density, that is, $\Delta\rho_{max}/\rho_0 \ll 1$. The Boussinesq approximation consists of the following two steps: (a) density ρ can be replaced by ρ_0, with the exception of the gravitational term, $(\rho - \rho_0)\mathbf{g}$ in the Navier-Stokes equation; (b) viscosity, thermal conductivity and molecular diffusivity are constant; (c) in the domain of variation of T and c, the density varies linearly with T and c, so that,

$$\rho = \rho_0 - \rho\beta_T(T - T_0) - \rho\beta_c(c - c_0), \tag{19.1.4}$$

where β_T and β_c are the coefficients of thermal and solutal expansion, respectively,

$$\beta_T = -\frac{1}{\rho}\left(\frac{\partial\rho}{\partial T}\right)_{T=T_0; c=c_0} \quad \text{and} \quad \beta_c = -\frac{1}{\rho}\left(\frac{\partial\rho}{\partial c}\right)_{T=T_0; c=c_0}, \tag{19.1.5}$$

and are assumed to be constant. Normally, $\beta_T > 0$, so that ρ decreases as T increases, while β_c can be positive or negative: indicatively, $\beta_c > 0$ when the

---

[1]Note that the shear stress and the heat and mass fluxes depend also on viscosity, thermal conductivity and molecular diffusivity, which, like density, depends on temperature and solute concentration. However, these dependencies are quite less pronounced than for density.

[2]J.V. Boussinesq (1842–1929) was a French physicist.

solute has a larger density than the fluid, otherwise $\beta_c < 0$. Obviously, the linearization (19.1.4) is not always possible. For example, in the case of the air updraft
described in the Introduction of this chapter, considering air as an ideal gas we have
$\beta_T = 1/T_0$, and assuming a $\Delta T = 10\,°C$ thermal excursion around ambient conditions (i.e., $T_0 \cong 300\,K$), we see that β_T changes by 3 % so that, in this case, the
Boussinesq approximation will provide results that are valid within a few percent.
On the other hand, if we consider water near the freezing point, β_T changes very
rapidly with temperature: in fact, $\beta_T < 0$ when $0 < T_0 < 4\,°C$, while $\beta_T > 0$ when
$T_0 > 4\,°C$, so that the linearization (19.1.4) is not valid in this case.

In the following, we will consider the case of a pure fluid, i.e., without solutes,
where ρ depends only on temperature (assuming small thermal excursions) through
the following linear relation,

$$\rho_0/\rho = 1 + \beta_T(T - T_0). \tag{19.1.6}$$

Substituting Eq. (19.1.6) into (19.1.3) the following Navier-Stokes equation is
obtained,

$$\frac{D\mathbf{v}}{Dt} = \frac{\partial \mathbf{v}}{\partial t} + \mathbf{v} \cdot \nabla\mathbf{v} = -\frac{1}{\rho_0}\nabla P - \beta_T(T - T_0)\mathbf{g} + \nu\nabla^2\mathbf{v}, \tag{19.1.7}$$

where $\nu = \mu/\rho_0$ is the kinematic viscosity, that is assumed to be constant and
evaluated at a reference temperature T_0. This equation must be coupled with the
continuity equation and the heat conservation equation,

$$\nabla \cdot \mathbf{v} = 0; \quad \frac{DT}{Dt} = \frac{\partial T}{\partial t} + \mathbf{v} \cdot \nabla T = \alpha\nabla^2 T, \tag{19.1.8a, b}$$

where fluid compressibility and viscous dissipation have been neglected.

We stress again that the Boussinesq approximation consists of assuming that the
only term where temperature variations are accounted for is the gravitational term in
the Navier-Stokes equation. At first sight, this may seem wrong, because in many
instances viscosity varies with temperature more rapidly than density. However,
one must consider that the driving force of the fluid flow, that is the reason why the
fluid moves at all, is the buoyancy term: without it, there would be no flux. In
addition, in free convection, the viscous term is generally very small, compared to
buoyancy, as we will see in Sect. 19.3, so that at leading order it is correct to
approximate the viscous term by assuming a constant viscosity.

19.2 Free Convection in a Vertical Channel

This is one of the very few problems in free convention that admits an exact analytical
solution, although, it must be admitted, its importance is mainly didactical. Consider a
channel of width 2 h and indefinite height, whose walls are kept at different temperatures,

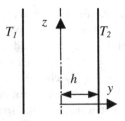

Fig. 19.1 Free convection in a vertical channel

i.e., T_1 on the left and $T_2 = T_1 + \Delta T$ on the right, with $\Delta T > 0$ (see Fig. 19.1). The fluid that is brought in contact with the cold wall will tend to descend, while the fluid on the other side of the channel will tend to ascend. Therefore, as we have assumed that the channel is infinitely long, the fluid velocity, $\mathbf{v}$, is uni-directional along the z-axis and will depend only on y, that is $\mathbf{v} = v(y)\mathbf{e}_z$, where $\mathbf{e}_z$ is a unit vector in the z-direction. The same considerations hold also for the temperature profile, so that $T = T(y)$; therefore, since there is transversal convection, heat will be transported along the y-direction only by conduction, and therefore the temperature profile will be linear. In fact, Eqs. (19.1.8a, b) gives:

$$\frac{\partial^2 T}{\partial y^2} = 0; \quad T(-h) = T_1; \quad T(h) = T_2 \Rightarrow \widetilde{T}(\eta) = \frac{T - T_m}{\Delta T} = \frac{1}{2}\eta, \quad (19.2.1)$$

where $T_m = (T_1 + T_2)/2$ is the mean temperature and $\eta = y/h$ is a dimensionless distance. Now, let us consider the momentum balance equation (19.1.7). First of all, as it is always the case in uni-directional flow fields, the inertial forces are identically null. In addition, since both $\mathbf{g}$ and $\mathbf{v}$ are directed along z, the pressure, p, is constant along y and depends only on z, i.e., $p = p(z)$. Therefore, assuming T_m as the reference temperature and choosing $V_v = v/h$ as a characteristic velocity due to viscous forces (it is the only possible scaling, since inertial forces are null), Eq. (19.1.7) yields,

$$\frac{d^2 \tilde{v}}{d\eta^2} = -\frac{1}{2} Gr\,\eta + \frac{d\tilde{P}}{d\tilde{z}}. \quad (19.2.2)$$

Here, we have defined the dimensionless variables $\tilde{v} = v/V_v$, $\widetilde{P} = P/(\mu V_v/h)$, and $\tilde{z} = z/h$, while

$$Gr = \frac{h^3 g \beta_T \Delta T}{v^2} = \frac{h^3 g (\Delta \rho/\rho)}{v^2} \quad (19.2.3)$$

is the Grashof number,[3] expressing the ratio between gravitational and viscous forces. Solving Eq. (19.2.2) with boundary conditions $\tilde{v}(\pm 1) = 0$ and considering

---

[3]Named after Franz Grashof (1826–1893), a German engineer.

Fig. 19.2 Cubic flow fields

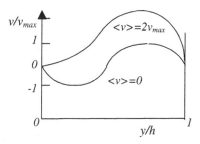

that the pressure gradient is constant,[4] we obtain the following velocity profile (see Fig. 19.2):

$$\tilde{v}(\eta) = \frac{1}{12}Gr(\eta - \eta^3) - \frac{1}{2}(d\tilde{P}/d\tilde{z})(1 - \eta^2). \qquad (19.2.4)$$

As one would expect, due to the linearity of the problem, the velocity is the sum of the effects induced by two driving forces, namely the constant pressure gradient and the zero-mean buoyancy force. The former induces a Poiseuille-type parabolic flow field, while the latter consists of a cubic velocity profile. Therefore, considering that

$$\langle \tilde{v} \rangle = \frac{1}{2}\int_{-1}^{1} \tilde{v}(\eta)d\eta = -\frac{1}{3}(d\tilde{P}/d\tilde{z}), \qquad (19.2.5)$$

we see that dP/dz can be determined from the net total flow rate, that is the difference between the upward and the downward flow fields. Now, if we assume that in reality the channel has a finite height, so that the same fluid flowing upward will eventually flow in the opposite direction, we may assume that there is no net flow, that is $\langle \tilde{v} \rangle = 0$, and therefore the pressure gradient must be null, i.e., $dP/dz = 0$. This is reasonable, because from its definition (19.1.2), the dynamic pressure already includes the effect of the *mean* gravitational term. Therefore we obtain:

$$\tilde{v}(\eta) = \frac{1}{12}Gr(\eta - \eta^3) = \frac{3\sqrt{3}}{2}\tilde{v}_{max}(\eta - \eta^3), \qquad (19.2.6)$$

where $\tilde{v}_{max} = hv_{max}/v = Gr/18\sqrt{3}$ is the maximum velocity, corresponding to $\eta = 1/\sqrt{3}$, while the minimum velocity, $-\tilde{v}_{max}$, corresponds to $\eta = -1/\sqrt{3}$. In dimensional terms, that means:

$$v_{max} = 0.032Gr(v/h). \qquad (19.2.6a)$$

---

[4]As we have seen in Sect. 5.1, this is a consequence of the translational symmetry of the problem.

Clearly, if the channel is open, we will have an imposed pressure difference between its two ends, thus inducing a mean flow $\langle v \rangle$. Then, from Eqs. (19.2.4) and (19.2.5) we find (see Fig. 19.2):

$$v(\eta) = \frac{3}{2}\left[\sqrt{3}v_{max}(\eta - \eta^3) + \langle v \rangle(1 - \eta^2)\right]. \tag{19.2.7}$$

As we have observed at the start, this problem is rather artificial, because by confining the fluid to flow within a pipe we have assumed the velocity field to be uni-directional, thus imposing that the inertial forces are identically null. In general, however, the geometries encountered in free convection problems are more complex, and inertial forces cannot be dealt with so easily.

19.3 Scaling of the Problem

Consider the natural convection of a fluid, with an unperturbed temperature T_∞, flowing past a macroscopic object of size L and temperature $T_0 = T_\infty + \Delta T$. From the Navier-Stokes equation (19.1.7), we see that buoyancy can be balanced by either viscous forces or by inertia, depending whether the Reynolds number is small or large, respectively. In the first case, denoting by V_v the characteristic viscous velocity, we obtain:

$$\nu V_v/L^2 = g\beta_T\Delta T \Rightarrow V_v = \frac{L^2 g\beta_T\Delta T}{\nu}. \tag{19.3.1}$$

This is the scaling of the problem that we have studied in Sect. 19.2, imposing $Gr = 1$ in Eq. (19.2.3), so that $V_v = \nu/h$, and also assuming that the pressure gradient balances the viscous forces, so that we obtain a characteristic pressure, $P_v = \mu V_v/L$. Finally, defining the dimensionless quantities,

$$\tilde{\mathbf{v}} = v/V_v; \quad \tilde{T} = (T - T_\infty)/\Delta T; \quad \tilde{P} = P/(\mu V_{cv}/L); \quad \tilde{\mathbf{r}} = \mathbf{r}/L, \tag{19.3.2}$$

the Navier-Stokes equation (19.1.7) at steady state becomes,

$$Gr\,\tilde{\mathbf{v}} \cdot \tilde{\nabla}\tilde{\mathbf{v}} = -\tilde{\nabla}\tilde{P} - \tilde{T}\mathbf{e}_g + \tilde{\nabla}^2\tilde{\mathbf{v}}, \tag{19.3.3}$$

where $\tilde{\nabla} = L\nabla$, $\mathbf{e}_g$ is a unit vector directed along the gravitational field and Gr is the Grashof number (19.2.3),

$$Gr = \frac{L^3 g\beta_T\Delta T}{\nu^2} = \frac{L^3 g(\Delta\rho/\rho)}{\nu^2}, \tag{19.3.4}$$

expressing the ratio between gravitational and viscous forces. This scaling applies in two cases, when either the inertial term is identically zero, as in the uni-directional problem of the previous Section, or when $Gr \ll 1$. In general, however, on one hand the unidirectional case is quite artificial and, on the other hand, in all practical cases the Grashof number is quite large. In fact, as we see in Problem 19.2, we can write Gr in the following way:

$$Gr = \gamma L^3 \Delta T, \quad \text{where } \gamma = \beta_T g/\nu^2.$$

Here, γ is a property of the fluid at the T_0 reference temperature; for example, at $T_0 = 50\,°C$ we find $\gamma = 1.2 \times 10^{10} m^{-3} K^{-1}$ for water, while $\gamma = 1.0 \times 10^8 m^{-3} K^{-1}$ for air. Thus, we see that even for "small" bodies, with $L = 1$ cm, and assuming $\Delta T = 10\,°C$, the Grashof number is very large.

Let us consider now the other, much more important, scaling, by balancing buoyancy with the inertial forces in Eq. (19.1.7), thus defining a characteristic inertial velocity, V_c,

$$V_c^2/L = g\beta_T \Delta T \Rightarrow V_c = \sqrt{Lg\beta_T \Delta T}. \tag{19.3.5}$$

Therefore, we see that the Grashof number can be interpreted as the square of the Reynolds number based on the inertial velocity, V_c:

$$Gr = \frac{L^3 g \beta_T \Delta T}{\nu^2} = \left(\frac{V_c L}{\nu}\right)^2. \tag{19.3.6}$$

Accordingly, the case $Gr \gg 1$ in free convention corresponds to $Re \gg 1$ in forced convection. In this case, since the characteristic pressure gradient will balance the inertial forces, we find the characteristic inertial pressure $P_c = \rho V_c^2$. Now, defining the dimensionless quantities,

$$\tilde{\mathbf{v}} = v/V_c; \quad \tilde{T} = (T - T_\infty)/\Delta T; \quad \tilde{P} = P/(\rho V_c^2); \quad \tilde{\mathbf{r}} = \mathbf{r}/L, \tag{19.3.7}$$

the Navier-Stokes at steady state becomes,

$$\tilde{\mathbf{v}} \cdot \tilde{\nabla}\tilde{\mathbf{v}} = -\tilde{\nabla}\tilde{P} - \tilde{T}\mathbf{1}_g + \frac{1}{Gr^{1/2}} \tilde{\nabla}^2 \tilde{\mathbf{v}}. \tag{19.3.8}$$

In the same way, the governing equations (19.1.8a, b) become, in dimensionless form,

$$\tilde{\nabla} \cdot \tilde{\mathbf{v}} = 0; \quad \tilde{\mathbf{v}} \cdot \tilde{\nabla}\tilde{T} = \frac{1}{Pr\ Gr^{1/2}} \tilde{\nabla}^2 \tilde{T}, \tag{19.3.9}$$

where $Pr = \nu/\alpha$ is the Prandtl number. Note that the following group,

$$Gr^{1/2}Pr = \left(\frac{V_cL}{\nu}\right)\left(\frac{\nu}{\alpha}\right) = \left(\frac{V_cL}{\alpha}\right) \qquad (19.3.10)$$

plays the same role in free convection as the Peclet number in forced convection.

19.4 The Boundary Layer in Free Convection

As we saw in Eq. (19.3.6), in free convention the Grashof number plays the same role of the square of the Reynolds number in forced convection. Therefore, we expect that when a fluid flows past an object by free convection at $Gr \gg 1$, a boundary layer will form at the surface. In fact, as we approach the body surface and the fluid velocity tends to zero, the magnitude of the inertial term decreases, while that of the viscous term increases, as the velocity profile becomes steeper and steeper, until, at the edge of the boundary layer, the two terms will have the same magnitude. Now, considering the four terms in Eq. (19.3.8), we see that the inertial term at the LHS and the first two terms on the RHS (i.e., the pressure gradient and the buoyancy term) have the same $O(1)$-magnitude. Therefore, imposing that these terms balance the last, viscous term on the RHS, we obtain the following expression for the boundary layer thickness, δ,

$$\frac{1}{Gr^{1/2}}\tilde{\nabla}^2\tilde{v} \approx \frac{1}{Gr^{1/2}}\frac{1}{\tilde{\delta}^2} \approx 1 \Rightarrow \tilde{\delta} = \frac{\delta}{L} \approx \frac{1}{Gr^{1/4}}. \qquad (19.4.1)$$

Considering that $Gr \approx Re^2$, Eq. (19.4.1) is basically identical to the expression $\delta/L \approx Re^{-1/2}$ in forced convection. The difference between free and forced convection is that in the forced convection of a fluid past a flat plate, pressure is uniform and therefore the profile of the fluid velocity near the wall is approximately linear. On the other hand, in free convection the buoyancy driving force induces a pressure gradient, which in turn generates a Poiseuille-like parabolic velocity profile near the wall.

Let us consider heat transport. From Eq. (19.3.9) we see that when $Pr\,Gr^{1/2} \gg 1$, i.e., the equivalent of $Pe \gg 1$ in forced convection, we have a thermal boundary layer of thickness, say, δ_T. As in forced convection, we must consider separately the two cases $Pr \ll 1$ and $Pr \gg 1$.

When $Pr \ll 1$, we expect that $\delta_T \gg \delta$, so that a fluid region at a distance δ_T from the wall lays outside the momentum boundary layer, where the fluid velocity in the longitudinal direction (i.e., parallel to the wall) is of $O(V_c)$. Therefore, balancing the two terms in Eq. (19.3.9), with $\tilde{v} \approx 1$ and $\tilde{\mathbf{v}} \cdot \tilde{\nabla}\tilde{T} \approx 1,$[5] we obtain,

---

[5]Remember that the temperature gradient is in the longitudinal direction and therefore it is of $O(1)$.

$$\frac{1}{Pr\, Gr^{1/2}} \frac{1}{\tilde{\delta}_T^2} \approx 1 \Rightarrow \tilde{\delta}_T = \frac{\delta_T}{L} \approx Gr^{-1/4} Pr^{-1/2}. \tag{19.4.2}$$

that satisfies the initial assumption $\delta_T \gg \delta$. So, in this case we have obtained the same result as in forced convention, with $Gr \approx Re^2$, namely $\delta_T/L \approx Re^{-1/2} Pr^{-1/2} \approx Pe^{-1/2}$.

When $Pr \gg 1$, we expect that $\delta_T \ll \delta$, so that a fluid region at a distance δ_T from the wall lays well inside the momentum boundary layer. There, as the inertial term must balance the diffusion term in the convection-diffusion equation, we obtain:

$$v_x \Delta T/L \approx \alpha \Delta T/\delta_T^2,$$

where, considering that the velocity profile is parabolic, we have: $v_x/V_c \approx (\delta_T/\delta)^2$. Finally, since $\delta/L \approx Gr^{-1/4}$, we find:

$$\frac{\delta_T}{L} \approx \frac{\delta}{L} Pr^{-1/4} \approx Gr^{-1/4} Pr^{-1/4} = Ra^{-1/4}, \tag{19.4.3}$$

where we have defined the following, so-called, *Rayleigh* number,

$$Ra = Gr\, Pr = \left(\frac{L^3 g \beta \Delta T}{\nu^2} \right) \left(\frac{\nu}{\alpha} \right) = \frac{L^3 g \beta \Delta T}{\nu \alpha}. \tag{19.4.4}$$

Now, let us calculate the heat flux at the wall, $J_{Uw} = -k(\partial T/\partial y)_w$, where y is the coordinate axis perpendicular to the wall and directed outward. We obtain: $J_{Uw} = h\Delta T$, where $h = k/\delta_T$ and therefore, defining the Nusselt number, $Nu = hL/k$, we obtain:

$$Nu = L/\delta_T = 1/\tilde{\delta}_T.$$

We conclude by summarizing the results of this Section as follows,

$$Nu = C_1 Gr^{1/4} Pr^{1/2} \quad \text{when } Gr \gg 1 \text{ and } Pr \ll 1;$$
$$Nu = C_2 Gr^{1/4} Pr^{1/4} \quad \text{when } Gr \gg 1 \text{ and } Pr \gg 1. \tag{19.4.5a, b}$$

Here, C_1 and C_2 are two $O(1)$ constant that can be determined from the knowledge of the temperature profile, and therefore depend on the geometry of the problem, such as the shape of the submerged object. For example, in the case of a flat vertical plate, there is an exact analytical solution, due to Le Fevre in 1956, who obtained $C_1 = 0.8005$ and $C_2 = 0.6703$, and where the characteristic distance L is the longitudinal vertical coordinate z, (as in the Blasius solution for forced convection).

It should be noted that, while the solution (19.4.5a) is valid only when $Pr \ll 1$, Eqs. (19.4.5b) applies also for moderate values of Pr (trivially, when $Pr \cong 1$, the dependence of Pr is not crucial any more). The reason for this unexpected agreement is the same as in forced convection, and has been discussed in Sect. 13.3.

Finally, it is worth noting that in free convection the process of heat transport is not linear; in particular, as ρ is temperature-dependent, the convective heat flux, $\rho c v \Delta T$, is not linear with ΔT, and therefore it is not surprising that the heat transfer coefficient, h, depends on ΔT. In fact, considering that $Gr \propto \Delta T$, we see that, when $Gr \gg 1$, we have, from Eqs. (19.4.5a, b), that $h \propto (\Delta T)^{1/4}$ and therefore,

$$J_U = h\Delta T \Rightarrow J_U \propto (\Delta T)^{5/4}. \tag{19.4.6}$$

19.5 Experimental Correlations

When a fluid at temperature T_∞ flows by free convention past a body at temperature T_0, the experimental measurements are in good agreement with the following simple relation,

$$Nu = C(Gr\,Pr)^n = C\,Ra^n, \tag{19.5.1}$$

where the constant C and n, together with the characteristic length L on which Gr is defined may vary from case to case, as indicated in the Table below.

Form	Ra	L	C	n
Vertical plate	10^4–10^9	Vertical height	0.58	0.25
Vertical cylinder	10^9–10^{13}	Axial length	0.02	0.4
Horizontal cylinder	10^4–10^9	Diameter	0.53	0.25
Sphere	10^4–10^9	Radius	0.53	0.25
Parallelepiped	10^4–10^9	$L = 1/(1/L_{horiz} + 1/L_{vert})$	0.53	0.25
Horizontal plate A × B	10^4–10^7	$L = (A + B)/2$	0.54	0.25
• Warmer upper surface	10^7–10^{11}		0.15	0.33
• Colder upper surface	10^5–10^{11}		0.58	0.20
Horizontal disk of diameter D: identical to the horizontal plate		$L = 0.9\,D$		

When natural convention takes place in a closed space, very few correlations are available. The case of a fluid confined between two vertical plates at different temperatures is not dissimilar from the problem that we have studied in Sect. 19.2 (see Fig. 19.3). Defining the Grashof and Nusselt number in terms of the distance between the plates, d, i.e., $Gr = d^3 g\beta\Delta T/\nu^2$ and $Nu = hd/k$, the two following correlations have been proposed, depending whether the fluid is a liquid or a gas:

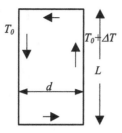

Fig. 19.3 Fluid confined between two vertical plates at different temperatures

- Liquid: $Nu = 0.42\,(Gr\,Pr)^{0.25}\,Pr^{0.012}\,(L/d)^{-0.3}$, valid when $10^4 < Ra < 10^7$ and when $10 < (L/d) < 40$.
- Gas: $Nu = 0.2\,(Gr\,Pr)^{0.25}\,(L/d)^{-0.11}$, valid when $10^3 < Ra < 10^5$ and when $10 < (L/d) < 40$.

Note that the typical value of the heat transfer coefficient for a gas in free convention is $h \approx 3\text{--}20$ W/m^2 K, and is therefore smaller than the typical $h \approx 10\text{--}100$ W/m^2 K in forced convection. On the other hand, in liquids we have $h \approx 100\text{--}500$ W/m^2 K, both in natural and forced convection.

19.6 Heat Transfer with Phase Transition

Phase transition, and in particular the processes of condensation and boiling, are very important because they use the large energy that is absorbed (or released) during the phase change (i.e., the latent heat) to cool (or heat) a fluid. Since these processes are deeply affected by the density differences between the two phases, they can be considered as particular cases of natural convection phenomena.

19.6.1 Film Condensation on a Vertical Plate

Condensation occurs when a saturated vapor comes in contact with a surface whose temperature is below the saturation temperature. Normally, a film of condensate is formed on the surface, and the thickness of this film increases with increase in extent of the surface. This is called *film* (or film-type) *condensation*. Another type of condensation, called *dropwise condensation*, occurs when the wall is not uniformly wetted by the condensate, with the result that the condensate appears in many small droplets at various points on the surface. In the following, we will consider film condensation, as it is far more common and more dependable,

Fig. 19.4 Film condensation
on a vertical plate

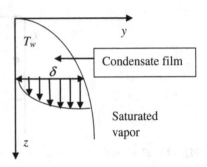

following the analysis that was first performed by Nusselt (and, in fact, it is generally called Nusselt's theory) (Fig. 19.4).

Consider a vapor at temperature T_g and pressure P_g, flowing past a vertical wall of width W. Assume that the wall is kept at a temperature T_w lower than the saturation temperature of the vapor (i.e., the temperature where, at that given pressure P_g, the vapor undergoes a vapor-liquid phase transition), so that

$$T_w = T_g - \Delta T < T^{sat}(P_g).$$

Then, a liquid film of thickness $\delta(z)$ will form at the wall surface that, under gravity, flows down the plate. In quasi stationary conditions, the velocity profile is the Poiseuille profile,

$$v(y) = \frac{g}{2v}(2\delta y - y^2), \tag{19.6.1}$$

where v is the fluid kinematic viscosity. Here, we have assumed a hydrostatic pressure drop, $\Delta P/L = \rho g$, where ρ is the liquid density, neglecting the much smaller vapor density. Now, define the mass flow rate per unit thickness, Γ,

$$\dot{m} = \int_0^\delta \rho v W dy = \Gamma W; \quad \Gamma = \frac{\rho g}{3v}\delta^3. \tag{19.6.2}$$

The heat exchanged through the film of thickness δ due to conduction (as there is no convection along the transversal y-direction) must be equal to the heat absorbed by condensation, that is,

$$d\dot{Q} = \Delta h_{lg} d\dot{m} = k\frac{\Delta T}{\delta} W dz, \tag{19.6.3}$$

where Δh_{lg} is the latent heat of vaporization, $d\dot{m}$ is the increase in mass flow rate due to the condensation occurring within dz and k is the coefficient of thermal conductivity. Substituting Eq. (19.6.2) into (19.6.3), we obtain:

$$\frac{d\Gamma}{dz} = \frac{k}{\Delta h_{\mathrm{lg}}} \frac{\Delta T}{\delta} = \frac{\rho g}{v} \delta^2 \frac{d\delta}{dz} \Rightarrow \int_0^{\delta} \delta^3 d\delta = \frac{kv}{\rho g \Delta h_{\mathrm{lg}}} \Delta T \int_0^z dz. \tag{19.6.4}$$

Hence, integrating and readjusting we find:

$$\delta(z) = \left[\frac{4kv\Delta T}{\rho g \Delta h_{\mathrm{lg}}}\right]^{1/4}. \tag{19.6.5}$$

Therefore, defining a heat transfer coefficient as:

$$d\dot{Q} = Wdzh(z)\Delta T; \quad h(z) = \frac{k}{\delta(z)} \propto \frac{1}{z^{1/4}}, \tag{19.6.6}$$

we obtain the Nusselt equation for the mean heat transfer coefficient, h_L,

$$h_L = \frac{1}{L} \int_0^L h(z)dz = \frac{4}{3} h(L) = 0.943 \left[\frac{\rho g k^3 \Delta h_{\mathrm{lg}}}{vL\Delta T}\right]^{1/4}. \tag{19.6.7}$$

Thus, we find the mean Nusselt number,

$$Nu = \frac{h_L L}{k} = 0.943 \left[\frac{\rho g \Delta h_{\mathrm{lg}}}{vk\Delta T} L^3\right]^{1/4} \tag{19.6.8}$$

and the total heat exchanged,

$$\dot{Q}_{tot} = h_L(LW)\Delta T. \tag{19.6.9}$$

Note that, as always in free convection, the heat transfer process is not linear and, in fact, we find: $h \propto (\Delta T)^{-1/4}$ and therefore, $\dot{Q}_{tot} \propto J_U = h\Delta T \propto (\Delta T)^{3/4}$. Also, note that, as usual, the Nusselt number is inversely proportional to the film thickness, that is,

$$Nu = \frac{h_L L}{k} = \frac{4}{3} \frac{h(L)L}{k} = \frac{4}{3} \frac{L}{\delta(L)}. \tag{19.6.10}$$

Let us calculate the total mass flow rate of condensate per unit width,

$$\Gamma_L = \Gamma(L) = \frac{\rho g}{3v} \delta(L)^3, \quad \text{or,} \quad \Gamma_L = \frac{Q_{tot}/W}{\Delta h_{\mathrm{lg}}}, \tag{19.6.11}$$

obtaining,

$$\Gamma_L = \frac{h_L L \Delta T}{\Delta h_{\mathrm{lg}}} \tag{19.6.12}$$

The Reynolds number is:

$$Re = \frac{4\Gamma_L}{\mu} = \frac{4}{3}\frac{g}{v^2}\delta(L)^3 = \frac{4}{3}Gr\frac{\delta(L)^3}{L^3}, \tag{19.6.13}$$

where we have defined the Grashof number $Gr = gL^3/v^2$. Finally expressing $\delta(L)$ as a function of Re and substituting the result into Eq. (19.6.10), we obtain the following relation:

$$Nu = 1.47\, Gr^{1/3} Re^{-1/3}. \tag{19.6.14}$$

19.6.2 Boiling

19.6.2.1 Pool Boiling

Pool boiling is what we experience when we place a pot filled with water on a hot plate and let the water boil. In general, pool boiling consists of the process occurring when a heating surface is surrounded by a relatively large body of fluid, which is initially stagnant and then is agitated only by natural-convection and bubble motion. The process is characterized by the typical bi-logarithmic boiling curve[6] that is shown in Fig. 19.5, representing heat flux as a function of the difference ΔT between the temperature T_w of the hot plate and the saturation temperature, T_{sat}. This curve is generally divided into four parts, representing four different regimes: natural convection, nucleate boiling, the transition region and film boiling.

For small values of the oversaturation ΔT, the curve is almost linear, with a 5/4 slope, indicating that there is no phase transition and the process is governed by *natural convection* [see Eq. (19.4.6)], with the Nusselt number, Nu, proportional to $Ra^{1/4}$, where Ra denotes the Rayleigh number. In this regime, the hot liquid is removed from the hot plate through natural convention, thus allowing colder liquid to replace it near the hot surface.

At point A, when ΔT reaches a few degrees (about 5 K), small bubbles (a few tens of micron size) start to appear on the surface of the hot plate; typically, they form preferentially in correspondence of the small crevices due to the roughness of the plate surface. Once formed, these bubbles continue to grow, until they detach by

---

[6]This curve was obtained experimentally by *Nukiyama* in 1934.

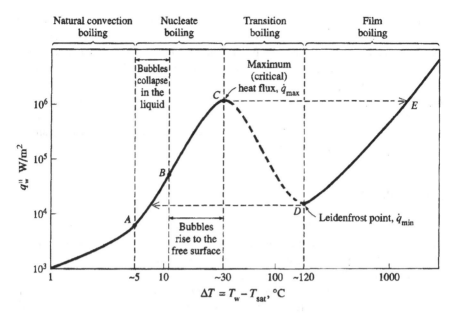

Fig. 19.5 Heat flux in pool boiling as a function of the wall temperature

gravity from the plate surface and move up in the surrounding liquid. In this, so-called, *nucleate boiling* regime, the boiling curve shown in Fig. 19.5 becomes much steeper, as the heat flux increases by two orders of magnitude as the over-saturation temperature ΔT goes from 5 to 30 K. In fact, it is observed that, as the heat exchanged grows, moving from point B to C, bubble formation becomes so intense that bubbles coalesce and form continuous vapor columns that rise from the bottom to the top of the pool. In this regime of isolated bubbles and vapor jets, which is still part of the *nucleate boiling* regime, the heat transfer is particularly efficient; in fact, typical values of the heat transfer coefficient h is as high as 10^4 W/m^2 K, that is at least two orders of magnitude larger than those encountered in single-phase convective heat transport. At this point, though, if ΔT is increased further, large regions of the plate surface will remain occasionally dry, i.e. they are not wetted by the liquid and remain in contact with a vapor film. This produces a so-called *sputtering* regime, with a drastic heat transfer reduction, caused by the low thermal conductivity of vapor compared to that of water (about 40 times smaller, in standard conditions). The flow regime corresponding to a ΔT varying between 30 and 120 K is not easily reproducible as it presents the hysteresis[7] shown in Fig. 19.5; this is why this region of the curve is named *transition region* and is represented with a broken line. The point C where the boiling curve reaches a

---

[7]When the heat flux is progressively increased from point A, we reach the *critical point* C and then jump to point E; at this point, if we decrease the heat flux from point E, we reach so so-called *Leidenfrost point* D and then jump back to the initial A-C part of the curve.

maximum corresponds to the *critical flux*, that is a condition that we should avoid at all cost. In fact, if we imagine to provide the system with a progressively increasing heat flux, once we reach the critical flux condition we jump from the point C of the boiling curve to the point E, corresponding to a much higher wall temperature (about 1500 °C). This is the so-called *burnout*, that in general causes the fusion of the heating metal surface. Finally, for even larger values of the oversaturation temperature, the wall is never wetted by the liquid and is in contact with a stable vapor film. In this so-called *film boiling* regime, the heat flux grows again as ΔT increases, due mainly to radiation.

19.6.2.2 Flow Boiling

Flow boiling denotes the boiling process undergone by a liquid flowing in a pipe, with an imposed heat flux at the wall.[8] This process is very similar to pool boiling, as shown in Fig. 19.6, where the different two-phase flow regimes and heat transfer regions of pool boiling are represented.

Starting from the entrance region, in the *natural convection* regime (region A) and the *subcooled boiling* regime (region B), the mean (i.e., bulk) temperature of the fluid lays below the saturation temperature, T_{sat}, resulting in the collapse of the bubbles as soon as they leave the hot surface.

Then, in region C, we enter the *saturated nucleate boiling* region, where the fluid bulk temperature is equal to the saturation temperature, resulting in net vapor generation. There, we observe the formation of bubbles that tend to grow and coalesce, until they occupy the whole cross section, forming a *slug flow* (region D).

Moving further up, we then encounter the *annular flow* regime (regions E and F), where the central part of the conduit is occupied by vapor, while the walls are wetted by the liquid. In this region, boiling at the wall surface is replaced by the evaporation taking place at the liquid-vapor interface. Consequently, the liquid film becomes progressively thinner, until it disappears completely at the *dryout point*. This point is analogous to the critical point in pool boiling, corresponding to a sharp drop of the heat transfer coefficient, with a consequent drastic increase of the wall temperature. Next, we observe a *drop flow*, that is a sort of fog, consisting of the residual drops being carried away by the vapor, until they, too, evaporate and the fluid becomes pure vapor.

All these different regimes are described by experimental correlations that can be found in most two-phase flow manuals.

---

[8]In fact, in most cases the tubes are placed inside a furnace, and so the heat flux is determined by the characteristics of the chemical reaction taking place in the combustion chamber.

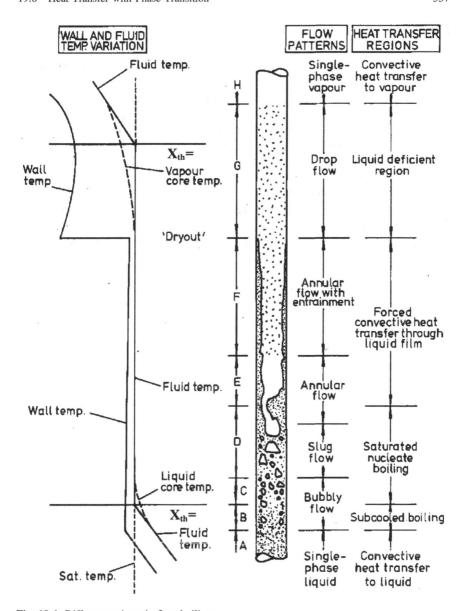

Fig. 19.6 Different regimes in flow boiling

19.7 Problems

19.1 What would we obtain if in Sect. 19.2 we chose a different reference temperature?

19.2 Calculate the maximum velocity of a fluid (consider air and water) confined between two vertical plates of infinite height, that are placed at a distance $d = 2$ cm from each other and are kept at temperatures 25 and 75 °C.

19.3 Air at atmospheric pressure is confined between two glass plates of length $L = 0.5$ m, that are placed vertically at a distance $d = 15$ mm from each other. Calculate the heat exchanged, assuming that the two plates are kept at temperatures 40 and 100 °C.

19.4 Repeat the dimensional analysis of Sect. 19.4, concerning the thermal boundary layer when $Gr \gg 1$ and $Pr \gg 1$, assuming that $\delta_T/\delta = Pr^{-a}$; $\tilde{v}_x \approx Pr^{-b}$, where a and b are positive constant to be determined by balancing the appropriate terms in Eqs. (19.3.8) and (19.3.9).

Chapter 20
Radiant Heat Transfer

Abstract Radiant heat transfer consists of the transfer of energy through the electromagnetic waves that are emitted by any material object as a consequence of its temperature. Unlike the other modality of energy transport, i.e., convection and diffusion, radiation does not need a medium, such as air or a metal, to propagate and, in fact, it can move across the void, as it happens with the solar energy reaching the earth surface. Also, radiation is much more dependent on temperature, compared to heat convection and diffusion. Accordingly, radiation is the dominant form of energy transport in furnaces, because of their high temperature, and in cryogenic insulation, because of the vacuum existing between particles. So, for example, gases in a combustion chamber lose more than 90 % of their energy by radiation. In this Section, we will consider the particular form of electromagnetic radiation that is connected to heat transfer, that is *thermal radiation*. Most energy of this type is in the infra-red region of the electromagnetic spectrum, although some of it is in the visible region, and should not be confused with other forms of electromagnetic radiation, from such as radio waves, X-rays, or gamma rays.

20.1 The Law of Stefan-Boltzmann

From an engineer's point of view, radiation is much simpler than convection, because the details of the electromagnetic field are unimportant: at the end, what we care about, practically, is only the temperature of the radiating surfaces by applying the Stefan-Boltzmann relation. Considering its importance, in the following this law is derived in detail.

Consider a closed box of volume V with perfectly elastic walls, containing $N = nV$ particles, where n is the particle number density. The pressure P exerted by these particles on the walls is equal to the force per unit area, that is the momentum transferred by the particles to the wall per unit time and per unit area. Now consider that (a) the momentum flux due to a single particle is qv, where q and v are the particle momentum and velocity, respectively; (b) at any instant of time only 1/6 of

© Springer International Publishing Switzerland 2015
R. Mauri, *Transport Phenomena in Multiphase Flows*,
Fluid Mechanics and Its Applications 112, DOI 10.1007/978-3-319-15793-1_20

the particles move in a direction orthogonal to a wall and directed towards it; (c) at each collision, a particle will transfer a momentum $2q$. Thus, we may conclude that,

$$P = \frac{1}{3}nqv. \tag{20.1.1}$$

When the particles are photons, v is the velocity of light, c, while q depends on the energy of a single photon, u_1, through the relation: $u_1 = qc$.[1] Then,

$$P = \frac{1}{3}n(u_1/c)c = \frac{1}{3}u, \tag{20.1.2}$$

where $u = nu_1$ is the specific (i.e., per unit volume) internal energy, $u = U/V$. Note that in the case of an ideal gas, with $q = mv$, Eq. (20.1.1) can be rewritten as $P = \frac{1}{3}nmv^2$, where v^2 is the mean square velocity. Then, since from the equipartition theorem $mv^2 = 3kT$, where k is the Boltzmann constant, we obtain the well-known equation of state for ideal gases, $P = nkT$, that is $P\tilde{V} = RT$, where $\tilde{V}$ is the molar volume, while $R = N_A k$ is the gas constant, with N_A denoting the Avogadro constant. Obviously, since the internal energy per unit volume of an ideal gas is $u = \frac{3}{2}nkT$, we see that $P = \frac{2}{3}u$, showing that the pressure of a photon gas is half that of an ideal gas with the same specific internal energy.

At this point, from the first law of thermodynamics we obtain,

$$dU = TdS - PdV \Rightarrow \left(\frac{\partial U}{\partial V}\right)_T = T\left(\frac{\partial P}{\partial T}\right)_V - P \tag{20.1.3}$$

where we have applied the Maxwell equation $(\partial S/\partial V)_T = (\partial P/\partial T)_V$. Now, considering that U is an extensive variable, we have: $U(T, V) = u(T)V$, stressing that u is a function of T only. Thus, Eq. (20.1.3), with $P = u/3$, becomes:

$$u = \frac{T}{3}\frac{du}{dT} - \frac{u}{3} \Rightarrow T\frac{du}{dT} = 4u,$$

obtaining the Stefan-Boltzmann law[2]

$$u = aT^4 = \frac{4\sigma}{c}T^4, \tag{20.1.4}$$

---

[1]This relation can be interpreted classically by associating to a photon a fictitious mass m, so that its energy (which is equal to hv, where h is the Planck constant and v the frequency) is $u_1 = mc^2$. Then, considering that $q = mc$, we obtain: $u_1 = qc$.

[2]A simple way to explain why $u \propto T^4$ is to consider that in any spatial direction the number of degrees of freedom is proportional to T. For example, if the photons have the same frequency, v, there will be kT/hv degrees of freedom on any direction. Thus, in 3D, the number of degrees of freedom will be $\propto T^3$ and any one of them will have an energy $kT/2$.

where a is the Stefan[3] energy-density constant. For sake of convenience [see Eq. (20.1.7) below], the Stefan constant is often redefined in terms of $\sigma = ac/4$ (see below).

Once u and P are known, we can determine any other thermodynamic quantity. For example, considering that $S(T, V) = s(T)V$ and applying the above-mentioned Maxwell equation, we obtain:

$$s = \frac{dP}{dT} = \frac{1}{3}\frac{du}{dT} = \frac{4}{3}\frac{u}{T} = \frac{16\sigma}{3c}T^3. \tag{20.1.5}$$

Thus, the enthalpy, h, the Helmholtz energy, a, and the Gibbs free energy, g, per unit volume are:

$$h = u + P = \frac{4}{3}u$$
$$a = u - Ts = -\frac{1}{3}u$$
$$g = a + P = 0.$$

This shows that the Gibbs free energy of a photon gas is null.

Note that, while in a photon gas we consider energies per unit volume, which turn out to be functions of T only, in an ideal gas the corresponding quantities are the energies per unit mass or mole. In fact, interpreting Eq. (20.1.3) as a molar relation and substituting the equation of state of an ideal gas, we find: $(\partial U/\partial V)_T = 0$, and then $U \propto T$.

Now we can determine the energy flux coming out of a cavity kept at a temperature T (see Fig. 20.1). Since radiation is isotropic, the total energy flux, equal to cu, is distributed uniformly in all directions, so that the energy flux that crosses the solid angle $do = 2\pi \sin\theta\, d\theta$ (i.e., a surface on a unit radius sphere forming an angle between θ and $\theta + d\theta$ with the normal to the cavity surface) is $uc \cos\theta\, do/4\pi$. Here we have considered that the net radiation flux will have a direction $\theta = 0$, and therefore the component $uc \sin\theta$ does not contribute. Finally, integrating over all possible angles θ we obtain:

$$J_U = \int_0^{\pi/2} \frac{uc}{4\pi}\cos(\theta)\, 2\pi\, \sin(\theta)\, d\theta = \frac{uc}{4}. \tag{20.1.6}$$

So, finally, we find another version of the law of Stefan-Boltzmann,

$$J_U = \sigma T^4. \tag{20.1.7}$$

---

[3]Named after Josef Stefan, a Slovenian physicist. σ is also called the Stefan-Boltzmann constant.

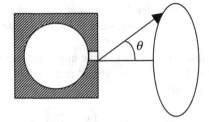

Fig. 20.1 Cavity radiation

20.1.1 Planck's Black Body Radiation Theory

The Stefan-Boltzmann law given in Eq. (20.1.4) connects the total thermal radiation energy per unit volume to the thermodynamic temperature T. Equation (20.1.4) was well known in the 19th century; in fact, Stefan's constant a was introduced into physics in 1879, long before any suggestions of quantum mechanics. However, it was immediately realized that classical physics could not justify Eq. (20.1.4) in a coherent way. In fact, considering that in a cavity of size L there will be an infinite number of resonating stationary waves with frequencies $v_n = nc/L$, where $n = 1,2...$, and, according to the equipartition theorem, each of these oscillations will have the same energy kT, then the total energy contained in the cavity should diverge. This is the famous *ultraviolet catastrophe* paradox, that stimulated a research effort culminating in Planck's theory of radiation.

Planck was interested into the spectrum of the thermal radiation emitted by a cavity, that is the radiation energy $du(v)$ whose frequencies are comprised between v and $v + dv$. Clearly, Eq. (20.1.4) provides no information regarding the radiation spectrum $du(v)$, although it imposes that its integral is fixed, as $u = \int_0^\infty du(v)$. Now, according to classical physics, $u(v) \propto v^3$ and therefore the integral of du should diverge [see Eq. (20.1.12) below]. To resolve this paradox, Planck realized that if we assume that energy can be exchanged only through discrete *quanta*, it becomes progressively more difficult to excite oscillators of increasing frequency, and therefore $u(v)$ must decrease sharply at large frequencies. At the end, he obtained the following so-called Planck distribution,

$$du(\omega) = \frac{\hbar}{\pi^2 c^3} \frac{\omega^3 d\omega}{e^{\hbar\omega/kT} - 1}, \tag{20.1.8}$$

where $\hbar = h/2\pi$, h is Planck's constant, and $\omega = 2\pi v$. The Planck distribution presents a maximum at the frequency ω_m, with

$$\hbar\omega_m = 2.822\,kT; \quad \text{hence:} \quad \lambda_m T = 2898\,\mu\text{m K}, \tag{20.1.9}$$

where $\lambda = 2\pi c/\omega = c/\nu$ is the wavelength. This is the *Wien displacement law*, showing that, if we increase the temperature of an object, the frequency of its emitted radiation increases proportionally.

Integrating the Planck distribution (20.1.8) over all frequencies,[4] and equating the result with Eq. (20.1.4), we obtain:

$$u = \int\limits_0^\infty du(\omega) = \frac{\pi^2 k^4}{15\hbar^3 c^3} T^4 = \frac{4\sigma}{c} T^4. \tag{20.1.10}$$

Therefore, we see that Stefan's constant, Planck's constant, Boltzmann's constant and the light speed are connected through the relation, $60\hbar^3 c^2 \sigma = \pi^2 k^4$. In general, this relation is used to determine an exact value of the Stefan constant, i.e.,

$$\sigma = \frac{\pi^2 k^4}{60\hbar^3 c^2} = 5.67 \times 10^{-5} \frac{g}{s^3 K^4} = 5.67 \times 10^{-8} \frac{W}{m^2 K^4}. \tag{20.1.11}$$

The classical (and diverging) result, which is based on the equipartition theorem, corresponds to the limit of the Planck distribution (20.1.8) when $\hbar\omega \ll kT$, obtaining,

$$du(\omega) = \frac{kT}{\pi^2 c^3} \omega^2 d\omega. \tag{20.1.12}$$

This is the *Rayleigh-Jeans distribution* which, if we assume it applicable at all frequencies, leads to the ultraviolet catastrophe paradox that we have mentioned above.

20.2 Emissivity and Absorptance

20.2.1 The Kirchhoff Law

In the previous Section, we have considered cavities that contain radiation by treating them as a photon gas at thermodynamic equilibrium with their walls. The properties that we have seen, in particular the Stefan-Boltzmann law and the Planck distribution, do not depend on the material that constitutes the wall. In fact, if we consider two cavities in contact with one another, as in Fig. 20.2, at equilibrium the flux of the radiation leaving each cavity must be equal to the flux entering, independently of the walls of the two cavities.

---

[4]Consider that: $\int_0^\infty \frac{x^3 dx}{e^x - 1} = \frac{\pi^4}{15}$.

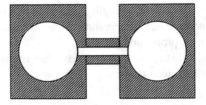

Fig. 20.2 Independence of the cavity radiation from the wall material

Now, define a material object, called *black body*, that absorbs all the radiation that reaches its surface, at any frequency. If we insert a black body in a cavity having the same temperature, being in a condition of thermal equilibrium, all the flux emitted by the cavity is absorbed by the black body, and, vice versa, all the flux emitted by the black body is absorbed by the cavity. Accordingly, a black body is characterized by a radiant heat flux, both emitted and absorbed, that equals the flux (20.1.7) of a cavity. Thus, a cavity is the simplest example of a black body.

In general, material objects do not behave as black bodies, as the radiation that reaches their surface is partly absorbed and partly reflected, or transmitted. Now, the fraction of the radiation that is absorbed is called *absorptance* (also called *absorptivity*) $a(v)$, which is then defined as the ratio between the absorbed radiant heat flux, $J_{Ua}(v)$, and the total incident flux $J_{Ui}(v)$, at a given frequency v, i.e.,

$$a(v) = J_{Ua}(v)/J_{Ui}(v). \qquad (20.2.1)$$

Since a black body is defined as an object that absorbs all the incident radiation at all frequencies (i.e., with $a = 1$), the absorptance can also be defined as the ratio between the absorbed radiant heat flux, $J_{Ua}(v)$, and the radiant heat flux that is absorbed (or emitted) by a back body in the same condition, $J_{Ubb}(v)$. In the same way, we can define the *emissivity* $e(v)$ as the ratio between the emitted radiant heat flux, $J_{Ue}(v)$, and the radiant heat flux that is emitted by a back body in the same condition, $J_{Ubb}(v)$. In summary:

$$a(v) = \frac{J_{Ua}(v)}{J_{Ubb}(v)}; \quad e(v) = \frac{J_{Ue}(v)}{J_{Ubb}(v)}, \qquad (20.2.2)$$

where $a(v)$ and $e(v)$, in general, depends on the radiation frequency, and we have considered that for a black body the radiant heats emitted and absorbed are equal to one another.

Now, suppose to insert an object in a cavity. When cavity walls and object have the same temperature, they are at mutual thermal equilibrium, and therefore the radiant heat fluxes absorbed and emitted by the object must be equal, that is $J_{Ua}(v) = J_{Ue}(v)$. Therefore, from the definition (20.2.2) we obtain the following

Kirchhoff law,[5] stating that the emissivity and the absorptivity of a surface are equal to one another:

$$e(v) = a(v). \tag{20.2.3}$$

In the following, we will drop the frequency dependence. That amounts to considering *gray bodies*, such that $e(v) = e \leq 1$. A black body is a particular case of gray bodies with $e = 1$.

20.2.2 The View Factor

Now we consider the case when several surfaces are present. Consider first the case of two black surfaces, S_1 and S_2, having temperatures T_1 and T_2, and areas A_1 and A_2. The heat power flux $\dot{Q}_{12}$, (heat per unit time) leaving S_1 and reaching S_2 (where it is completely absorbed, because we are considering black bodies) is $\dot{Q}_{12} = A_1 F_{12} J_{U1}$, where $J_{U1} = \sigma T_1^4$ is the radiant heat flux (heat per unit time and unit surface) leaving S_1, while F_{12} is the *view factor*, indicating the fraction of the radiation leaving S_1 which is intercepted by S_2. Therefore, $A_1 F_{12}$, called the *direct-exchange area*,[6] is the portion of the area A_1 from which the radiation intercepting 2 comes from, or, equivalently, it is the part of S_1 that "sees" S_2; naturally, the direct-exchange area is a purely geometric feature. Conversely, the heat power emitted by surface S_2 and absorbed by S_1 is $\dot{Q}_{21} = A_2 F_{21} J_{U2}$, where $A_2 F_{21}$ is the direct-exchange area representing the portion of S_2 that "sees" S_1. At equilibrium, when the two surfaces have the same temperature, the two heat powers must be equal to each other, that is $\dot{Q}_{21} = \dot{Q}_{12}$ and $J_{U1} = J_{U2}$, and therefore we find that the two direct-exchange areas are equal to each other, that is $A_1 F_{12} = A_2 F_{21}$. In general, we obtain the following *reciprocity relation*,

$$A_i F_{ik} = A_k F_{ki}. \tag{20.2.4}$$

When the two surfaces have different temperatures, the net heat power exchanged between the two black surfaces is:

$$\dot{Q}_{net} = (J_{U1} - J_{U2})A_1 F_{12} = (J_{U1} - J_{U2})A_2 F_{21}. \tag{20.2.5}$$

---

[5]Named after Gustav Robert Georg Kirchhoff (1824–1887), a German physicist and mathematician.
[6]Sometimes it is indicated simply as 12.

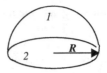

Fig. 20.3 The radiant heat flux exchanged between a hemisphere and a plane

The view factor can be determined out of purely geometric reasoning, finding[7]

$$F_{12} = \frac{1}{\pi A_1} \iint \frac{\cos(\theta_1)\cos(\theta_2)}{r_{12}^2} dA_1 dA_2, \qquad (20.2.6)$$

where $\mathbf{r}_{12}$ is the vector connecting the areas dA_1 and dA_2, with $r_{12} = |\mathbf{r}_{12}|$, while θ_1 and θ_2 are the angles between $\mathbf{r}_{12}$ and, respectively, the vectors orthogonal to dA_1 and dA_2. Note that this relation identically satisfies the equality (20.2.4).

Based on the view factor definition, the sum of all the view factors of a given body equals unity, that is,

$$\sum_{j=1}^{n} F_{ij} = 1, \qquad (20.2.7)$$

where we have assumed that the radiation leaving surface i can be intercepted by n surrounding surfaces. Note that, in general, one must consider also the view factors F_{ii}, indicating the radiation that leaves i and is intercepted again by i, as it happens in the case of concave geometries (see the example below).

Example: Consider the radiant heat flux exchanged between a hemisphere (surface 1) and the plane (surface 2) represented in Fig. 20.3. Here, all the radiation that leaves 2 is intercepted by 1, so that $F_{21} = 1$; consequently, since $F_{22} + F_{21} = 1$, we see that $F_{22} = 0$, indicating that no part of the radiation leaving 2 is intercepted by 2. Now consider that reciprocity relation: $A_1 F_{12} = A_2 F_{21}$, where $A_1 = 2\pi R^2$ and $A_2 = \pi R^2$, so that $F_{12} = F_{21}/2 = \frac{1}{2}$. Consequently, since $F_{11} + F_{12} = 1$, we may conclude that $F_{11} = \frac{1}{2}$, indicating that half of the radiation leaving the hemisphere is intercepted again by the hemisphere, while the other half ends up on the plane.

20.2.3 Example: Radiation in a Furnace Chamber

A simplified sketch of a furnace chamber is shown in Fig. 20.4. It presents a flame (i.e., a heat source) on a surface 1, with temperature T_1, and a heat exchanger (i.e., a

---

[7]See for example Bird, Stewart and Lightfoot, *Transport Phenomena*, Sect. 14.4. The values of the view factor for the most common geometries can be found in Perry, *Handbook of Chemical Engineering*.

Fig. 20.4 A furnace chamber

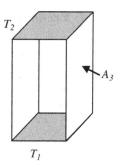

heat sink) on a surface 2, with temperature T_2, while the lateral surfaces 3 are radiatively adiabatic, with a negligible heat absorbed. Assuming that all surfaces behave as black bodies, the net heat power leaving any surface i (with $i = 1, 2, 3$) is [see Eq. (20.2.5)]:

$$\dot{Q}_i = \sigma A_i \sum_{j=1}^{3} F_{ij}\left(T_i^4 - T_j^4\right). \tag{20.2.8}$$

These are 3 equations, with $\dot{Q}_3 = 0$ and where T_1 and T_2 are known. Therefore, knowing all the view factors, we can determine the three unknown, $\dot{Q}_1$, $\dot{Q}_2$ and T_3 solving the following equations:

$$\dot{Q}_1 = \sigma A_1 \left[F_{12}\left(T_1^4 - T_2^4\right) + F_{13}\left(T_1^4 - T_3^4\right)\right]$$
$$\dot{Q}_2 = \sigma A_2 \left[F_{21}\left(T_2^4 - T_1^4\right) + F_{23}\left(T_2^4 - T_3^4\right)\right];$$
$$0 = \sigma A_3 \left[F_{31}\left(T_3^4 - T_1^4\right) + F_{32}\left(T_3^4 - T_2^4\right)\right]$$

Now, considering that, out of symmetry, $F_{31} = F_{32}$, from the last equation we find T_3 as:

$$T_3^4 = \frac{1}{2}\left(T_1^4 + T_2^4\right). \tag{20.2.9}$$

We could find the same result summing the first two equations and imposing that all the heat emitted by 1 is absorbed by 2, so that $\dot{Q}_1 = -\dot{Q}_2$, with $A_1 = A_2 = A$ and $F_{13} = F_{23}$. Finally, we obtain:

$$\dot{Q}_1 = \sigma A \left(F_{12} + \frac{1}{2}F_{1L}\right)\left(T_1^4 - T_2^4\right). \tag{20.2.10}$$

Obviously, as $T_1 > T_2$, we see that $\dot{Q}_1 > 0$, that is a net radiant heat leaves the flame and is absorbed by the heat exchanger.

20.2.4 Exchange of Radiant Heat Between Gray Bodies

The radiant heat flux leaving a gray body, $J_{U,out}$, is equal to the sum of the heat flux emitted by the body, $e\sigma T^4$, and the heat flux reflected, $(1-a)J_{U,in}$, where $J_{U,in}$ is the incident radiant heat flux, while e and a denote, respectively, the emissivity and the absorptance of the body which, in agreement with the Kirchhoff law, are equal to each other. Thus,

$$J_{U,out} = e\sigma T^4 + (1-e)J_{U,in}. \tag{20.2.11}$$

For a black body, $e = 1$ and we find again Eq. (20.1.7). Thus, the net radiant heat flux leaving the gray body, i.e., $J_{U,net} = J_{U,out} - J_{U,in}$, is:

$$J_{U,net} = e\left(\sigma T^4 - J_{U,in}\right). \tag{20.2.12a}$$

At equilibrium, $J_{U,out} = J_{U,in}$, and therefore $J_{U,in} = \sigma T^4$, indicating that the temperature of all the surrounding object must be the same as that of the body. Sometimes, it is more convenient to rewrite Eq. (20.2.12a, 20.2.12b) by substituting Eq. (20.2.11), obtaining,

$$J_{U,net} = \frac{e}{1-e}\left(\sigma T^4 - J_{U,out}\right). \tag{20.2.12b}$$

Therefore, the net heat power exchanged between the two surfaces (from 1 to 2), $\dot{Q}_{net,1-2}$, can be written in the following ways,

$$
\begin{aligned}
\dot{Q}_{net,1-2} &= \frac{e_1 A_1}{1-e_1}\left(\sigma T_1^4 - J_{U,out,1}\right) \\
&= \frac{e_2 A_2}{1-e_2}\left(J_{U,out,2} - \sigma T_2^4\right) = A_1 F_{12}\left(J_{U,out,1} - J_{U,out,2}\right).
\end{aligned}
$$

Hence, eliminating $J_{U,out,1}$ and $J_{U,out,2}$, we obtain:

$$\dot{Q}_{net,1-2} = \frac{\sigma\left(T_1^4 - T_2^4\right)}{\frac{1-e_1}{e_1 A_1} + \frac{1}{A_1 F_{12}} + \frac{1-e_2}{e_2 A_2}}. \tag{20.2.13}$$

The denominator in Eq. (20.2.13) is a thermal resistance, equal to the sum of three resistances in series: the first and the third are the resistances to heat emission and absorbance due to the grayness of the bodies (in fact, these resistances are null for black bodies, when $e = 1$), while the second thermal resistance depends on the view factor, indicating that only a portion of the radiation that leaves 1 actually will reach 2, and viceversa. For black bodies, this result reduces to Eq. (20.2.5).

Example: Consider a cubic oven, with 80 cm side, where 5 walls are kept at 1000 °C, while the sixth wall (that is, the oven door) is open on the surrounding, at

300 K. Assuming that the emissivity of the oven walls is 0.95, calculate the heat power leaving the oven door.

The heat is exchanged between the gray surface of the oven (surface 1, with area $A_1 = 3.20$ m^2) and the black surface consisting of the oven door (surface 2, with area $A_2 = 0.64$ m^2). The heat power exchanged, $\dot{Q}_{net,1-2}$, is given by Eq. (20.2.13), with $e_1 = 0.95$ and $e_2 = 1$, while $A_1 F_{12}$ can be obtained from the reciprocity relation $A_1 F_{12} = A_2 F_{21}$, where $F_{21} = 1$. At the end, we find:

$$\dot{Q}_{net,12} = \frac{5.67 \times 10^{-8}(\text{W/m}^2\,\text{K}^4) \times (1273^4 - 300^4)\text{K}^4}{\frac{1-0.95}{0.95 \times 3.2\,\text{m}^2} + \frac{1}{1 \times 0.64\,\text{m}^2}} = 94\ \text{KW}.$$

20.3 Radiation and Conduction

When a solid body is heated by radiation, its inside temperature satisfies the heat equation,

$$\frac{\partial T}{\partial t} = \alpha \nabla^2 T, \qquad (20.3.1)$$

where $\alpha = k/\rho c$ is thermal diffusivity, with k, ρ and c denoting, respectively, the coefficient of heat conduction, the density and the specific heat of the object. At the wall (indicated with the suffix w), the heat flux equals the radiant heat, so that, for an object with emissivity e, the following boundary conditions can be written:

$$-k\frac{\partial T}{\partial z}\bigg|_w = J_{Uw} = e\sigma F_{12}\left(T_w^4 - T_e^4\right), \qquad (20.3.2)$$

where T_w and T_e are the temperatures of the wall and of the emitting body, respectively. Obviously, thermal equilibrium is reached when the body temperature is uniform (and therefore there is no conduction) and equal to T_e (so that the radiant heat is also null). In general, far from thermal equilibrium, the solution of the problem is quite difficult. However, we have seen (see Sect. 9.4) that the problem simplifies considerably when the Suasi Steady State (QSS) approximation can be applied. In this case, that requires that the Biot number is very small, i.e.,

$$Bi = \frac{hL}{k} \approx \frac{e\sigma F_{12} T_e^3 L}{k} \ll 1, \qquad (20.3.3)$$

where L is a characteristic linear dimension of the object, while h is the heat transfer coefficient, defined from Eq. (20.3.2) as follows:

$$J_{Uw} = h(T_w - T_e), \quad \text{with} \quad h = e\sigma F_{12}\frac{(T_w^4 - T_e^4)}{(T_w - T_e)} = e\sigma F_{12}(T_w + T_e)(T_w^2 + T_e^2).$$

$$(20.3.4)$$

When the QSS approximation is valid, the temperature of the object is uniform, i.e., $T_w = T$. Then, a simple heat balance gives,

$$\frac{d}{dt}(\rho c V T) = A F_{12} e\sigma(T_e^4 - T^4),$$

$$(20.3.5)$$

where V and S are the volume and external surface of the object, respectively. Now, defining the dimensionless variables $\Theta = T/T_e$ and $\tau = t\alpha/L^2$, with $L = V/A$, Eq. (20.3.5) becomes:

$$\frac{d\Theta}{d\tau} = Bi(1 - \Theta^4),$$

$$(20.3.6)$$

subjected to the initial condition $\Theta(\tau = 0) = \Theta_0 = T_0/T_e$. The solution of this problem is:

$$Bi\ \tau = \frac{1}{4}\ln\frac{1 + \Theta}{1 - \Theta}\Big|_{\Theta_0}^{\Theta} + \frac{1}{2}\tan^{-1}\Theta\Big|_{\Theta_0}^{\Theta}.$$

$$(20.3.7)$$

In particular, when $\Theta_0 \ll 1$ and the time τ is short enough, so that $\Theta \ll 1$, then the last term on the RHS of Eq. (20.3.5) can be neglected and the solution reduces to:

$$\Theta = \Theta_0 + Bi\ \tau.$$

$$(20.3.8)$$

20.4 Example: The Design of a Solar Panel

On a clear sunny day, the solar radiation reaches the earth surface with a radiant heat flux of about $J_{U,sol} = 1.14$ kW/m^2. First of all, let us calculate the maximum temperature that a plate can reach assuming that (a) the plate is placed horizontally and the lower face is insulated; (b) the plate behaves as a black body; (c) it does not dissipate heat by convection; (d) it radiates towards the sky, which in turn behaves as a black body at 0 K. Then, we obtain:

$$J_{U,sol} - \sigma T_w^4 = 0 \Rightarrow T_w = \left(\frac{1.14 \times 10^3\ \text{W/m}^2}{5.67 \times 10^{-8}\ \text{W/m}^2\ \text{K}^4}\right)^{1/4} = 376\ \text{K} = 103\ °\text{C}. \quad (20.4.1)$$

At this temperature, assuming that the air temperature is $T_a = 20\ °$C, the effect of free convection cannot be neglected and the heat balance equation becomes,

$$J_{U,sol} - h_{fc}(T_w - T_a) - \sigma T_w^4 = 0 \Rightarrow T_w \cong 70\,°C. \qquad (20.4.2)$$

Here, h_{fc} denotes the heat transfer coefficient due to free convection, that can be determined from Eq. (19.5.1):

$$Nu = CRa^n, \quad \text{with} \quad C = 0.15 \quad \text{and} \quad n = 0.33, \qquad (20.4.3)$$

where $Nu = h_{fc}L/k$ is the Nusselt number, while $Ra = Gr\,Pr$ is the Rayleigh number. Here, k is air thermal conductivity and L the plate size; in addition, $Gr = \gamma L^3 \Delta T$ is the Grashof number and $Pr = \nu/\alpha$ is the Prandtl number, with $\gamma = \beta g/\nu^2$, where β is the thermal expansion coefficient, ν the kinematic viscosity, and α the air thermal diffusivity, while $\Delta T = T_w - T_a$ is the difference between the temperature of the wall and that of the air.

Generally, the solution can be obtained iteratively. For example, start assuming that the wall temperature is 80 °C while the air has a 50 °C temperature (i.e., the mean value between 80 and 20 °C): we see that air at 50 °C has $\gamma = 10^8 \mathrm{m}^{-3}\,\mathrm{K}^{-1}$ and $Pr = 0.7$. So, with a plate of size $L = 1$ m, we obtain:

$$Ra = 0.42 \times 10^{10} \text{ and therefore } Nu = 242, \text{ that is } h_{fc} = 7.3\ \mathrm{W/m^2\,K}, \quad (20.4.4)$$

where we have considered that $k = 0.03$ W/mK. Note that h_{fc} does not depend on the plate size L. Substituting this value of h_{fc} into Eq. (20.4.2), we obtain: $T_w \cong 70\,°C$. At this point, we can already stop, since the corrections that we would obtain at the next iteration are very small.

If we want to decrease the heat losses due to convection, we could cover the plate with a glass, so that the air confined between plate and glass could act as a buffer. Actually, when the glass is very close to the plate surface, convection could be neglected in the buffer region, and would be confined in the region outside of the glass, so that the convective heat flux could be estimated using the same method that we have described before. At the end, we find that the glass has a temperature of about 30 °C, so that the temperature difference ΔT appearing in the Rayleigh number is 10 °C, instead of 60 °C, as we had before. So, at the end, we find a heat transfer coefficient $h_{fc} = 4.1\ \mathrm{W/m^2K}$, that is about half the one that we had previously, leading to a plate temperature $T_w = 87\,°C$.

20.5 Problems

20.1 A thin silicon wafer, of 150 mm diameter and 1 mm thickness, and initially at $T_0 = 800$ K, is placed in a void chamber. The chamber walls are kept at 300 K and have an $e_1 = 0.9$ emissivity. How long will it take the silicon wafer to reach a 350 K temperature, assuming that it has the following properties:

$$e_2 = 0.6; k = 0.15 \text{ W/cm K}; c = 0.7 \text{ J/g K}; \rho = 2.3 \text{ g/cm}^3.$$

20.2   Determine the entropy change of a black body as a function of the variations of temperature and volume.

20.3   Based on the result of the previous problem, determine: (a) the heat that must be provided to the radiation contained in a cavity to increase its volume at constant temperature; (b) how the volume of a cavity changes as a function of its temperature during an adiabatic transformation.

Chapter 21
Antidiffusion

Abstract In Chap. 14 we have seen that the constitutive relation of the material flux of a chemical species in a mixture should be expressed in terms of the chemical potential gradient of that species. This statement can be easily justified considering the analogy between mass and heat transport: as thermal equilibrium is characterized by a uniform temperature, it is natural that when this condition is perturbed, the system will tend to return to its equilibrium state by inducing a thermal flux from hot to cold regions, that is proportional to the temperature gradient. Therefore, as chemical equilibrium is characterized by a uniform chemical potential, it is natural to assume that a system will tend to return to its equilibrium state by inducing a material flux that is proportional to the (negative) chemical potential gradient (see R. Mauri, *Non-equilibrium Thermodynamics in Multiphase Flows*, Springer, 2013). In this chapter, some of the consequences of this statement will be studied, starting in Sect. 21.1 with some elementary considerations about mixture thermodynamics. Then, in Sects. 21.2 and 21.3, we describe van der Waals' theory of chemical stability and phase transition, applying these results to simple symmetric and regular binary mixtures in Sect. 21.4. Finally, in Sects. 21.5 and 21.6 diffusive fluxes are modeled in terms of chemical potential gradients.

21.1 The Chemical Potential

Any change in the internal energy, $U = U(S, V, N_i)$ of a multicomponent system can be expressed as,

$$dU = TdS - PdV + \sum_{i=1}^{n} \mu_i dN_i, \quad \text{with } \mu_i = (\partial U/\partial N_i)_{S,V,N_{j \neq i}},$$

where T and S are temperature and entropy, P and V are pressure and volume, N_i and μ_i are the number of moles and the chemical potential of the ith component. This relation expresses the first law of thermodynamics, showing that the homogeneity of the intensive quantities, T, P and μ_i, describe the state of thermodynamic

© Springer International Publishing Switzerland 2015   353
R. Mauri, *Transport Phenomena in Multiphase Flows*,
Fluid Mechanics and Its Applications 112, DOI 10.1007/978-3-319-15793-1_21

equilibrium. In particular, a uniform T describes a state of thermal equilibrium, a uniform P a state of mechanical equilibrium, and uniform μ_i's describe a state of chemical equilibrium.

Now, define the Gibbs free energy as $G = U - TS + PV$, so that,

$$dG = -SdT + VdP + \sum_{i=1}^{n} \mu_i dN_i.$$

Since we are interested in modeling a process, namely molecular diffusion, that takes place in isothermal and isobaric conditions, we see that the free energy G is very convenient to use, as it describes the energy change at constant T and P. In the following, we will always assume that T and P are kept constant, even if we do not indicate it explicitly. So, the above equation can be written as

$$dG = \sum_{i=1}^{n} \mu_i dN_i. \tag{21.1.1}$$

Accordingly, we see that the chemical potential μ_i of a homogeneous mixture composed of n species is defined as the variation of the free energy of the mixture, $G(T, P, \mathbf{N})$, due to an increase of N_i at constant temperature, T, pressure, P, and number of moles, N_j, of all the other components, that is,

$$\mu_i(T, P, \mathbf{x}) = \left(\frac{\partial G}{\partial N_i}\right)_{T,P,N_{j\neq i}} = \left(\frac{\partial (Ng)}{\partial N_i}\right)_{T,P,N_{j\neq i}}. \tag{21.1.2}$$

Here, $N = \sum_{i=1}^{n} dN_i$ is the total number of moles, and $g = G/N$ denotes the molar free energy.

Note that since both G and N_i are extensive variables (that is, they are proportional to the mass, or number of moles, of the system), then both g and μ_i are intensive variables and therefore they are independent of N and are functions of the molar fractions $x_i = N_i/N$ (and, in addition, of T and P as well). That means that g and μ_i depend on the $(n - 1)$ independent variables $x_1, x_2, \ldots, x_{n-1}$ (consider that $x_n = -\sum_{i=1}^{n-1} x_i$, since the sum of all the molar fractions must be equal to unity).

21.1.1 The Gibbs-Duhem Relation

The free energy G, like any extensive quantity, satisfies the relation,[1]

---

[1]This is one of the many *Euler* equations.

$$G = \sum_{i=1}^{n} N_i \left(\frac{\partial G}{\partial N_i}\right) = \sum_{i=1}^{n} N_i \mu_i. \tag{21.1.3}$$

This relation can be easily demonstrated mathematically. Physically, imagine to extract from a homogeneous mixture a fraction k of its volume (and of mass and moles as well). From Eq. (21.1.1), the variation of the free energy due to this operation is $\Delta G = \sum \mu_i \Delta N_i$, where $\Delta G = kG$ and $\Delta N_i = kN_i$, hence we obtain Eq. (21.1.3). From this simple reasoning, we see that the chemical potential μ_i represents the free energy of the ith component within the mixture. In thermodynamics, such quantities are denoted as *partial molar properties*; other examples consists of the *partial molar volume*, the *partial molar entropy*, etc.

Now, differentiating Eq. (21.1.3), we find $dG = \sum_{1}^{n} (\mu_i dN_i + N_i d\mu_i)$; thus, applying Eq. (21.1.2) we obtain:

$$\sum_{i=1}^{n} N_i d\mu_i = 0. \tag{21.1.4}$$

This is the so-called *Gibbs-Duhem* equation, stating that chemical potentials depend on each other and therefore cannot be chosen arbitrarily.

21.1.2 Binary Mixtures

For binary mixtures, since g, μ_1 and μ_2 depend only on $\phi \equiv x_1$ (and on T and P, as well), Eqs. (21.1.1)–(21.1.4) assume a particularly simple form. In fact, considering that

$$\mu_1(\phi) = \left(\frac{\partial[(N_1 + N_2)g(\phi)]}{\partial N_1}\right)_{N_2} = g + N \frac{dg}{d\phi}\frac{d\phi}{dN_1}$$

and

$$N\phi = N_1 \quad \Rightarrow \quad N\,d\phi = (1 - \phi)dN_1,$$

we obtain:

$$\mu_1(\phi) = g(\phi) + \left(\frac{dg}{d\phi}\right)(1 - \phi),$$
$$\mu_2(\phi) = g(\phi) - \left(\frac{dg}{d\phi}\right)\phi, \tag{21.1.5}$$

Fig. 21.1 Graphic
interpretation of Eq. (21.1.5)

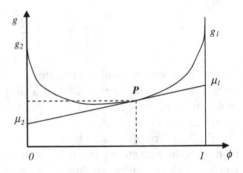

where the second equation is derived from the first one by replacing x_1 with x_2, and vice versa. From these equations we see that the Gibbs-Duhem relation is identically satisfied,

$$\phi \frac{d\mu_1}{d\phi} + (1 - \phi) \frac{d\mu_2}{d\phi} = 0. \tag{21.1.6}$$

In addition, summing and subtracting Eq. (21.1.5) we obtain:

$$g = \phi \mu_1 + (1 - \phi) \mu_2. \tag{21.1.7}$$

and

$$\mu = \mu_1 - \mu_2 = \frac{dg}{d\phi}. \tag{21.1.8}$$

Equation (21.1.7) coincides with (21.1.3), while Eq. (21.1.8) shows that the quantity that is thermodynamically conjugated to the molar fraction x_1 of component 1 is the chemical potential difference $\mu = \mu_1 - \mu_2$ (and not μ_1 alone).

Equations (21.1.5) have an immediate graphic interpretation, as shown in Fig. 21.1, where the curve $g(\phi)$ is represented (remind that T and P are assumed to be constant). Here, the values of $\mu_1(\phi)$ and $\mu_2(\phi)$ are the intercepts of the tangent line in the point $P = (\phi, g(\phi))$, respectively on the axes $\phi = 1$ and $\phi = 0$. Note that, as expected, $\mu_1(\phi = 1)$ coincides with the free energy g_1 of the pure component 1, and analogously $\mu_2(\phi = 0) = g_2$.

21.2 Chemical Stability

Consider an isolated system composed of the mixture we are studying, together with a heat reservoir having the same temperature T, a mechanical reservoir having the same pressure P and a series of material reservoirs, with which the mixture can exchange heat, work and matter. The second law of thermodynamics states that a spontaneous transformation of an isolated system requires that its entropy σ does not decrease. Therefore, as the exchanges of work and matter do not involve any entropy variation, we obtain, applying the first law of thermodynamics,

$$\delta\sigma = \delta S - \frac{\delta Q}{T} = \delta S - \frac{1}{T}\left(\delta U + P\delta V - \sum \mu_k \delta N_k\right) \geq 0, \qquad (21.2.1)$$

where $\delta\sigma$ is the total entropy change, δS is the entropy variation of the mixture, δQ is the heat provided to the mixture, δU and δV are the variations of internal energy and volume of the mixture, $\mu_k = \partial G/\partial N_k$ is the chemical potential of the species k, and δN_k is the change in the number of moles of the component k of the mixture. Now, when Eq. (21.2.1) is applied to closed systems (i.e., with $\delta N_k = 0$), we can show that the conditions of thermal and mechanical stability require that the specific heat, c_V, and the coefficient of isothermal compressibility, κ_T, are positive quantities. As for chemical stability, we see that for open systems, at constant T and P, Eq. (21.2.1) becomes:

$$-T\delta\sigma = \delta G - \sum \mu_k \delta N_k \leq 0, \qquad (21.2.2)$$

where $G = U - TS + PV$, showing that, at equilibrium, $\delta G|_{T,P} - \sum \mu_k \delta N_k$ must be minimum. Now, starting from an equilibrium state, let us introduce a virtual variation δN_k, keeping all the other $N_{i \neq k}$ (together with T and P) constant. Substituting the following expansion,

$$\delta G = \left(\frac{\partial G}{\partial N_i}\right)_{N_{k \neq i}} \delta N_i + \frac{1}{2}\left(\frac{\partial^2 G}{\partial N_i^2}\right)_{N_{k \neq i}} (\delta N_i)^2 + \cdots,$$

into Eq. (21.2.2), and considering that $\delta\sigma < 0$, since we are moving away from equilibrium, from the definition (21.1.2) we obtain:

$$\left(\frac{\partial^2 G}{\partial N_i^2}\right)_{N_{k \neq i}} \geq 0 \quad \Rightarrow \quad \left(\frac{\partial \mu_i}{\partial N_i}\right)_{N_{k \neq i}} \geq 0. \qquad (21.2.3)$$

Therefore, we conclude that the condition of chemical equilibrium imposes that the free energy is a concave function of the concentration of anyone of its component or, equivalently, the chemical potential of the component i is a monotonic increasing function of the concentration of the same component. In particular, for binary mixtures, considering Eq. (21.1.8), the stability conditions can be written in the following equivalent forms:

$$\delta g - \mu\delta\phi \leq 0; \quad \frac{d^2 g}{d\phi^2} \geq 0; \quad \frac{d\mu}{d\phi} \geq 0. \qquad (21.2.4)$$

This stability condition can be easily interpreted on the plot of Fig. 21.2, where the molar free energy of mixing, $\Delta g = g(x_1) - (x_1 g_1 + x_2 g_2)$, is represented as a function of ϕ, presenting a convex part between the inflection points c and d. These are called *spinodal* points, with compositions ϕ_s^α and ϕ_s^α. Now, consider a mixture having a concentration ϕ^*, with $\phi_s^\alpha < \phi^* < \phi_s^\beta$: if the mixture were homogeneous, it would have a free energy corresponding to point f. However, from the first of the relations (21.2.4), we know that the free energy of a system with a fixed composition ϕ^* will tend to a minimum; therefore, the equilibrium state will correspond to point g, representing two coexisting phases, α and β, indicated by points b and e, with concentrations ϕ_e^α and ϕ_e^β.

This graphical interpretation confirms that, since at equilibrium the chemical potentials of two coexisting phases are equal to each other, i.e., $\mu^\alpha = \mu^\beta$, then from (21.1.8) we have: $(dg/d\phi)^\alpha = (dg/d\phi)^\beta$, showing that the slopes of the tangents to the free energy curve at the two equilibrium points must be equal to each other. In summary,

$$\mu^\alpha = \mu^\beta \quad \Rightarrow \quad \left(\frac{dg}{d\phi}\right)^\alpha = \left(\frac{dg}{d\phi}\right)^\beta = \frac{g^\alpha - g^\beta}{\phi^\alpha - \phi^\beta}. \tag{21.2.5}$$

We conclude that a mixture, at a given temperature and pressure, is thermodynamically unstable whenever, in a certain concentration interval, $d^2 g / d\phi^2 < 0$, i.e., its free energy (at constant T and P) is a convex function of concentration. It means that, within that concentration interval, the mixture separates into two coexisting phases, whose compositions can be determined through Eq. (21.2.5).

The relative quantities of the two phases are determined through the so-called *lever rule*. In fact, if we have N^* moles of a mixture having composition ϕ^*, indicating by N^α and N^β the number of moles of phase α and β, respectively, a simple mass balance gives:

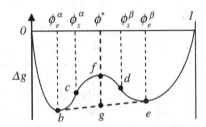

Fig. 21.2 Free energy of mixing as a function of the mixture composition

$$x_1^* N^* = x_1^\alpha N^\alpha + x_1^\beta N^\beta;$$
$$x_2^* N^* = x_2^\alpha N^\alpha + x_2^\beta N^\beta,$$
(21.2.6)

Then, eliminating N^* from the first equation, we obtain the *lever rule*:

$$\frac{N^\alpha}{N^\beta} = \frac{\phi^\beta - \phi^*}{\phi^* - \phi^\beta}.$$
(21.2.7)

Now, it is quite evident that the Gibbs free energy g, the composition ϕ, and the chemical potential difference $\mu = (\partial g / \partial \phi)_{T,P}$ of a binary mixture play the same role as the Helmholtz free energy f, the specific volume $\tilde{v}$ and the pressure $P = -(\partial f / \partial \tilde{v})_T$ of a single-component fluid. In fact, taking the derivative of the Gibbs free energy of Fig. 21.2, we obtain the $(\mu - \phi)$ diagram for binary mixtures that is identical (apart from the sign) to the $(P - \tilde{v})$ diagram of a single-component fluid, as indicated in Fig. 21.3. Here, as in Fig. 21.2, points b and e correspond to the states of equilibrium of two coexisting phases, while points c and d are the spinodal points. The spinodal points are the inflection points in the $(g - \phi)$ diagram, and here are defined such that $d\mu/d\phi = 0$ at constant T and P. Accordingly, in the region $\phi_s^\alpha < \phi^* < \phi_s^\beta$, the chemical potential difference μ decreases with ϕ and therefore corresponds to unstable states. As for the equilibrium points, they are characterized by the condition (21.2.5). Thus, as phase transition takes place at constant temperature, pressure and chemical potential difference, it can be represented as a horizontal isotherm isobaric segment in the $(\mu - \phi)$ diagram of Fig. 21.3. The chemical potential and the concentrations at equilibrium can be determined defining a generalized potential, $\Phi = g - \mu\phi$, with $d\Phi = -sdT + vdP - \phi d\mu$. Considering that at equilibrium $dT = dP = d\mu = 0$, so that the generalized potentials of the two phases are equal to each other, i.e., $\Phi^\alpha = \Phi^\beta$, we obtain:

$$\Phi_{Th}^\beta - \Phi_{Th}^\alpha = \int_b^e d\Phi_{Th} = 0 \quad \Rightarrow \quad \int_b^e \phi d\mu_{Th} = \int_b^e \phi \left(\frac{d\mu_{Th}}{d\phi}\right)_{T,P} d\phi = 0, \quad (21.2.8)$$

where we have considered that the phase transition is isothermal and isobaric. From a geometrical point of view, this relation manifests the equality between the shaded area of Fig. 21.3 (Maxwell's rule), where the points b and e correspond to the equilibrium points of the two phases at that temperature and pressure, with compositions ϕ_e^α and ϕ_e^β.

Fig. 21.3 P-v diagram of
single-component
fluids or μ-φ diagram of
binary mixtures

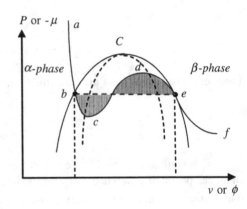

21.3 The Critical Point

For each temperature and pressure, a $(g - \phi)$-diagram can be drawn, with two
corresponding equilibrium points and two spinodal points. In particular, this would
lead to a $(\mu - T)$-diagram, representing, at each temperature, the equilibrium
chemical potential difference. The same analysis could be extended to $(\mu - P)$-
diagrams also, although we should consider that, for liquid mixtures, pressure is
rather unimportant and therefore its influence will be considered only implicitly in
the following. Now, just like in a $(P - T)$-diagram for a single component fluid, in
the $(\mu - T)$-diagram for binary mixtures we see that the phase equilibrium curve
stops at the *critical point*, characterized by a critical temperature T_C and a critical
chemical potential difference, μ_C. As the critical point is approached from below
(i.e. with two coexisting phases), the difference between the composition of the two
phases decreases, until it vanishes altogether at the critical point. At higher temperatures, $T > T_C$, the differences between the two phases vanish altogether and the
system is always in a single phase.

Accordingly, near the critical point, we obtain:

$$0 = \mu_C(T, \phi) = \mu_C(T, \phi + \delta\phi)$$
$$\Rightarrow 0 = \left(\frac{\partial\mu}{\partial\phi}\right)_{T_C} (\delta\phi) + \frac{1}{2}\left(\frac{\partial^2\mu}{\partial\phi^2}\right)_{T_C} (\delta\phi)^2 + \cdots,$$

where we have considered that, since the two phases are at equilibrium with each
other, they have the same chemical potential (in addition to having the same
pressure and temperature). At this point, dividing by $\delta\phi$ and letting $\delta\phi \to 0$, we see
that at the critical point we have:

$$(\partial\mu/\partial\phi)_{T_C} = 0. \tag{21.3.1}$$

Note that this condition is the limit case of the inequality (21.2.3), i.e., $(\partial\mu/\partial\phi)_T \geq 0$, which manifests the internal stability of any two-phase system. In addition, since near an equilibrium point [cf. Eq. (21.2.2)], $\delta g - \mu\delta\phi > 0$, expanding δg in a power series of $\delta\phi$, with constant T and P, we obtain:

$$\delta g = \left(\frac{\partial g}{\partial\phi}\right)_T (\delta\phi) + \frac{1}{2!}\left(\frac{\partial^2 g}{\partial\phi^2}\right)_T (\delta\phi)^2 + \frac{1}{3!}\left(\frac{\partial^3 g}{\partial\phi^3}\right)_T (\delta\phi)^3 + \frac{1}{4!}\left(\frac{\partial^4 g}{\partial\phi^4}\right)_T (\delta\phi)^4 + \cdots$$

Finally, considering that $(\partial g/\partial\phi)_T = \mu$, and that at the critical point $(\partial^2 g/\partial\phi^2)_T = 0$, we obtain:

$$\frac{1}{3!}\left(\frac{\partial^2\mu}{\partial\phi^2}\right)(\delta\phi)^3 + \frac{1}{4!}\left(\frac{\partial^3\mu}{\partial\phi^3}\right)(\delta\phi)^4 + \cdots > 0.$$

Since this equality must be valid for any value (albeit small) of $\delta\phi$ (both positive and negative), we obtain:

$$\left(\frac{\partial^2\mu}{\partial\phi^2}\right)_{T_C} = 0, \quad \text{and} \quad \left(\frac{\partial^3\mu}{\partial\phi^3}\right)_{T_C} > 0. \tag{21.3.2}$$

Therefore, the critical point corresponds to a horizontal inflection point in the $\mu - \phi$-diagram, which means that,

$$\left(\frac{\partial^2 g}{\partial\phi^2}\right)_{T_C} = 0, \quad \text{and} \quad \left(\frac{\partial^3 g}{\partial\phi^3}\right)_{T_C} = 0. \tag{21.3.3}$$

Below the critical point, the mixture separates into two phases having the same chemical potential, so that, at each temperature (and pressure) we can determine the equilibrium composition of the mixture, i.e.,

$$\mu^\alpha = \mu^\beta \quad \Rightarrow \quad \left(\frac{\partial g}{\partial\phi}\right)_T^\alpha = \left(\frac{\partial g}{\partial\phi}\right)_T^\beta \quad \Rightarrow \quad (\phi_e^\alpha, \phi_e^\beta). \tag{21.3.4}$$

When we represent the curve $T = T(\phi_e)$, we obtain the *equilibrium curve* (also denoted as *miscibility curve*) of Fig. 21.4, representing all points b and e of Figs. 21.2 and 21.3.[2] So, if from a single-phase state h the mixture is quenched to the unstable state f, lying inside the equilibrium curve, the mixture will phase

---

[2]In Fig. 21.4 we have assumed that the mixture presents an upper critical point. Actually, there are also cases with a lower critical point, or even cases presenting both an upper and a lower critical point.

Fig. 21.4 Equilibrium and
spinodal curves

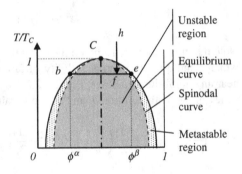

separate into the two coexisting states, represented by points b and e. In Fig. 21.4,
we have also represented the, so called, *spinodal curve*, defined as the locus $T =
T(\phi_s)$ of all points satisfying the condition,

$$\left(\frac{\partial \mu}{\partial \phi}\right)_T = \left(\frac{\partial^2 g}{\partial \phi^2}\right)_T = 0 \quad \Rightarrow \quad \left(\phi_s^\alpha, \phi_s^\beta\right). \tag{21.3.5}$$

All points lying outside the region encompassing the equilibrium curve represent
homogeneous, single-phase mixtures in a state of stable equilibrium, while all points
lying inside that region represent systems in a state of non equilibrium, which tend to
separate into two phases. In turn, all points lying in the region inside the spinodal
curve are unstable, that is any infinitesimal perturbation can trigger the phase transition process, while all points sandwiched between the equilibrium and the spinodal
curves represent metastable systems, i.e. mixtures that need a finite activation energy
to phase separate. In single-phase fluids, metastable states correspond, for example,
to undercooled water: this state can exists (i.e., its existence does not violate the
second law of thermodynamics), but, eventually, the water will freeze.

21.4 Example: Binary Symmetric Mixtures

Consider two fluids, 1 and 2. The molar Gibbs free energy is the sum of an ideal
part, g^{id}, and a so-called *excess* part, g^{ex}. The former depends on the entropy
increase due to the mixing between two ideal gases, and is therefore present in any
mixture, with,

$$g^{id} = g_1 x_1 + g_2 x_2 + RT[x_1 \ln(x_1) + x_2 \ln(x_2)],$$

where g_1 and g_2 are the molar free energies of the pure components. On the other
hand, g^{ex} depends on the non-ideality of the mixture and can vary enormously from
mixture to mixture. Here, we consider the simplest form for g^{ex}, namely the one-
parameter Margules correlation,

$$g^{ex} = RT\Psi x_1 x_2,$$

where $\Psi = \Psi(T)$. The variation of the free energy due to mixing at constant T (and P) is then:

$$\Delta g = g - [g_1 x_1 + g_2 x_2] = RT[x_1 \ln(x_1) + x_2 \ln(x_2) + \Psi x_1 x_2]. \quad (21.4.1)$$

The ideal (also referred to as *entropic*) part of Δg is always negative and concave, indicating that an ideal mixture always favors mixing. On the other hand, the non ideal (also referred to as *entalpic*) part of Δg can be positive or negative, depending on the sign of $\Psi(T)$, and therefore can accelerate or contrast mixing. In fact, considering that for an isothermal mixing,

$$\Delta g = \Delta h - T \Delta s,$$

where h is the molar enthalpy, and $\Delta s = -R(x_1 \ln x_1 + x_2 \ln x_2)$ for an ideal mixture of ideal gases, it can be shown[3] that

$$\Delta h \propto RT\Psi \propto F_{11} + F_{22} - 2F_{12},$$

where F_{ij} are the attractive forces between components i and j. Thus, Δh, and therefore Ψ, is positive when the attractive forces between molecules of the same species are larger than those between molecules of different species. A particularly important case corresponds to regular mixtures, defined as mixtures of van der Waals fluids. In this case, as the excess entropy and the excess volume are null, Δh must be independent of T and therefore $\Psi \propto 1/T$.

From Eqs. (21.1.8) and (21.4.1) we have:

$$\mu = \frac{dg}{d\phi} = (g_1 - g_2) + RT\left[\ln\left(\frac{\phi}{1-\phi}\right) + \Psi(1 - 2\phi)\right]. \quad (21.4.2)$$

and

$$\frac{d\mu}{d\phi} = \frac{d^2g}{d\phi^2} = RT\left[\frac{1}{\phi(1-\phi)} - 2\Psi\right]. \quad (21.4.3)$$

From the condition (21.2.4) of thermodynamic stability, i.e., $d^2g/d\phi^2 > 0$, we obtain:

$$\phi^2 - \phi + \frac{1}{2\Psi} \geq 0. \quad (21.4.4)$$

---

[3]See I.S. Sandler, *Chemical and Engineering Thermodynamics*, 3rd ed., Chap. 7, Wiley, New York (1999).

This inequality is always satisfied when,

$$\Psi \leq 2. \tag{21.4.5}$$

Therefore, $\Psi = 2$ corresponds to the critical point, i.e., $\Psi(T_C) = 2$, so that for regular mixtures, $\Psi = 2T_C/T$. When $T > T_C$, then $\Psi < 2$ and the $(\Delta g - \phi)$-curve is concave, indicating that the mixture is miscible at any composition (see also Fig. 21.4). Instead, when $T < T_C$, then $\Psi > 2$ and the free energy of the mixture is represented by the curve in Fig. 21.5. Here, we see that mixtures with composition ϕ^* corresponding to the point f, will phase separate. The composition of the two coexisting phases (points b and e) can be easily determined, due to the symmetry of the curve, imposing that the derivative is zero, i.e.,

$$\Delta\mu = \left(\frac{dg}{d\phi}\right)_T = 0$$

$$\Rightarrow \frac{d}{d\phi}[\phi \ln(\phi) + (1 - \phi) \ln(1 - \phi) + \Psi\phi(1 - \phi)] = 0,$$

hence:

$$\ln\frac{\phi_e}{1 - \phi_e} + \Psi(1 - 2\phi_e) = 0. \tag{21.4.6}$$

Therefore, Eq. (21.4.6) represents the equilibrium curve, while the spinodal curve corresponds to Eq. (21.4.4),

$$\phi_s^2 - \phi_s + \frac{1}{2\Psi} = 0. \tag{21.4.7}$$

At the critical point, when $\Psi = 2$, we find that the two compositions at equilibrium and the two spinodal composition coincide, with $\phi_e = \phi_s = 1/2$, as one would expect out of symmetry. In general, when $\Psi > 2$, Eqs. (21.4.6) and (21.4.7) have 2 solutions. In particular, in the vicinity of the critical point, defining,

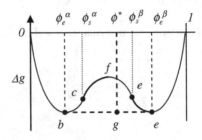

Fig. 21.5 Free energy of a binary mixture

$$\tilde{\psi} = \frac{1}{2}(\Psi - 2); \quad \tilde{u} = 2\phi - 1, \tag{21.4.8}$$

we find:

$$\tilde{u}_e = \pm\sqrt{3\tilde{\psi}}, \quad \text{and} \quad \tilde{u}_s = \pm\sqrt{\tilde{\psi}}. \tag{21.4.9}$$

21.5 Molecular Diffusion in Binary Symmetric Mixtures

In Eq. (21.1.8) we have seen that the chemical potential difference, $\mu = \mu_1 - \mu_2$, is thermodynamically conjugated to ϕ. Accordingly, it is reasonable to model molecular diffusion assuming that the molar flux of component 1, $\mathbf{J}_\phi$, is proportional to the gradient of the chemical potential difference, i.e.,

$$\mathbf{J}_\phi = -D(\phi)\nabla\mu, \tag{21.5.1}$$

where $D(\phi)$ is a composition-dependent diffusion coefficient. This constitutive relation is subjected to a constrain; it must reduce to the usual Fick law in the dilute limit, i.e. when $\phi \to 0$ and $\phi \to 0$. Now, since $\nabla\mu = (d\mu/d\phi)\nabla\phi$, and considering Eq. (21.4.3), we see that the molar flux tends to diverge in the dilute limit, unless we choose an appropriate form for $D(\phi)$. The simplest choice is the following:

$$\mathbf{J}_\phi = -\frac{D}{RT}\phi(1 - \phi)\nabla\mu, \tag{21.5.2}$$

where D is the molecular diffusivity, that here we consider to be composition-independent.

From the Gibbs-Duhem relation (21.1.6), we see that

$$\phi\nabla\mu_1 + (1 - \phi)\nabla\mu_2 = 0, \quad \text{so that} \quad \nabla\mu = \frac{1}{1 - \phi}\nabla\mu_1.$$

Therefore, we obtain for the molar flux of component 1 the following equivalent constitutive relation,

$$\mathbf{J}_\phi = -\frac{D}{RT}\phi\nabla\mu_1. \tag{21.5.3}$$

Substituting Eq. (21.4.3) into (21.5.2) we obtain:

Fig. 21.6 D^* as a function
of ϕ

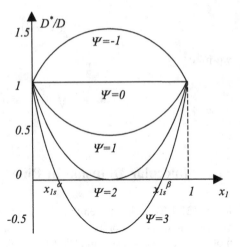

$$\mathbf{J}_\phi = -\frac{D}{RT}\phi(1-\phi)\frac{d\mu}{d\phi}\nabla\phi \quad \Rightarrow \quad \mathbf{J}_\phi = -D^*\nabla\phi, \qquad (21.5.4a)$$

where

$$D^* = D[1 - 2\Psi\phi(1-\phi)] \qquad (21.5.4b)$$

is the effective diffusion coefficient. Inverting ϕ with $(1-\phi)$ in Eq. (21.5.4a), we see that (a) the diffusivity of species 1 into 2 equals the diffusivity of species 2 into 1, as it should be, and (b) the flux of species 2 is opposite to the flux of species 1, that is $J_2 = -J_1$, showing that these are really diffusive fluxes, with no convective components. In addition, for ideal or dilute mixtures, i.e. when either $\Psi = 0$, $\phi \ll 1$ or $[1-\phi] \ll 1$, we obtain that $D^* = D$ and therefore Eq. (21.5.4a) reduces to the Fick law.

When we plot D^* as a function of ϕ, we see in Fig. 21.6 that for $\Psi < 0$, that is when particles tend to have near molecules of the other species, effective diffusivity is larger than molecular diffusion. Instead, when $\Psi > 0$, particles of the same species tend to attract each other, thereby decreasing the mixing process and the effective diffusivity. This tendency to demix, due to the enthalpic term in the expression (21.4.1), is contrasted by the entropic tendency to mix, that is intrinsic in all mixtures. Thus, there is a threshold value, i.e., $\Psi_C = 2$, where the two tendency exactly balance each other. Then, as $\Psi > 2$ there is a region of negative diffusion, corresponding to the convex region in the $(\Delta g - \phi)$-curve, where $d^2g/d\phi^2 < 0$. In that region, the mixture will phase separate through a process where species will diffuse from regions of small to large concentrations. This is what we call *anti-diffusion*.

21.6 Non-ideal Mixtures

In general, non-ideal binary mixtures are described through the activity coefficients γ_1 and γ_2, defined as the non-ideal part of the chemical potential. Therefore, we obtain:

$$g^{ex} = RT(x_1 \ln \gamma_1 + x_2 \ln \gamma_2), \tag{21.6.1}$$

which is the analogous of Eq. (21.1.7). Now, summing Eq. (21.6.1) to g^{id} and then applying (21.1.8) we obtain:

$$\mu = \frac{dg}{d\phi} = (g_1 - g_2) + RT \ln\left(\frac{\phi\gamma_1}{(1-\phi)\gamma_2}\right). \tag{21.6.2}$$

where we have considered that $\ln \gamma_i$, like μ_i, satisfies the Gibbs-Duhem relation (21.1.6):

$$x_1 \frac{d\ln\gamma_1}{dx_1} + x_2 \frac{d\ln\gamma_2}{dx_1} = 0 \quad \Rightarrow \quad x_1 \frac{d\ln\gamma_1}{dx_1} = x_2 \frac{d\ln\gamma_2}{dx_2}. \tag{21.6.3}$$

When $\ln\gamma_1 = \Psi(1-\phi)^2$ and $\ln\gamma_2 = \Psi\phi^2$, we recover the case that we have studied in the previous section.

Now, consider the constitutive relation (21.5.2) and obtain:

$$\mathbf{J}_\phi = -\frac{D}{RT}\phi(1-\phi)\frac{d\mu}{d\phi}\nabla\phi \quad \Rightarrow \quad \mathbf{J}_\phi = -D^*\nabla\phi, \tag{21.6.4}$$

where, applying the Gibbs-Duhem relation (21.6.3),

$$D^* = D\left[1 + \phi\frac{d\ln\gamma_1}{d\phi}\right] = D\left[1 + \frac{d\ln\gamma_1}{d\ln\phi}\right]. \tag{21.6.5}$$

Note that we find again that the diffusive flux of species 2 and of species 1 are opposite to each other, i.e.,

$$\mathbf{J}_2 = -D^*\nabla x_2, \tag{21.6.6}$$

where,

$$D^* = D\left[1 + x_2\frac{d\ln\gamma_2}{dx_2}\right] = D\left[1 + x_1\frac{d\ln\gamma_1}{dx_1}\right]. \tag{21.6.7}$$

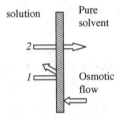

solution   Pure
                solvent

2

1   Osmotic
                flow

Fig. 21.7 Example of osmotic flow

21.7 Osmotic Flow

The osmotic flow is perhaps the most common example of antidiffusion, as it occurs whenever a membrane separates a solution from its solvent. As an example, consider the case when the membrane is permeable to the solvent (species 2), but is impermeable to the solute (species 1), as shown in Fig. 21.7. Then, the chemical potential of the solvent dissolved within the solution is lower than its value for the pure solvent; therefore, the solvent will move from the pure solvent to the solution (i.e., from right to left in Fig. 21.7). One could say that, in agreement with Le Chatelier's rule, the solvent tends to equilibrate the solute concentration on the two sides of the membrane: as the solute cannot move from left to right, then the solvent will move in the opposite direction. As a result, the pressure inside the solution will progressively increase, with a consequent decrease of the solvent flow, until equilibrium is reached. At this point, the pressure difference between the two sides of the membrane is called *osmotic pressure*, $\Delta\Pi$.

The relation between the osmotic pressure and the concentration of the solute in the solution can be determined by imposing that at equilibrium the chemical potential of the solvent must be the same on the two sides of the membrane.[4] Then we have:

$$\mu_2^{solv}(T,P) = \mu_2^{solut}(T,P+\Delta\Pi,x_1), \tag{21.7.1}$$

where $\mu_2^{solv} = g_2$ is the free energy of the pure solvent. In addition, for ideal solutions (e.g., in the dilute case),

$$\mu_2^{soluz}(T,P+\Delta\Pi,x_1) = g_2(T,P+\Delta\Pi) + RT\ln x_2 \tag{21.7.2}$$

and

$$g_2(T,P+\Delta\Pi) = g_2(T,P) + \tilde{V}_2\Delta\Pi, \tag{21.7.3}$$

---

[4]Note that the condition (21.2.5) of chemical equilibrium can be applied only to the solvent, which is free to move across the membrane.

where $\tilde{V}_2 = (\partial g_2/\partial P)_T$ is the molar volume of the solvent. In the dilute case, ln $(1 - x_1) \cong -x_1$, and we obtain the *van 't Hoff's equation*,

$$\Delta\Pi = \frac{RT}{\tilde{V}_2}x_1 = RTc_1. \tag{21.7.4}$$

where c_1 is the molar concentration of the solute (i.e., moles of solute per unit volume).

In general, when a pressure difference, ΔP, is applied between the two sides of a membrane of thickness d, the solvent mean velocity, v, will follow the Darcy equation, provided that the osmotic pressure is added (or subtracted) to ΔP, i.e.,

$$v = \frac{\kappa}{\mu d}(\Delta P \pm \sigma\Delta\Pi), \tag{21.7.5}$$

where κ is the membrane permeability, while σ is the *coefficient of reflection*, with $\sigma = 1$ when the membrane is permeable to the solvent and impermeable to the solute, and $\sigma = 0$ when the membrane is equally permeable to both solvent and solute.

Note: The osmotic pressure is often very large. For example, suppose to dissolve 3.5 g of salt (NaCl) into 96.5 g of water at 15 °C. Then, we easily find $x_1 = 0.011$, corresponding to about half the salt concentration at saturation. Finally, applying Eq. (21.7.4), we obtain:

$$\Delta\Pi = RT\,x_1/\tilde{V}_2 = 8.314 \times 288 \times 0.011/(18 \times 10^{-6}) \cong 1.46\,\text{MPa} = 14.6\,\text{atm},$$

where we have considered that the molar volume of water is $\tilde{V}_2 \cong 18\,\text{mL/mole}$.

Chapter 22
Stationary Diffusion

Abstract At steady state, pure diffusion is described by two equations, namely the Laplace equation and Stokes equation. The former, $\nabla^2 f = \frac{\partial}{\partial r_i} \frac{\partial f}{\partial r_i} = 0$, describes the stationary temperature (or concentration) profile, in the absence of any heat convection. The Stokes equation, $\nabla p = \mu \nabla^2 \mathbf{v}; \quad \nabla \cdot \mathbf{v} = 0$, describes the steady state velocity and pressure profiles at low Reynolds number conditions (that is, again, in the absence of any convective effects) of an incompressible fluid. In this chapter, we study some important properties of these problems, all centered upon the behavior of harmonic functions, as described in Sect. 22.1. Then, in Sect. 22.2, these properties are applied to determine the general solution of the Stokes equation. In particular, the uniform flow past a sphere is studied, finding the celebrated Stokes law expressing the drag force in terms of the unperturbed fluid velocity.

22.1 Harmonic Functions

A twice continuously differentiable function $f(\mathbf{r})$ of the position vector $\mathbf{r}$ is a *harmonic function* when it satisfies the Laplace equation,

$$\nabla^2 f = \frac{\partial}{\partial r_i} \frac{\partial f}{\partial r_i} = 0, \tag{22.1.1}$$

with appropriate boundary conditions. Examples of harmonic functions are the temperature and concentration fields at steady state due to diffusion only, i.e. without convection.

Harmonic functions are very regular. For example, according to the *mean value property*, the mean value of a harmonic function on the surface of a sphere is equal to its value at the center. So, if we choose the origin of the reference frame at the center of the sphere, we have[1]:

---

[1]The symbol $d^2\mathbf{r}$ indicates an elementary area on the surface of a circle of radius r, and so $d^2\mathbf{r} = r^2 \sin\theta \, d\theta \, d\phi$ in spherical coordinates. Therefore, $d^2\mathbf{n}$ refers to a sphere of unit radius.

© Springer International Publishing Switzerland 2015

R. Mauri, *Transport Phenomena in Multiphase Flows*,
Fluid Mechanics and Its Applications 112, DOI 10.1007/978-3-319-15793-1_22

$$f(0) = \frac{1}{4\pi r^2} \oint_r f(\mathbf{r}) d^2\mathbf{r}. \tag{22.1.2}$$

To prove this theorem, let us show first that the RHS of Eq. (22.1.2) does not depend on r. We see that:

$$\frac{d}{dr}\left(\frac{1}{r^2}\oint_r f(\mathbf{r})d^2\mathbf{r}\right) = \frac{d}{dr}\left(\oint_1 f(r\mathbf{n})d^2\mathbf{n}\right) = \oint_1 \mathbf{n}\cdot\nabla f(r\mathbf{n})d^2\mathbf{n},$$

where $\mathbf{n} = \mathbf{r}/r$ is a unit vector and we have considered that:

$$\partial f(\mathbf{r})/\partial r = \partial f(\mathbf{r})/\partial\mathbf{r}\cdot\partial\mathbf{r}/\partial r = \nabla f\cdot\mathbf{n},$$

Therefore we obtain:

$$\frac{1}{r^2}\oint_r \mathbf{n}\cdot\nabla f(\mathbf{r})d^2\mathbf{r} = \frac{1}{r^2}\int_r \nabla^2 f(\mathbf{r})d^3\mathbf{r} = 0,$$

where we have considered that $\nabla^2 f = 0$. Now, since we have seen that the RHS of Eq. (22.1.2) does not depend on r, taking the limit as $r \to 0$ and expanding $f(\mathbf{r})$ in Taylor series, we obtain:

$$\lim_{r\to 0}\frac{1}{4\pi r^2}\oint_r [f(0) + \mathbf{r}\cdot\nabla f(0) + \cdots]d^2\mathbf{r},$$

which yields the mean value theorem, Eq. (22.1.2). Now, integrating Eq. (22.1.2) we obtain:

$$\int_r 4\pi r^2 f(0)dr = \int_r \left[\oint_r f(\mathbf{r})d^2\mathbf{r}\right]dr,$$

and therefore we see that the mean value theorem can be expressed also in the following form:

$$f(0) = \frac{1}{\frac{4}{3}\pi r^3}\int_r f(\mathbf{r})d^3\mathbf{r}. \tag{22.1.3}$$

This shows that the value of a harmonic function at the center of a sphere is equal to its mean value inside the sphere.

An important corollary of the mean value property is that a harmonic function f defined in a volume V cannot have maximum or minimum points inside the volume; such points must necessarily lay on the external surface S of the volume.[2] From this corollary and from the continuity of f, follows that if f has a constant value on a closed surface S, then it has that same value at all points within S. A Faraday cage is the most important application of this corollary.

22.1.1 Decaying Harmonics

A decaying harmonic is a function that satisfies the Laplace equation, with vanishing boundary conditions at infinity. The fundamental decaying harmonics is $f(\mathbf{r}) = 1/r$; in addition, since the Laplace equation is linear, any gradient of $1/r$ is also a harmonic function. Thus, in general, decaying harmonic functions can be expressed as linear combinations of $1/r$ and of all its gradients. These latter are tensor harmonic functions, named vector harmonics, that can be written in invariant notation, i.e. in a form that is independent of the coordinate system and involves only the position vector, $\mathbf{r}$, and its length, $r = |\mathbf{r}|$. Thus, decaying harmonic functions are the 3D generalization of the 1D functions $f(x)$, that admit a power series expansion, $f(x) = \sum_{n=1}^{\infty} C_n x^{-n}$, for $x < a$. The complete solution of the Laplace equation should also include the growing harmonic functions, namely $\mathbf{r}$ and all its inverse gradients, as can be found in specialized texts.[3]

Among the decaying harmonic functions, the first to be considered is $1/r$, which is often referred to as a monopole and represents the Green function, or propagator, of the Laplace equation, i.e. it is the solution of the equation $\nabla^2 T = \delta(\mathbf{r})$. The monopole is the temperature field generated by an energy impulse, $\dot{Q}$, placed at the origin, that is a point source which radiates equally well in all directions. In fact, since at steady state the energy that crosses per unit time any closed surface that includes the origin is constant and equal to $\dot{Q}$, for a sphere of radius a we have,

$$\dot{Q} = \oint_{r=a} \mathbf{n} \cdot \mathbf{J}_U d^2 \mathbf{r}_S; \quad \text{with} \quad \mathbf{J}_U = -k \nabla T, \tag{22.1.4}$$

where $\mathbf{r}_S$ is a position vector located on the sphere surface, $\mathbf{n} = \mathbf{r}_S/a$ is the outer unit vector, k is the heat conductivity, and $\mathbf{J}_U$ is the heat flux, so we easily find that a temperature distribution $T = 1/r$ induces a heat flux $\dot{Q} = 4\pi k$. Accordingly, the monopole temperature distribution is often indicated as:

$$T(\mathbf{r}) = \dot{Q} T^{(m)}(\mathbf{r}), \quad \text{with} \quad T^{(m)}(\mathbf{r}) = \frac{1}{4\pi k} \frac{1}{r}, \tag{22.1.5}$$

showing that the strength of the monopole is directly related to the total heat flux.

---

[2]In fact, if at a point P inside the volume V the harmonic function f had a maximum, then there would be a small region around P where $f < f(P)$, thus contradicting the mean value theorem.

[3]See R. Mauri, *Non-equilibrium Thermodynamics in Multiphase Flows*, Springer (2013).

Due to the linearity of the Laplace equation, all the gradients of the singular fundamental solution, r^{-1}, will be (singular) solutions as well. So, for example, ∇r^{-1}, which is often referred to as a dipole distribution, is the solution of the equation $\nabla \nabla^2 T = \nabla \delta(\mathbf{r})$, and therefore it represents the temperature field generated by two monopole sources of infinitely large, equal, and opposite strengths, separated by an infinitesimal distance d, located at the origin[4]; therefore, as one monopole generates the same energy that is absorbed by the other, the dipole has no net energy release. In the same way, we can define a quadrupole, as two opposing dipoles, an octupole, as two opposing quadrupole, and so on Thus, in general, any decaying temperature field is determined as a linear combination of decaying vector harmonics, which are defined (after multiplication by convenient constants) by the following nth order tensor functions,

$$\mathbf{H}^{-(n+1)}(\mathbf{r}) = \frac{(-1)^n}{1 \times 3 \times 5 \times \cdots (2n-1)} \underbrace{\nabla\nabla\nabla\ldots\nabla}_{n\text{ times}}\left(\frac{1}{r}\right), \qquad (22.1.6)$$

for $n = 1, 2,\ldots$ In particular,

$$H^{-1}(\mathbf{r}) = -\frac{1}{r}; \quad H_i^{-2}(\mathbf{r}) = \frac{r_i}{r^3}; \quad H_{ij}^{-3}(\mathbf{r}) = \frac{r_i r_j}{r^5} - \frac{\delta_{ij}}{3r^3}. \qquad (22.1.6a)$$

Concretely, we find the so-called *multipole expansion* as follows:

$$T(\mathbf{r}) = \sum_{n=0}^{\infty} \mathbf{C}_n(\cdot)^n \mathbf{H}^{-(n+1)}(\mathbf{r}), \qquad (22.1.7)$$

where $\mathbf{C}_n$ are nth order constant tensors, representing the strength of the monopole, dipole, quadrupole, etc., to be determined by satisfying the boundary conditions and, above all, taking advantage of all the symmetries of the problem.

For example, suppose that we want to determine the temperature distribution around a sphere induced by an imposed ΔT, i.e., $T = T_\infty$, as $r \to \infty$ and $T = T_\infty + \Delta T$ at the surface $r = a$ of the sphere. Then, the temperature difference field $(T - T_\infty)$ satisfies the Laplace equation with vanishing boundary conditions at infinity, and therefore it can be expressed as the multipole expansion (22.1.7). Due to the linearity of the problem, $(T(\mathbf{r}) - T_\infty)$ must be proportional to the driving force ΔT, and therefore only the zeroth order harmonics will enter the solution, so that $(T(\mathbf{r}) - T_\infty) = \lambda \Delta T/r$, with $\lambda = a$ after matching the boundary condition at $r = a$. At the end, we find the well-known solution,

$$T(\mathbf{r}) = T_\infty + \Delta T(a/r), \qquad (22.1.8)$$

---

[4]In fact, the dipole strength is the product of the strength of the two opposite impulses by their distance.

This solution can also be written as $(T(\mathbf{r}) - T_\infty) = \dot{Q} T^{(m)}(\mathbf{r})$, with $\dot{Q} = 4\pi ka\Delta T$, showing that the temperature distribution around a sphere induced by an imposed ΔT is determined uniquely by a monopole at the center of the sphere, due to a total heat flux $\dot{Q} = 4\pi ka\Delta T$.

A more complex case is when we impose a constant temperature gradient, G_i, at infinity, i.e. $T(\mathbf{r}) = T_\infty(\mathbf{r}) = G_i r_i$ as $r \to \infty$, while $T(r = a) = 0$. Note that the unperturbed temperature distribution far from the sphere satisfies identically the Laplace equation, as it must. Then, the temperature field $(T - T_\infty)$ satisfies the Laplace equation, with vanishing boundary conditions at infinity, and therefore it can be expressed as the multipole expansion (22.1.7). Due to the linearity of the problem, $(T - T_\infty)$ must be proportional to the driving force G_i, and therefore only the first-order harmonics, $\mathbf{H}^{(-2)}$, will enter the solution, so that $(T(\mathbf{r}) - T_\infty) = \lambda \mathbf{G} \cdot \mathbf{r}/r^3$, with $\lambda = -a^3$ after matching the boundary condition at $r = a$. At the end we find:

$$T(\mathbf{r}) = (\mathbf{G} \cdot \mathbf{r})(1 - a^3/r^3). \tag{22.1.9}$$

Thus, in this case we find a dipole distribution around the sphere, with no net heat flow.

22.2 Creeping Flow

Creeping flow, also called Stokes flow, is a type of fluid regime for incompressible fluids, where convective inertial forces are small compared with viscous forces, and therefore the Reynolds number is small. The equations of motion for creeping flow are the Stokes equations, consisting of the stationary Navier-Stokes equations, where the inertial terms are neglected, i.e.,

$$\nabla p = \mu \nabla^2 \mathbf{v}; \quad \nabla \cdot \mathbf{v} = 0. \tag{22.2.1a, b}$$

Here, $\mathbf{v}(\mathbf{r})$ and $p(\mathbf{r})$ are the velocity and pressure field, while μ is the constant fluid viscosity.

Taking the divergence of (22.2.1a) and considering Eq. (22.2.1b) (i.e., the velocity field is solenoidal), we see that the pressure field is a harmonic function, i.e., it satisfies the Laplace equation,

$$\nabla^2 p = 0. \tag{22.2.2}$$

In addition, it is easy to verify that the velocity field can be written as,

$$\mathbf{v}(\mathbf{r}) = \frac{1}{2\mu} \mathbf{r} p(\mathbf{r}) + \mathbf{u}(\mathbf{r}), \tag{22.2.3}$$

where $\mathbf{u}(\mathbf{r})$ is a harmonic vector function, called *homogeneous velocity*, with,

$$\nabla^2 \mathbf{u} = 0. \tag{22.2.4}$$

Obviously, only three of the four harmonic functions defining p and $\mathbf{v}$ are independent, as the condition that the velocity field is divergence-free must be implemented, obtaining:

$$\nabla \cdot \mathbf{u} = -\frac{1}{2\mu}(3p + \mathbf{r} \cdot \nabla p). \tag{22.2.5}$$

Therefore, we have seen that the solutions of the Stokes equations are determined in terms of harmonic functions, in one case for a scalar field (i.e., the pressure) and in the other for a vector field (i.e., the homogeneous velocity).

22.2.1 Stokeslet

The simplest application of these procedure arises when we calculate the propagator, or Green function, of the Stokes equation, that is the velocity and pressure fields induced in an unbounded and otherwise quiescent fluid by a point body force located at the origin, $\mathbf{f}(\mathbf{r}) = \mathbf{F}\delta(\mathbf{r})$, where $\mathbf{F}$ is a force that is applied to the fluid. Then, both the pressure p and the homogeneous velocity $\mathbf{u}$ are decaying harmonic function, proportional to $\mathbf{F}$, i.e.,

$$p = \lambda F_k \frac{r_k}{r^3}, \tag{22.2.6a}$$

and,

$$u_i = \frac{F_k}{2\mu}\left[\lambda'\frac{\delta_{ik}}{r} + \lambda''\left(\frac{r_i r_k}{r^5} - \frac{\delta_{ik}}{3r^3}\right)\right]. \tag{22.2.6b}$$

Now, since λ'' has the units of a square length, considering that there is no characteristic dimension in this problem. It must be: $\lambda'' = 0$. Consequently, imposing that the velocity field is divergence free, so that Eq. (22.2.5) is satisfied, we see that $\lambda = \lambda'$, and therefore we obtain:

$$v_i = \frac{F_k}{2\mu}\lambda\left(\frac{\delta_{ik}}{r} + \frac{r_i r_k}{r^3}\right). \tag{22.2.7}$$

The value of λ can be determined imposing that the total force applied to the fluid located outside a sphere of radius a must be equal to $\mathbf{F}$. Thus, proceeding as for Eq. (22.1.6), we find:

$$\oint_{r=a} n_i T_{ij} d^2 \mathbf{r}_S = -F_i, \tag{22.2.8}$$

where $\mathbf{r}_S$ is a position vector located on the sphere surface, $\mathbf{n} = \mathbf{r}_S/a$ is the outer unit vector and the minus sign reflects the fact that $\mathbf{F}$ is a force that is applied to the fluid. Here, T_{ij} is the stress tensor,

$$T_{ij} = -p\delta_{ij} + \mu\left(\frac{\partial v_i}{\partial r_k} + \frac{\partial v_k}{\partial r_i}\right) = -3\lambda F_k \frac{r_i r_j r_k}{r^5}. \tag{22.2.9}$$

Now, considering that

$$n_i T_{ij}\big|_{r=a} = -3\lambda F_k \frac{r_j r_k}{a^4},$$

together with the following identity[5]

$$\oint_{r=a} r_i r_k d^2 \mathbf{r}_S = \frac{4}{3}\pi a^4 \delta_{ik},$$

we finally obtain: $\lambda = 1/4\pi$. Therefore, the Stokeslet has the following pressure and velocity field:

$$p(\mathbf{r}) = F_k P_k^{(s)}(\mathbf{r}), \quad \text{with } P_k^{(s)}(\mathbf{r}) = \frac{1}{4\pi}\frac{r_k}{r^3}, \tag{22.2.10a}$$

$$v_i(\mathbf{r}) = F_k V_{ik}^{(s)}(\mathbf{r}), \quad \text{with } V_{ik}^{(s)}(\mathbf{r}) = \frac{1}{8\pi\mu r}\left(\delta_{ik} + \frac{r_i r_k}{r^2}\right), \tag{22.2.10b}$$

The multiplier $\mathbf{V}^{(s)}(\mathbf{r})$ is called *Oseen tensor*, describing the disturbance to an unperturbed flow due to a point force.

The Stokeslet plays in creeping flow the same role that is played in heat conduction by the monopole $1/r$ solution. Thus, the disturbance to the flow field due to the presence of a submerged object can be expressed as a multipole expansion in terms of the gradients of the Stokeslet, quite similar to Eq. (22.1.7).

---

[5]When $i \neq k$, the integral is zero and therefore it must be proportional to δ_{ik}. Then, multiplying both members by δ_{ik} and considering that $\delta_{ik}\delta_{ik} = 3$, we easily verify the identity.

22.2.2 Uniform Flow Past a Sphere

In this case, the unperturbed velocity field is uniform, with $\mathbf{v}(\mathbf{r}) = \mathbf{U}$ as $r \to \infty$, while the sphere is kept fixed at the origin, i.e. $\mathbf{v}(\mathbf{r}) = \mathbf{0}$ at $r = a$. Then, proceeding as before, we obtain the same expressions (22.2.6a),

$$p = \lambda a \mu U_k \frac{r_k}{r^3}, \tag{22.2.11a}$$

and

$$v_i = U_i + \frac{1}{2} a U_k \left[\lambda \left(\frac{\delta_{ik}}{r} + \frac{r_i r_k}{r^3} \right) + \lambda'' \left(\frac{r_i r_k}{r^5} - \frac{\delta_{ik}}{3r^3} \right) \right]. \tag{22.2.11b}$$

where $\lambda = \lambda'$, so that the divergence-free condition for the velocity field, i.e., Eq. (22.2.5), is satisfied. Note that now $\lambda'' \neq 0$, since we have a characteristic dimension a. Imposing that $\mathbf{v}(\mathbf{r}) = \mathbf{0}$ at $r = a$, we find: $\lambda = -\lambda''/a^2 = -3/2$. So, at the end, we find:

$$p = -\frac{3}{2} a \mu U_k \frac{r_k}{r^3}, \tag{22.2.12a}$$

and

$$v_i(\mathbf{r}) = U_i - \frac{3}{4} a U_k \left[\left(\frac{\delta_{ik}}{r} + \frac{r_i r_k}{r^3} \right) - a^2 \left(\frac{r_i r_k}{r^5} - \frac{\delta_{ik}}{3r^3} \right) \right]. \tag{22.2.12b}$$

At large r, the flow perceives only a point force $\mathbf{F}$, so that the pressure and the velocity fields must reduce to the Stokeslet (22.2.10a). Therefore, $F_k/4\pi = 3/2 a \mu U_k$, i.e.,

$$\mathbf{F} = 6\pi \mu a \mathbf{U}. \tag{22.2.13}$$

This is the Stokes law, establishing the drag force exerted on a sphere by a uniform fluid flow.

Note that Eq. (22.2.12b) can be expressed in terms of the following singular solutions of the Stokes equation:

$$\mathbf{v}(\mathbf{r}) = \mathbf{U} - \mathbf{F}_0 \cdot \left[\mathbf{V}^{(s)}(\mathbf{r}) + \mathbf{V}^{(d)}(\mathbf{r}) \right], \tag{22.2.14}$$

where $\mathbf{F}_0$ is the Stokes drag force (22.2.13), $\mathbf{V}^{(s)}$ is the Oseen tensor (22.2.10b), while $\mathbf{V}^{(d)}$ is defined as:

$$\mathbf{V}^{(d)}(\mathbf{r}) = \frac{a^2}{6}\nabla^2\mathbf{V}^{(s)}(\mathbf{r}) = \frac{a^2}{8\pi\mu r^3}\left(\frac{\mathbf{rr}}{r^2} - \frac{\mathbf{I}}{3}\right) = \frac{a^2}{8\pi\mu}\mathbf{H}^{(-3)}(\mathbf{r}). \qquad (22.2.15)$$

Here, $\mathbf{H}^{-3}$ is the harmonic tensorial function (22.1.6a) with $n = 2$, which is referred to as a potential doublet, since it is identical to a doublet in potential flow. From a different point of view, imposing that Eq. (22.2.14) satisfies the Stokes equation, we see that $\nabla^2\mathbf{V}^{(d)} = \mathbf{0}$, and therefore the pressure field associated with the potential doublet is identically zero, that is, $\mathbf{P}^{(d)} = \mathbf{0}$. Consequently, according to the d'Alambert paradox, the drag force exerted by the potential doublet (as well as by any potential flow) is equal zero.

22.2.3 Faxen's Law

From Eq. (22.2.14) we see that $\mathbf{V}^{(d)}$ is the disturbance to the unperturbed flow (that here consists of a Stokeslet), $\mathbf{v}^\infty(\mathbf{r}) = \mathbf{U} - \mathbf{F}_0 \cdot \mathbf{V}^{(s)}(\mathbf{r})$, due to the presence of a sphere. In fact, Eq. (22.2.14) can be reformulated as:

$$\mathbf{v}(\mathbf{r}) = \mathbf{U} - \mathbf{F} \cdot \mathbf{V}^{(s)}(\mathbf{r}), \qquad (22.2.16)$$

with,

$$\mathbf{F} = 6\pi\mu a\left(1 + \frac{a^2}{6}\nabla^2\right)\mathbf{v}^\infty(\mathbf{0}). \qquad (22.2.17)$$

This result is generally referred to as *Faxen's law*, stating that the force exerted on a rigid sphere by a known unperturbed flow field can be calculated directly, without actually solving the flow field problem, and it depends only on the undisturbed velocity at the center of the sphere and its Laplacian. As we saw, this is equivalent to saying that the net force is determined only by a Stokeslet and a potential doublet located at the center of the sphere (i.e., the origin, in this case).

In the previous analysis, we have assumed that the flow field vanishes at infinity, or, equivalently, that the velocity of the sphere is null. In general, denoting by $\mathbf{V}$ the velocity of a sphere centered in the origin, Faxen's law can be written as:

$$\mathbf{V} = \left(1 + \frac{a^2}{6}\nabla^2\right)\mathbf{v}^\infty(\mathbf{0}) - \frac{\mathbf{F}}{6\pi\mu a}. \qquad (22.2.18)$$

Here we see that the velocity of a neutrally buoyant sphere, i.e., with $\mathbf{F} = \mathbf{0}$, immersed in a linear flow field, coincides with the unperturbed fluid velocity at its center. However, this is not true in general. For example, the velocity of a neutrally buoyant sphere immersed in a Poiseuille flow field is lower than the unperturbed fluid velocity at its center.

An identical relation can also be written for the temperature, T, and the total heat flux, $\dot{Q}$, of a sphere with infinite thermal conductivity (so that temperature is constant inside the sphere), immersed in medium with unperturbed temperature T^∞, obtaining:

$$T = T^\infty(0) + \frac{\dot{Q}}{4\pi ka}. \qquad (22.2.19)$$

Here we have considered that the force $\mathbf{F}$ exerted on the sphere in Eq. (22.2.18) is replaced by the net heat flux $\dot{Q}$ released by the sphere in Eq. (22.2.19), and that the temperature field is harmonic, so that $\nabla^2 T^\infty = 0$.

Similar relations can be written in terms of the angular velocity of a sphere and the torque exerted on it.

Appendix A
Properties of Pure Components at 1 atm

Fluid	ρ (kg/m^3)	μ (10^{-2} g/cm*s)	ν (10^{-2} cm^2/s)	σ^a (dyn/cm)	k (W/m*K)	c_p (J/Kg*K)	α (10^{-2} cm^2/s)
Gas at 300 K							
Air	1.16	0.0185	15.9	0 (misc.)	0.0263	1010	22.5
NH$_3$	0.692	0.0103	14.8	0 (misc.)	0.0246	2298	15.5
CO$_2$	1.789	0.0149	8.40	0 (misc.)	0.0166	852	10.9
CH$_4$	0.644	0.0111	17.3	0 (misc.)	0.0342	2240	23.7
N$_2$	1.12	0.0178	15.10	0 (misc.)	0.0259	1040	22.1
H$_2$	0.0819	0.00896	109	0 (misc.)	0.182	14320	155
O$_2$	1.31	0.0200	11.6	0 (misc.)	0.027	911	22.6
Liquid at 300 K							
Acetone	782	0.331	0.423	24	0.169	2180	0.0991
Water	988	1.002	1.014	73	0.600	4180	0.143
Water, 100 °C	958	0.279	0.0291	59	0.670	4220	0.168
Ethanol	802	1.05	1.31	22.5	0.168	2460	0.0853
Methanol	785	0.53	0.675	23	0.200	2480	0.103
Benzene	881	0.58	0.658	29	0.144	1730	0.0945
Glycerin	1260	1490	1200	63	0.287	2380	0.95
Mercury	1350	1.51	0.114	435	0.00858	139	4.56
Olive oil	916	84	91.7	35	–	–	–
Oil (SAE-5 W-30)	860	96.3	112	36.5	0.138	1850	0.0867
Oil SAE-10 W-30)	872	108	124	35	0.136	1840	0.0855
Oil (Castor)	970	986	1016	35	0.145	2161	0.069
Propanol	803	1.72	2.14	24	0.154	2477	0.0774
Solid at 300 K							
Aluminum	2702	–	–	–	236	902	97
Chrome	7160	–	–	–	95	451	29
Cupper	8933	–	–	–	401	385	116

(continued)

© Springer International Publishing Switzerland 2015
R. Mauri, *Transport Phenomena in Multiphase Flows*,
Fluid Mechanics and Its Applications 112, DOI 10.1007/978-3-319-15793-1

Fluid	ρ (kg/m^3)	μ (10^{-2} g/cm*s)	v (10^{-2} cm^2/s)	σ^a (dyn/cm)	k (W/m*K)	c_p (J/Kg*K)	α (10^{-2} cm^2/s)
Steel (inox)	7900	–	–	–	14	477	4.0
Iron	7870	–	–	–	83	440	23
Uranium	19070	–	–	–	27	116	12
Brick	1600	–	–	–	0.7	840	0.52
Carbon	1370	–	–	–	0.24	1260	0.14
Clay	1500	–	–	–	1.4	880	1.1
Sand	1500	–	–	–	0.3	800	0.25
Glass for windows	2700	–	–	–	0.84	800	0.39
Glass Pyroceram	2600	–	–	–	4.1	810	1.9
Ice	920	–	–	–	2.2	2000	1.2
Polystyrene	50	–	–	–	0.025	–	–
Cork	160	–	–	–	0.043	1900	0.14
Granite	2640	–	–	–	3.0	800	1.4
Human skin	–	–	–	–	0.37	–	–
Wood (oak)	600	–	–	–	0.17	2400	0.12
Wood (compressed)	550	–	–	–	0.12	1200	0.18
Wool	200	–	–	–	0.038	–	–

^{a}At an interface with air

Appendix B
Viscosity and Surface Tension of Selected Fluids

Fluids a 27 °C	Viscosity		Srf. Tens.[a] (dyn/cm)
	Shear rate (s^{-1})	Viscosity (cP)	
Newtonian liquids			
Water	225–450	1.0	73
Oil (sunflower)	22–45	42	33.5
Oil (olive)	22–45	39	33.5
Soy sauce	225–450	3.11	53
White vinegar	225–450	0.82	56
Anti-cough syrup	90–450	17.5	47
Spic and span	90–450	7.2	31
Coca Cola (classic)	90–450	1.4	50
Whole milk	225–450	2	48
Skimmed milk	225–450	1.3	50
Shear-thinning liquids			
"Mylanta" anti-acid	22	50	29
	45	43	
"Dawn" dish detergent	2	185	24
	11	158	
"J.&J." Baby Shampoo	1	364	32
	2.3	321	
"Coppertone" sun lotion	4.5	75	33
	11	72	
	22	67	

[a]At an interface with air

© Springer International Publishing Switzerland 2015
R. Mauri, *Transport Phenomena in Multiphase Flows*,
Fluid Mechanics and Its Applications 112, DOI 10.1007/978-3-319-15793-1

Appendix C
Conversion Factors

Constant

R gas universal constant = 8.31 J/(g × mol × K) = 1544 ft × lbf/(lb × mol × °F)

N Avogadro number = 6.02 10^{23} mol^{-1}

Length	Volume
1 ft = 12 in = 0.3048 m 1 in = 0.0254 m 1 mi(mile) = 1.609 km 1 yd(yard) = 3 ft = 0.9144 m	1 L(litro) = 10^3 cm^3 1 gal(U.S. gallon) = 3.785 L = 3785 cm^3 = 231 in^3 = 0.1337 ft^3
Mass	**Density**
1 lbm (pound-mass) = 16 oz = 0.45359 kg = 7000 grains 1 oz (ounces) = 1/16 lbm = 28.35 g 1 ton = 2000 lb = 907 kg	1 lb/ft^3 = 16.02 kg/m^2 1 g/cm^3 = 10^3 kg/m^3
Force	**Pressure**
1 dyne = 10^{-5} N 1 lbf (pound-force) = 4.448 N 1 poundal = 0.138 N	1 bar = 10^5 Pa = 10^5 N/m^2 = 100 kPa 1 atm (atmosphere) = 101.325 kPa 1 psi = 6.895 kPa 1 mm Hg (Torr) = 133.3224 Pa
Energy	**Power**
1 erg = 1 dyne * cm = 10^{-7} J 1 Btu (British Thermal Units) = 1054 J 1 cal (calorie) = 4.1868 J 1 kWh (kilowattore) = 1 kW × h = 3.6 × 10^6 J	1 hp (horsepower) = 0.746 kW 1 Btu/s = 1.41 hp = 1.054 kW 1 ft × lbf/s = 1.356 W
Thermal conductivity	**Specific heat**
1 Btu/(h × ft × °F) = 1.731 W/(m × K) 1 cal/(s × cm × K) = 418.4 W/(m × K)	1 Btu/(lb × °F) = 4190 J/(Kg × K) cal/(g × K) = 4190 J/(Kg × K)
Diffusivity and kinematic viscosity	**Dynamic viscosity**
1 ft^2/s = 0.0929 m^2/s 1 ft^2/h = 2.58 × 10^{-5} m^2/s	1 cP (centipoise) = 10^{-3} kg/(m × s) 1 poise = 1 g/(cm × s) = 0.1 kg/(m × s) 1 lb/(ft × s) = 1.49 kg/(m × s)
Surface tension	**Heat transfer coefficient**
1 lbf/ft = 14.59 N/m	1 Btu/(ft^2 × h × °F) = 5.68 W/(m^2 × K) 1 cal/(cm^2 × s × K) = 4.184 × 10^4 W/(m^2 × K)

© Springer International Publishing Switzerland 2015

R. Mauri, *Transport Phenomena in Multiphase Flows*,
Fluid Mechanics and Its Applications 112, DOI 10.1007/978-3-319-15793-1

Appendix D
Governing Equations

D.1 Cartesian Coordinates x, y, z

Components of the stress tensor:

$$T_{xx} = -p + 2\mu\frac{\partial v_x}{\partial x} \quad \left| \quad T_{xy} = T_{yx} = \mu\left(\frac{\partial v_x}{\partial y} + \frac{\partial v_y}{\partial x}\right) \quad \right| \quad T_{xz} = T_{zx} = \mu\left(\frac{\partial v_x}{\partial z} + \frac{\partial v_z}{\partial x}\right)$$
$$T_{yz} = T_{zy} = \mu\left(\frac{\partial v_y}{\partial z} + \frac{\partial v_z}{\partial y}\right) \quad \left| \quad T_{yy} = -p + 2\mu\frac{\partial v_y}{\partial y} \quad \right| \quad T_{zz} = -p + 2\mu\frac{\partial v_z}{\partial z}$$

The continuity equation and the Navier-Stokes equations are:

$$\frac{\partial v_x}{\partial x} + \frac{\partial v_y}{\partial y} + \frac{\partial v_z}{\partial z} = 0$$

$$\frac{Dv_x}{Dt} = \frac{\partial v_x}{\partial t} + v_x\frac{\partial v_x}{\partial x} + v_y\frac{\partial v_x}{\partial y} + v_z\frac{\partial v_x}{\partial z} = -\frac{1}{\rho}\frac{\partial P}{\partial x} + \nu\left(\frac{\partial^2 v_x}{\partial x^2} + \frac{\partial^2 v_x}{\partial y^2} + \frac{\partial^2 v_x}{\partial z^2}\right),$$

$$\frac{Dv_y}{Dt} = \frac{\partial v_y}{\partial t} + v_x\frac{\partial v_y}{\partial x} + v_y\frac{\partial v_y}{\partial y} + v_z\frac{\partial v_y}{\partial z} = -\frac{1}{\rho}\frac{\partial P}{\partial y} + \nu\left(\frac{\partial^2 v_y}{\partial x^2} + \frac{\partial^2 v_y}{\partial y^2} + \frac{\partial^2 v_y}{\partial z^2}\right),$$

$$\frac{Dv_z}{Dt} = \frac{\partial v_z}{\partial t} + v_x\frac{\partial v_z}{\partial x} + v_y\frac{\partial v_z}{\partial y} + v_z\frac{\partial v_z}{\partial z} = -\frac{1}{\rho}\frac{\partial P}{\partial z} + \nu\left(\frac{\partial^2 v_z}{\partial x^2} + \frac{\partial^2 v_z}{\partial y^2} + \frac{\partial^2 v_z}{\partial z^2}\right).$$

The heat equation is:

$$\frac{DT}{dt} = \frac{\partial T}{\partial t} + v_x\frac{\partial T}{\partial x} + v_y\frac{\partial T}{\partial y} + v_z\frac{\partial T}{\partial z} = \alpha\left(\frac{\partial^2 T}{\partial x^2} + \frac{\partial^2 T}{\partial y^2} + \frac{\partial^2 T}{\partial z^2}\right) + \frac{\dot{q}}{\rho c_p}.$$

In the following, we report the above equations using cylindrical r, ϕ, z, and spherical, r, θ, ϕ, coordinates, where:

$$x = r\cos\phi; \quad y = r\sin\phi; \quad z = z$$

© Springer International Publishing Switzerland 2015
R. Mauri, *Transport Phenomena in Multiphase Flows*,
Fluid Mechanics and Its Applications 112, DOI 10.1007/978-3-319-15793-1

and

$$x = r \sin\theta \cos\phi; \quad y = r \sin\theta \sin\phi; \quad z = r \cos\theta.$$

D.2 Cylindrical Coordinates r, ϕ, z

Components of the stress tensor:

$$T_{rr} = -p + 2\mu \frac{\partial v_r}{\partial r} \quad\bigg| \quad T_{r\phi} = T_{\phi r} = \mu\left(\frac{1}{r}\frac{\partial v_r}{\partial \phi} + \frac{\partial v_\phi}{\partial r} - \frac{1}{r}v_\phi\right) \quad\bigg| \quad T_{rz} = T_{zr} = \mu\left(\frac{\partial v_r}{\partial z} + \frac{\partial v_z}{\partial r}\right)$$

$$T_{\phi z} = T_{z\phi} = \mu\left(\frac{\partial v_\phi}{\partial z} + \frac{1}{r}\frac{\partial v_z}{\partial \phi}\right) \quad\bigg| \quad T_{\phi\phi} = -p + 2\mu\left(\frac{1}{r}\frac{\partial v_\phi}{\partial \phi} + \frac{1}{r}v_r\right) \quad\bigg| \quad T_{zz} = -p + 2\mu\frac{\partial v_z}{\partial z}$$

The continuity equation and the Navier-Stokes equations are:

$$\frac{1}{r}\frac{\partial(rv_r)}{\partial r} + \frac{1}{r}\frac{\partial v_\phi}{\partial \phi} + \frac{\partial v_z}{\partial z} = 0$$

$$\frac{Dv_r}{Dt} = \frac{\partial v_r}{\partial t} + v_r\frac{\partial v_r}{\partial r} + \frac{v_\phi}{r}\frac{\partial v_r}{\partial \phi} - \frac{v_\phi^2}{r} + v_z\frac{\partial v_r}{\partial z} = -\frac{1}{\rho}\frac{\partial P}{\partial r} + v\left[\frac{\partial}{\partial r}\left(\frac{1}{r}\frac{\partial}{\partial r}(rv_r)\right) + \frac{1}{r^2}\frac{\partial^2 v_r}{\partial \phi^2} - \frac{2}{r^2}\frac{\partial v_\phi}{\partial \phi} + \frac{\partial^2 v_r}{\partial z^2}\right],$$

$$\frac{Dv_\phi}{Dt} = \frac{\partial v_\phi}{\partial t} + v_r\frac{\partial v_\phi}{\partial r} + \frac{v_\phi}{r}\frac{\partial v_\phi}{\partial \phi} + \frac{v_\phi v_r}{r} + v_z\frac{\partial v_\phi}{\partial z} = -\frac{1}{\rho}\frac{1}{r}\frac{\partial P}{\partial \phi} + v\left[\frac{\partial}{\partial r}\left(\frac{1}{r}\frac{\partial}{\partial r}(rv_\phi)\right) + \frac{1}{r^2}\frac{\partial^2 v_\phi}{\partial \phi^2} + \frac{2}{r^2}\frac{\partial v_r}{\partial \phi} + \frac{\partial^2 v_\phi}{\partial z^2}\right],$$

$$\frac{Dv_z}{Dt} = \frac{\partial v_z}{\partial t} + v_r\frac{\partial v_z}{\partial r} + \frac{v_\phi}{r}\frac{\partial v_z}{\partial \phi} + v_z\frac{\partial v_z}{\partial z} = -\frac{1}{\rho}\frac{\partial P}{\partial z} + v\left[\frac{1}{r}\frac{\partial}{\partial r}\left(r\frac{\partial v_z}{\partial r}\right) + \frac{1}{r^2}\frac{\partial^2 v_z}{\partial \phi^2} + \frac{\partial^2 v_z}{\partial z^2}\right].$$

The heat equation is:

$$\frac{DT}{dt} = \frac{\partial T}{\partial t} + v_r\frac{\partial T}{\partial r} + \frac{v_\phi}{r}\frac{\partial T}{\partial \phi} + v_z\frac{\partial T}{\partial z} = \alpha\left[\frac{1}{r}\frac{\partial}{\partial r}\left(r\frac{\partial T}{\partial r}\right) + \frac{1}{r^2}\frac{\partial^2 T}{\partial \phi^2} + \frac{\partial^2 T}{\partial z^2}\right] + \frac{\dot{q}}{\rho c_p}.$$

D.3 Spherical Coordinates r, θ, ϕ

Components of the stress tensor:

$$T_{rr} = -p + 2\mu\frac{\partial v_x}{\partial x} \quad\bigg| \quad T_{r\theta} = T_{\theta r} = \mu\left(\frac{1}{r}\frac{\partial v_r}{\partial \theta} + \frac{\partial v_\theta}{\partial r} - \frac{1}{r}v_\theta\right)$$

$$T_{r\phi} = T_{\phi r} = \mu\left(\frac{\partial v_\phi}{\partial r} + \frac{1}{r\sin\theta}\frac{\partial v_r}{\partial \phi} - \frac{v_\phi}{r}\right) \quad\bigg| \quad T_{\theta\theta} = -p + 2\mu\left(\frac{1}{r}\frac{\partial v_\theta}{\partial \theta} + \frac{1}{r}v_r\right)$$

$$T_{\theta\phi} = T_{\phi\theta} = \mu\left(\frac{1}{r\sin\theta}\frac{\partial v_\theta}{\partial \phi} + \frac{1}{r}\frac{\partial v_\phi}{\partial \theta} - \frac{v_\phi\cos\theta}{r\sin\theta}\right) \quad\bigg| \quad T_{\phi\phi} = -p + 2\mu\left(\frac{1}{r\sin\theta}\frac{\partial v_\phi}{\partial \phi} + \frac{v_r}{r} + \frac{v_\theta\cos\theta}{r\sin\theta}\right)$$

The continuity equation and the Navier-Stokes equations are:

$$\frac{1}{r^2}\frac{\partial(r^2 v_r)}{\partial r} + \frac{1}{r\sin\theta}\frac{\partial v_\phi}{\partial\phi} + \frac{1}{r\sin\theta}\frac{\partial(v_\theta\sin\theta)}{\partial\theta} = 0$$

$$\frac{Dv_r}{Dt} = \frac{\partial v_r}{\partial t} + v_r\frac{\partial v_r}{\partial r} + \frac{v_\theta}{r}\frac{\partial v_r}{\partial\theta} + \frac{v_\phi}{r\sin\theta}\frac{\partial v_r}{\partial\phi} - \frac{v_\theta^2 + v_\phi^2}{r} = -\frac{1}{\rho}\frac{\partial P}{\partial r}$$
$$+ v\left[\frac{1}{r^2}\frac{\partial^2(r^2 v_r)}{\partial r^2} + \frac{1}{r^2\sin\theta}\frac{\partial}{\partial\theta}\left(\sin\theta\frac{\partial v_r}{\partial\theta}\right) + \frac{1}{r^2\sin^2\theta}\frac{\partial^2 v_r}{\partial\phi^2}\right],$$

$$\frac{Dv_\theta}{Dt} = \frac{\partial v_\theta}{\partial t} + v_r\frac{\partial v_\theta}{\partial r} + \frac{v_\theta}{r}\frac{\partial v_\theta}{\partial\theta} + \frac{v_\phi}{r\sin\theta}\frac{\partial v_\theta}{\partial\phi} + \frac{v_r v_\theta}{r} - \frac{v_\phi^2\cot\theta}{r} = -\frac{1}{\rho}\frac{1}{r}\frac{\partial P}{\partial\theta}$$
$$+ v\left[\frac{1}{r^2}\frac{\partial}{\partial r}\left(r^2\frac{\partial v_\theta}{\partial r}\right) + \frac{1}{r^2}\frac{\partial}{\partial\theta}\left(\frac{1}{\sin\theta}\frac{\partial}{\partial\theta}(v_\theta\sin\theta)\right) + \frac{1}{r^2\sin^2\theta}\frac{\partial^2 v_\theta}{\partial\phi^2}\right.$$
$$\left. + \frac{2}{r^2}\frac{\partial v_r}{\partial\theta} - \frac{2\cos\theta}{r^2\sin^2\theta}\frac{\partial v_\phi}{\partial\phi}\right],$$

$$\frac{Dv_\phi}{Dt} = \frac{\partial v_\phi}{\partial t} + v_r\frac{\partial v_\phi}{\partial r} + \frac{v_\theta}{r}\frac{\partial v_\phi}{\partial\theta} + \frac{v_\phi}{r\sin\theta}\frac{\partial v_\phi}{\partial\phi} + \frac{v_r v_\phi}{r} + \frac{v_\theta v_\phi\cot\theta}{r} = -\frac{1}{\rho}\frac{1}{r\sin\theta}\frac{\partial P}{\partial\phi}$$
$$+ v\left[\frac{1}{r^2}\frac{\partial}{\partial r}\left(r^2\frac{\partial v_\phi}{\partial r}\right) + \frac{1}{r^2}\frac{\partial}{\partial\theta}\left(\frac{1}{\sin\theta}\frac{\partial}{\partial\theta}(v_\phi\sin\theta)\right)\right.$$
$$\left. + \frac{1}{r^2\sin^2\theta}\frac{\partial^2 v_\phi}{\partial\phi^2} + \frac{2}{r^2\sin\theta}\frac{\partial v_r}{\partial\phi} + \frac{2\cos\theta}{r^2\sin^2\theta}\frac{\partial v_\theta}{\partial\phi}\right].$$

The heat equation is:

$$\frac{DT}{dt} = \frac{\partial T}{\partial t} + v_r\frac{\partial T}{\partial r} + \frac{v_\theta}{r}\frac{\partial T}{\partial\theta} + \frac{v_\phi}{r\sin\theta}\frac{\partial T}{\partial\phi}$$
$$= \alpha\left[\frac{1}{r^2}\frac{\partial}{\partial r}\left(r^2\frac{\partial T}{\partial r}\right) + \frac{1}{r^2\sin\theta}\frac{\partial}{\partial\theta}\left(\sin\theta\frac{\partial T}{\partial\theta}\right) + \frac{1}{r^2\sin^2\theta}\frac{\partial^2 T}{\partial\phi^2}\right] + \frac{\dot{q}}{\rho c_p}.$$

Appendix E
Balance Equations (Eulerian Approach)

In this Section, we want to study the transport of a physical quantity F, namely mass ($F = M$), momentum ($F = Q_1, Q_2, Q_3$) and energy ($F = U$), by considering the balance equation in an elementary fixed volume dV, as indicated in Fig. E.1,

$$\frac{dF}{dt} = \dot{F}_{in} - \dot{F}_{out} + \dot{F}_{gen}. \tag{E.1}$$

Here, dF/dt is the temporal variation of F in dV, $\dot{F}_{in}$ and $\dot{F}_{out}$ represent the amount of F entering or exiting dV, while $\dot{F}_{gen}$ is a source term, that is the amount of F generated in dV. Now, let us define the density f of F per unit mass, i.e., $(dF) = \rho f (dV)$, where ρ is the density. In addition, from the definition of flux, J, as the amount of F crossing a section of unit area per unit time, we have: $\dot{F}_{in} = J_{in}S_{in}$ and $\dot{F}_{out} = J_{out}S_{out}$, where S is the area of the cross section. Also, define the amount of F generated per unit mass as $\dot{f}_{gen} = \dot{F}_{gen}/(\rho V)$. Finally, when Eq. (E.1) is applied to the volume $dV = dx_1 dx_2 dx_3$ shown in Fig. E.1, we obtain:

$$\frac{\partial}{\partial t}(\rho f)dx_1 dx_2 dx_3 = J_1(x_1)dx_2 dx_3 - J_1(x_1 + dx_1)dx_2 dx_3 + J_2(x_2)dx_1 dx_3 - J_2(x_2 + dx_2)dx_1 dx_3$$
$$+ J_3(x_3)dx_1 dx_2 - J_3(x_3 + dx_3)dx_1 dx_2 + \rho \dot{f}_{gen}dx_1 dx_2 dx_3.$$

Fig. E.1 Balance of any physical quantity in an elementary volume

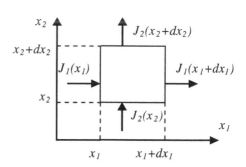

© Springer International Publishing Switzerland 2015

391

R. Mauri, *Transport Phenomena in Multiphase Flows*,
Fluid Mechanics and Its Applications 112, DOI 10.1007/978-3-319-15793-1

Now, expanding with Taylor,

$$J(x + dx) = J(x) + \frac{dJ}{dx} dx + O(dx)^2,$$

find:

$$\frac{\partial}{\partial t}(\rho f) = -\left(\frac{\partial J_1}{\partial x_1} + \frac{\partial J_2}{\partial x_2} + \frac{\partial J_3}{\partial x_3}\right) + \rho \dot{f}_{gen}, \qquad (E.2)$$

that is,

$$\frac{\partial}{\partial t}(\rho f) + \nabla \cdot \mathbf{J} = \rho \dot{f}_{gen}, \qquad (E.3)$$

where $\nabla \cdot \mathbf{J}$ is the divergence of the vector $\mathbf{J} = (J_1, J_2, J_3)$.

Let us consider now a few particular cases.

E.1 Conservation of Mass (for One-phase, One-component Fluid)

In this case:

(i) $f = 1$.
(ii) $\mathbf{J} = \mathbf{J}_M = \rho \mathbf{v}$, where $\mathbf{v}$ is the fluid velocity.
(iii) $\dot{f}_{gen} = 0$, because mass cannot be created.

Therefore, we find:

$$\frac{\partial \rho}{\partial t} + \nabla \cdot (\rho \mathbf{v}) = 0. \qquad (E.4)$$

Obviously, for incompressible fluids ρ is constant and the continuity equation becomes:

$$\frac{\partial v_1}{\partial x_1} + \frac{\partial v_2}{\partial x_2} + \frac{\partial v_3}{\partial x_3} = \nabla \cdot \mathbf{v} = 0, \qquad (E.5)$$

indicating that the velocity field is solenoidal, that is divergence-free.

E.2 Conservation of Momentum (for Newtonian Incompressible Fluids)

Let us consider the balance equation for the momentum along the x_1-direction. Then,

(i) $f = v_1$.

(ii) The momentum flux along the x_1-direction is the sum of a convective part, equal to $\rho v_1 \mathbf{v}$ and a diffusive part that, for Newtonian fluids, equal to $-\mu \nabla v_1$. Therefore:

$$J_1 = \rho v_1 v_1 - \mu \partial v_1 / \partial x_1; \quad J_2 = \rho v_1 v_2 - \mu \partial v_1 / \partial x_2$$
$$J_3 = \rho v_1 v_3 - \mu \partial v_1 / \partial x_3.$$

(iii) $\rho \dot{f}_{gen}$ represents the forces (that is, the momentum rate of change with time) per unit volume that are exerted on the fluid along the x_1-direction. Therefore, it is equal to the sum of a volumetric force, ρg_1, and a pressure force. This latter is equal to the difference between the pressure forces exerted on the two sides of the volume element sketched in Fig. E.1, that is $[P(x_1) - P(x_1 + dx_1)]dx_2 dx_3$. Therefore, expanding with Taylor, we find: $\rho \dot{f}_{gen} = \rho g_1 - \partial P / \partial x_1$.
Finally, assuming constant viscosity, the momentum balance equation along the x_1-direction is:

$$\frac{\partial(\rho v_1)}{\partial t} + \frac{\partial}{\partial x_1}(\rho v_1 v_1) + \frac{\partial}{\partial x_2}(\rho v_2 v_1) + \frac{\partial}{\partial x_3}(\rho v_3 v_1) + \frac{\partial P}{\partial x_1}$$
$$= \mu\left(\frac{\partial^2 v_1}{\partial x_1^2} + \frac{\partial^2 v_1}{\partial x_2^2} + \frac{\partial^2 v_1}{\partial x_3^2}\right) + \rho g_1.$$

Hence, assuming incompressibility, i.e., constant ρ, and applying Eq. (E.5), we obtain the Navier-Stokes equation,

$$\rho\left(\frac{\partial v_1}{\partial t} + v_1 \frac{\partial v_1}{\partial x_1} + v_2 \frac{\partial v_1}{\partial x_2} + v_3 \frac{\partial v_1}{\partial x_3}\right) + \frac{\partial P}{\partial x_1} = \mu\left(\frac{\partial^2 v_1}{\partial x_1^2} + \frac{\partial^2 v_1}{\partial x_2^2} + \frac{\partial^2 v_1}{\partial x_3^2}\right) + \rho g_1.$$

$$(E.6)$$

In vector form, we have:

$$\frac{\partial(\mathbf{v})}{\partial t} + \mathbf{v} \cdot \nabla \mathbf{v} + \frac{1}{\rho}\nabla P = \nu \nabla^2 \mathbf{v} + \mathbf{g}, \qquad (E.7)$$

where $\nabla^2 = \nabla \cdot \nabla$ is the Laplacian, and $\nu = \mu/\rho$ is the kinematic viscosity.

E.3 Conservation of Energy (Heat Equation for Incompressible Fluids)

In this case,

(i) $f = c\Delta T$, where c is the specific heat (for incompressible fluids, the specific heat at constant volume is equal to that at constant pressure), while ΔT is the fluid temperature, measured with respect to a reference temperature.

(ii) The heat flux is the sum of a convective term, $\rho c \Delta T \mathbf{v}$, and a diffusive component, $-k\nabla T$, where we have used the Fourier constitutive relation. Therefore,

$$J_1 = \rho c \Delta T v_1 - k\partial T/\partial x_1; \quad J_2 = \rho c \Delta T v_2 - k\partial T/\partial x_2;$$
$$J_3 = \rho c \Delta T v_3 - k\partial T/\partial x_3.$$

(iii) $\rho \dot{f}_{gen}$ represents the heat power generated per unit volume $\rho \dot{f}_{gen} = \dot{q}$.

Finally, assuming constant thermal conductivity k, we obtain:

$$\rho c \left(\frac{\partial T}{\partial t} + \frac{\partial (v_1 \Delta T)}{\partial x_1} + \frac{\partial (v_2 \Delta T)}{\partial x_2} + \frac{\partial (v_3 \Delta T)}{\partial x_3} \right) = k \left(\frac{\partial^2 T}{\partial x_1^2} + \frac{\partial^2 T}{\partial x_2^2} + \frac{\partial^2 T}{\partial x_3^2} \right) + \dot{q}.$$

Substituting the incompressibility condition (E.5), we finally obtain the heat equation,

$$\frac{\partial T}{\partial t} + \mathbf{v} \cdot \nabla T = \alpha \, \nabla^2 T + \frac{\dot{q}}{\rho c_p}, \tag{E.8}$$

where $\alpha = k/c$ is the coefficient of thermal diffusivity.

E.4 Conservation of Chemical Species (for Incompressible Fluids with Constant Total Concentration)

In this case:

(i) $f = c_A$, where c_A is the mass or molar concentration of species A.

(ii) The material (mass- or molar-based) flux of A is the sum of a convective term, $c_A \mathbf{v}$, and a diffusive component, $-cD\nabla x_A$, where $\mathbf{v}$ is the mean (mass or molar) velocity, c is the total concentration, $x_A = c_A/c$, is the molar or mass fraction of A, and D is the diffusion coefficient, appearing in the Fick constitutive relation. Therefore,

$$J_1 = c_A v_1 - cD\partial x_A/\partial x_1; \quad J_2 = c_A v_2 - cD\partial x_A/\partial x_2$$
$$J_3 = c_A v_3 - cD\partial x_A/\partial x_3.$$

(iii) $\rho \dot{f}_{gen}$ represents the number of moles of A generated per unit time and volume, due to chemical reaction, $\rho \dot{f}_{gen} = R_A$.

Finally we obtain:

$$\frac{\partial c_A}{\partial t} + \frac{\partial}{\partial x_1}\left(v_1^* c_A\right) + \frac{\partial}{\partial x_2}\left(v_2^* c_A\right) + \frac{\partial}{\partial x_3}\left(v_3^* c_A\right)$$
$$= D\left[\frac{\partial}{\partial x_1}\left(c\frac{\partial x_A}{\partial x_1}\right) + \frac{\partial}{\partial x_2}\left(c\frac{\partial x_A}{\partial x_2}\right) + \frac{\partial}{\partial x_3}\left(c\frac{\partial x_A}{\partial x_3}\right)\right] + R_A.$$

Therefore, assuming that c is constant and substituting the incompressibility condition (E.5), we finally obtain the following balance equation for chemical species,

$$\frac{\partial c_A}{\partial t} + \mathbf{v}^* \cdot \nabla c_A = D\nabla^2 c_A + R_A. \tag{E.9}$$

Appendix F
Introduction to Linear Algebra

F.1 Tensor and Vector Representation

General definitions

Consider three mutually orthogonal unit vectors $\mathbf{e}_1$, $\mathbf{e}_2$, $\mathbf{e}_3$, defining a system of Cartesian axes. Any vector $\mathbf{v}$ can be written as a linear combination of $\mathbf{e}_1$, $\mathbf{e}_2$, $\mathbf{e}_3$, that is,

$$\mathbf{v} = v_1\mathbf{e}_1 + v_2\mathbf{e}_2 + v_3\mathbf{e}_3 = \sum_{i=1}^{3} v_i\mathbf{e}_i, \tag{F.1.1}$$

where v_i are the component of $\mathbf{v}$. Geometrically, a vector is a directed segment and the components v_i are its projections along the Cartesian axes. Often, we have adopted the *Einstein summation convention*, that implies summation over a set of indexed terms in a formula, thus achieving notational brevity, i.e.,

$$\mathbf{v} = \sum_{i=1}^{3} v_i\mathbf{e}_i \equiv v_i\mathbf{e}_i. \tag{F.1.2}$$

Equation (F.1.1) can be generalized, defining a second-order tensor as follows,

$$\mathbf{T} = T_{11}\mathbf{e}_1\mathbf{e}_1 + T_{12}\mathbf{e}_1\mathbf{e}_2 + \cdots + T_{33}\mathbf{e}_3\mathbf{e}_3 = \sum_{i,j=1}^{3} T_{ij}\mathbf{e}_i\mathbf{e}_j = T_{ij}\mathbf{e}_i\mathbf{e}_j. \tag{F.1.3}$$

Further generalization to n-th order tensors is obvious.

Vectors and tensors can also be represented as strings and matrices,

$$\mathbf{v} = (v_1, v_2, v_3); \quad \mathbf{T} = \begin{pmatrix} T_{11} & T_{12} & T_{13} \\ T_{21} & T_{22} & T_{23} \\ T_{31} & T_{32} & T_{33} \end{pmatrix}. \tag{F.1.4}$$

© Springer International Publishing Switzerland 2015
R. Mauri, *Transport Phenomena in Multiphase Flows*,
Fluid Mechanics and Its Applications 112, DOI 10.1007/978-3-319-15793-1

Sums among vectors and tensors

We can sum (and subtract) vector and tensors as follows,

$$\mathbf{v} = \mathbf{u} + \mathbf{w} \quad \Leftrightarrow \quad v_i = u_i + w_i; \qquad \mathbf{T} = \mathbf{S} + \mathbf{U} \quad \Leftrightarrow \quad T_{ij} = S_{ij} + U_{ij}. \quad \text{(F.1.5)}$$

The transpose of a tensor

Define the transpose of a given tensor $\mathbf{T}$, indicated as $\mathbf{T}^+$, as the tensor corresponding to the transpose matrix, where rows and columns are exchanged, that is,

$$T_{ij}^+ = T_{ji}. \qquad \text{(F.1.6)}$$

Symmetric and antisymmetric tensors

A tensor is symmetric when $\mathbf{T} = \mathbf{T}^+$, that is $T_{ij} = T_{ij}^+ = T_{ji}$, while, on the contrary, it is anti-symmetric when $\mathbf{T} = -\mathbf{T}^+$, that is $T_{ij} = -T_{ij}^+ = -T_{ji}$. Therefore, a symmetric tensor $\mathbf{T}^{(s)}$ is composed of 6 independent elements, while an anti-symmetric tensor has all the diagonal elements equal to zero (that is, $T_{ii}^{(a)} = 0$) and is composed of only 3 independent elements. In matricial form, $\mathbf{T}^{(s)}$ and $\mathbf{T}^{(a)}$ have the following structures,

$$\mathbf{T}^{(s)} = \begin{pmatrix} T_{11}^{(s)} & T_{12}^{(s)} & T_{13}^{(s)} \\ T_{12}^{(s)} & T_{22}^{(s)} & T_{23}^{(s)} \\ T_{13}^{(s)} & T_{23}^{(s)} & T_{33}^{(s)} \end{pmatrix} \quad \text{and} \quad \mathbf{T}^{(a)} = \begin{pmatrix} 0 & T_{12}^{(a)} & T_{13}^{(a)} \\ -T_{12}^{(a)} & 0 & T_{23}^{(a)} \\ -T_{13}^{(a)} & -T_{23}^{(a)} & 0 \end{pmatrix}. \quad \text{(F.1.7)}$$

Tensor decomposition

Any tensor $\mathbf{T}$ can always be decomposed as the sum of a symmetric and an anti-symmetric tensors,

$$T_{ij} = \frac{1}{2}(T_{ij} + T_{ji}) + \frac{1}{2}(T_{ij} - T_{ji}) = T_{ij}^{(s)} + T_{ij}^{(a)}, \qquad \text{(F.1.8)}$$

with $T_{ij}^{(s)} = \frac{1}{2}(T_{ij} + T_{ji})$ and $T_{ij}^{(a)} = \frac{1}{2}(T_{ij} - T_{ji})$, that is,

$$\mathbf{T} = \underbrace{\begin{pmatrix} T_{11} & \frac{1}{2}(T_{12} + T_{21}) & \frac{1}{2}(T_{13} + T_{31}) \\ \frac{1}{2}(T_{12} + T_{21}) & T_{22} & \frac{1}{2}(T_{23} + T_{32}) \\ \frac{1}{2}(T_{13} + T_{31}) & \frac{1}{2}(T_{23} + T_{32}) & T_{33} \end{pmatrix}}_{\mathbf{T}^{(s)}}$$

$$+ \underbrace{\begin{pmatrix} 0 & \frac{1}{2}(T_{12} - T_{21}) & \frac{1}{2}(T_{13} - T_{31}) \\ \frac{1}{2}(T_{21} - T_{12}) & 0 & \frac{1}{2}(T_{23} - T_{32}) \\ \frac{1}{2}(T_{31} - T_{13}) & \frac{1}{2}(T_{32} - T_{23}) & 0 \end{pmatrix}}_{\mathbf{T}^{(a)}}$$

Dyadics

A *dyadic* is a second order tensor, formed by juxtaposing pairs of vectors:

$$\mathbf{uv} = u_i v_j \mathbf{e}_i \mathbf{e}_j = \begin{pmatrix} u_1 v_1 & u_1 v_2 & u_1 v_3 \\ u_2 v_1 & u_2 v_2 & u_2 v_3 \\ u_3 v_1 & u_3 v_2 & u_3 v_3 \end{pmatrix}. \tag{F.1.9}$$

From Eq. (F.1.3) we see that a tensor can always be written as a combination of elementary dyadics.

Unit tensor

The unit tensor, $\mathbf{I}$, is defined as:

$$\mathbf{I} = \delta_{ij} \mathbf{e}_i \mathbf{e}_j = \mathbf{e}_i \mathbf{e}_i = \begin{pmatrix} 1 & 0 & 0 \\ 0 & 1 & 0 \\ 0 & 0 & 1 \end{pmatrix}, \tag{F.1.10}$$

where δ_{ij} is the so-called Kronecker's delta,

$$I_{ij} = \delta_{ij} = \begin{cases} 1 & \text{per } i = j; \\ 0 & \text{per } i \neq j. \end{cases} \tag{F.1.11}$$

The unit tensor is invariant to any rotation of the Cartesian axes. It can also be defined through its property (see below): $\mathbf{v}.\mathbf{I} = \mathbf{I}.\mathbf{v} = \mathbf{v}$.

Levi-Civita tensor

The Levi-Civita tensor $\boldsymbol{\varepsilon}$ is a third-order, antisymmetric tensor, defined so that $\varepsilon_{ijk} = 1$ when (ijk) is an even permutation of (123), $\varepsilon_{ijk} = -1$ when (ijk) is an odd permutation of (123), and $\varepsilon_{ijk} = 0$ in all the other cases, that is when there are two repeated indices, i.e.,

$$\boldsymbol{\varepsilon} = \varepsilon_{ijk} \mathbf{e}_i \mathbf{e}_j \mathbf{e}_k, \quad \text{where } \varepsilon_{ijk} = \begin{cases} +1 & \text{when } ijk = 123,\ 231 \text{ or } 312; \\ -1 & \text{when } ijk = 132,\ 213 \text{ or } 321; \\ 0 & \text{otherwise (when } i = j,\ j = k \text{ or } i = k) \end{cases} \tag{F.1.12}$$

The Levi-Civita tensor is also invariant to any rotation of the Cartesian axes. However, it changes sign under mirror reflection, when clockwise rotations are transformed into counter-clockwise rotations.

Inner product

The *inner product* (also called *scalar product*, or *dot product*) between two vectors
u and **v** is a scalar c, defined as:

$$c = \mathbf{u} \cdot \mathbf{v} = uv \cos \theta_{uv}, \tag{F.1.13}$$

where u and v are the magnitudes of **u** and **v**, while θ_{uv} is the angle between the two
vectors. Note that since u, v, and θ_{uv} are invariant to a rotation of the reference
frame, the inner product is also invariant.

In a Cartesian frame, we have:

$$\mathbf{e}_i \cdot \mathbf{e}_j = \delta_{ij}, \tag{F.1.14}$$

where the Kronecker delta is defined in (F.1.11). Then, we obtain:

$$\mathbf{v} \cdot \mathbf{e}_j = (v_1\mathbf{e}_1 + v_2\mathbf{e}_2 + v_3\mathbf{e}_3) \cdot \mathbf{e}_j = v_i(\mathbf{e}_i \cdot \mathbf{e}_j) = v_i\delta_{ij} = v_j, \tag{F.1.15}$$

showing that the j-th component of a vector equals indeed its projection along the
x_j-axis. Generalizing this result we obtain:

$$c = \mathbf{u} \cdot \mathbf{v} = u_i\mathbf{e}_j \cdot v_j\mathbf{e}_j = u_iv_j\delta_{ij} = u_iv_i = u_1v_1 + u_2v_2 + u_3v_3. \tag{F.1.16}$$

In the case of tensors we have:

$$\mathbf{u} = \mathbf{T} \cdot \mathbf{v} \quad \Rightarrow \quad u_i = T_{ij}v_j, \tag{F.1.17}$$

that is:

$$\mathbf{u} = \begin{pmatrix} u_1 \\ u_2 \\ u_3 \end{pmatrix} = \mathbf{T} \cdot \mathbf{v} = \begin{pmatrix} T_{11} & T_{12} & T_{13} \\ T_{21} & T_{22} & T_{23} \\ T_{31} & T_{32} & T_{33} \end{pmatrix} \cdot \begin{pmatrix} v_1 \\ v_2 \\ v_3 \end{pmatrix}$$
$$= \begin{pmatrix} T_{11}v_1 + T_{12}v_2 + T_{13}v_3 \\ T_{21}v_1 + T_{22}v_2 + T_{23}v_3 \\ T_{31}v_1 + T_{32}v_2 + T_{33}v_3 \end{pmatrix}.$$

Therefore, we see that a tensor can also be defined as the proportionality term
between two vectors. For example, in Eq. (5.3.2), i.e., $\mathbf{f}_n = \mathbf{T} \cdot \mathbf{e}_n$, we see that the
stress tensor **T** is the proportionality term between the force vector $\mathbf{f}_n$ exerted on a
surface and the unit vector $\mathbf{e}_n$ perpendicular to the surface. Therefore, since both $\mathbf{f}_n$
and $\mathbf{e}_n$ are vectors, we can conclude that **T** is a tensor, and therefore it can be
decomposed as in Eq. (F.1.4). In the same way we have:

$$\mathbf{S} \cdot \mathbf{T} = \mathbf{W} \quad \Rightarrow \quad S_{ij}T_{jk} = W_{ik}. \tag{F.1.18}$$

Note the inner product has been defined so that the row elements of the first term, $\mathbf{S}$, are multiplied by the column elements of the second component, $\mathbf{T}$.

Double inner product

As we have seen, the inner product between two vectors is a scalar, while the inner product between two tensor is a tensor. Now, define the double inner product as:

$$\left(\mathbf{e}_i\mathbf{e}_j\right)\!:\!\left(\mathbf{e}_k\mathbf{e}_\ell\right) = \delta_{i\ell}\delta_{jk} = \begin{cases} 1 & \text{if } i=\ell \text{ e } j=k, \\ 0 & \text{otherwise.} \end{cases} \tag{F.1.19}$$

Then, considering the decomposition (F.1.4) of a tensor, we obtain:

$$c = \mathbf{S}:\mathbf{T} = \left(S_{ij}\mathbf{e}_i\mathbf{e}_j\right):\left(T_{k\ell}\mathbf{e}_k\mathbf{e}_\ell\right) = \delta_{i\ell}\delta_{jk}S_{ij}T_{k\ell} = S_{ij}T_{ji}. \tag{F.1.20}$$

Invariants

An invariant is a scalar quantity that remains invariant when the reference frame is rotated. It must be obtained by multiplying a vector or a tensor by itself of by one of the isotropic tensors, $\mathbf{I}$ and $\boldsymbol{\varepsilon}$. For a vector $\mathbf{v}$, there is only one invariant, i.e. its magnitude v (or its square),

$$v^2 = v_1^2 + v_2^2 + v_3^2 = v_i v_i = \delta_{ij} v_i v_j = \mathbf{I}:\mathbf{vv}. \tag{F.1.21}$$

For a second-order tensor $\mathbf{T}$, there are three invariants. The first two are the followings:

$$\mathbf{I}:\mathbf{T} = T_{ii} = T_{11} + T_{22} + T_{33} = Tr(\mathbf{T}), \tag{F.1.22}$$

$$T^2 = \mathbf{T}:\mathbf{T} = T_{ij}T_{ji}. \tag{F.1.23}$$

$Tr(\mathbf{T})$ is the trace of the tensor $\mathbf{T}$, while T is its magnitude. The third invariant of a second-order tensor corresponds to its determinant, which is defined as,

$$\det(\mathbf{T}) = \frac{1}{6}\varepsilon_{ikm}\varepsilon_{j\ell n}T_{ij}T_{k\ell}T_{mn}. \tag{F.1.24}$$

Outer product

The *vector product* (also called *vector product*, or *cross product*) between two vectors $\mathbf{u}$ and $\mathbf{v}$ is a vector $\mathbf{w}$ defined as:

$$\mathbf{w} = \mathbf{u} \times \mathbf{v} = uv\sin\theta_{uv}\mathbf{e}_{uv}, \tag{F.1.25}$$

where θ_{uv} is the angle between $\mathbf{u}$ and $\mathbf{v}$, while $\mathbf{e}_{uv}$ is a unit vector normal to the plane containing $\mathbf{u}$ and $\mathbf{v}$ and pointing in the direction that a right-handed screw will move if turned from $\mathbf{u}$ to $\mathbf{v}$. Note that,

$$\mathbf{u} \times \mathbf{v} = -\mathbf{v} \times \mathbf{u}.$$

In addition, when $\mathbf{u}$ and $\mathbf{v}$ are perpendicular to one another, the axes $\mathbf{u}$, $\mathbf{v}$, and $(\mathbf{u} \times \mathbf{v})$ form a Cartesian reference frame.

Based on the definition (F.1.25), we have:

$$\begin{array}{llll} \mathbf{e}_1 \times \mathbf{e}_1 = \mathbf{0} & \mathbf{e}_1 \times \mathbf{e}_2 = \mathbf{e}_3 & \mathbf{e}_1 \times \mathbf{e}_3 = -\mathbf{e}_2 & \\ \mathbf{e}_2 \times \mathbf{e}_1 = -\mathbf{e}_3 & \mathbf{e}_2 \times \mathbf{e}_2 = \mathbf{0} & \mathbf{e}_2 \times \mathbf{e}_3 = \mathbf{e}_1 & \text{that is: } \mathbf{e}_i \times \mathbf{e}_j = \varepsilon_{ijk}\mathbf{e}_k, \\ \mathbf{e}_3 \times \mathbf{e}_1 = \mathbf{e}_2 & \mathbf{e}_3 \times \mathbf{e}_2 = -\mathbf{e}_1 & \mathbf{e}_3 \times \mathbf{e}_3 = \mathbf{0} & \end{array}$$

$$\text{(F.1.26)}$$

where ε_{ijk} is the Levi-Civita tensor (F.1.12). Therefore, using the decomposition (F.1.11), we obtain:

$$\mathbf{u} \times \mathbf{v} = u_i \mathbf{e}_i \times v_j \mathbf{e}_j = \varepsilon_{ijk} u_i v_j \mathbf{e}_k = \begin{pmatrix} u_2 v_3 - u_3 v_2 \\ u_3 v_1 - u_1 v_3 \\ u_1 v_2 - u_2 v_1 \end{pmatrix}. \qquad \text{(F.1.27)}$$

F.2 Vector Differential Operators

Gradient

The spatial derivatives of a field variable (e.g., velocity, temperature and concentration) are evaluated using the *gradient operator* ∇, defined as:

$$\nabla = \mathbf{e}_1 \frac{\partial}{\partial x_1} + \mathbf{e}_2 \frac{\partial}{\partial x_2} + \mathbf{e}_3 \frac{\partial}{\partial x_3} = \mathbf{e}_i \frac{\partial}{\partial x_i}. \qquad \text{(F.2.1)}$$

Practically, the ∇-operator behaves like a normal vector, although it does not commute with other operations (trivially, for example, $\mathbf{x} \cdot \nabla \neq \nabla \cdot \mathbf{x}$). The ∇-operator can be applied to both scalar and vector fields, that is:

$$\nabla f = \mathbf{e}_1 \frac{\partial f}{\partial x_1} + \mathbf{e}_2 \frac{\partial f}{\partial x_2} + \mathbf{e}_3 \frac{\partial f}{\partial x_3} = \mathbf{e}_i \frac{\partial f}{\partial x_i} = \begin{pmatrix} \dfrac{\partial f}{\partial x_1} & \dfrac{\partial f}{\partial x_2} & \dfrac{\partial f}{\partial x_3} \end{pmatrix}; \qquad \text{(F.2.2)}$$

$$\nabla \mathbf{v} = \mathbf{e}_i \frac{\partial \mathbf{v}}{\partial x_i} = \mathbf{e}_i \mathbf{e}_j \frac{\partial v_j}{\partial x_i} = \begin{pmatrix} \partial v_1/\partial x_1 & \partial v_2/\partial x_1 & \partial v_3/\partial x_1 \\ \partial v_1/\partial x_2 & \partial v_2/\partial x_2 & \partial v_3/\partial x_2 \\ \partial v_1/\partial x_3 & \partial v_2/\partial x_3 & \partial v_3/\partial x_3 \end{pmatrix}. \qquad \text{(F.2.3)}$$

Divergence

The *divergence* of a vector (or tensor) field equals the inner product of the gradient operator by the field vector,

$$\nabla \cdot \mathbf{v} = \left(\mathbf{e}_i \frac{\partial}{\partial x_i} \right) \cdot (v_j \mathbf{e}_j) = \frac{\partial v_i}{\partial x_i} = \frac{\partial v_1}{\partial x_1} + \frac{\partial v_2}{\partial x_2} + \frac{\partial v_3}{\partial x_3}; \qquad (F.2.4)$$

$$\nabla \cdot \mathbf{T} = \left(\mathbf{e}_i \frac{\partial}{\partial x_i} \right) \cdot (T_{jk} \mathbf{e}_j \mathbf{e}_k) = \frac{\partial T_{ik}}{\partial x_i} \mathbf{e}_k = \begin{pmatrix} \frac{\partial T_{11}}{\partial x_1} + \frac{\partial T_{21}}{\partial x_2} + \frac{\partial T_{31}}{\partial x_3} \\ \frac{\partial T_{12}}{\partial x_1} + \frac{\partial T_{22}}{\partial x_2} + \frac{\partial T_{32}}{\partial x_3} \\ \frac{\partial T_{13}}{\partial x_1} + \frac{\partial T_{23}}{\partial x_2} + \frac{\partial T_{33}}{\partial x_3} \end{pmatrix}. \qquad (F.2.5)$$

Curl

The *curl* of a vector (or tensor) field equals the cross product of the gradient operator by the field vector,

$$\nabla \times \mathbf{v} = \left(\mathbf{e}_i \frac{\partial}{\partial x_i} \right) \times (v_j \mathbf{e}_j) = \varepsilon_{ijk} \frac{\partial v_j}{\partial x_i} \mathbf{e}_k = \begin{pmatrix} \frac{\partial v_3}{\partial x_2} - \frac{\partial v_2}{\partial x_3} \\ \frac{\partial v_1}{\partial x_3} - \frac{\partial v_3}{\partial x_1} \\ \frac{\partial v_2}{\partial x_1} - \frac{\partial v_1}{\partial x_2} \end{pmatrix}. \qquad (F.2.6)$$

Laplacian

The *Laplacian* is the inner product of the gradient operator with itself,

$$\nabla^2 = \nabla \cdot \nabla = \frac{\partial}{\partial x_i} \frac{\partial}{\partial x_i} = \frac{\partial^2}{\partial x_1^2} + \frac{\partial^2}{\partial x_2^2} + \frac{\partial^2}{\partial x_3^2}. \qquad (F.2.7)$$

Material derivative

The *material derivative* (also called *substantial derivative*) is the time rate of change of a physical quantity referred to a material element that moves together with the fluid,

$$\frac{D}{Dt} = \frac{\partial}{\partial t} + \mathbf{v} \cdot \nabla = \frac{\partial}{\partial t} + v_i \frac{\partial}{\partial x_i}, \qquad (F.2.8)$$

where $\mathbf{v}$ is the fluid velocity. In 1D, the material derivative expresses the chain rule:

$$\frac{d}{dt} f[x(t), t] = \frac{\partial f}{\partial t} + \frac{\partial f}{\partial x} \frac{dx}{dt}.$$

F.3 Integral Theorems

Divergence theorem

In studying the transport of a physical quantity F, we often encounter the flux of F exiting (or entering) a control volume, V, by crossing its outer surface, S. As shown in Sect. 6.1, a balance of F leads to Eq. (6.1.2), that is, neglecting the source term.

$$-\frac{dF}{dt} = -\int_V \frac{\partial(\rho f)}{\partial t} dV = \oint_S J_n dS, \tag{F.3.1}$$

where J_n is the F-flux, i.e., the amount of F that crosses, per unit time, the unit area directed along the direction of a unit vector $\mathbf{e}_n$. Then, the Cauchy lemma, shown in Sect. 6.1, states that J_n is the projection of a vector flux, $\mathbf{J} = (J_1, J_2, J_3)$, along $\mathbf{e}_n$, that is,

$$J_n = \mathbf{e}_n \cdot \mathbf{J}. \tag{F.3.2}$$

In Eq. (F.3.1) we have defined $dF = \rho f dV$, where dV is the volume element, ρ is the mass density, and f is then the mass density of F.

Now, applying Eq. (F.3.1) to a volume $dV = dx_1 dx_2 dx_3$ (see Fig. F.1) we obtain:

$$(\mathbf{e}_n \cdot \mathbf{J}) dS = [J_1(x_1 + dx_1) - J_1(x_1)] dx_2 dx_3 + [J_2(x_2 + dx_2) - J_2(x_2)] dx_1 dx_3$$
$$+ [J_3(x_3 + dx_3) - J_3(x_3)] dx_1 dx_2.$$

Thus, expanding with Taylor,

$$J(x + dx) = J(x) + \frac{dJ}{dx} dx + O(dx)^2, \tag{F.3.3}$$

we may conclude:

$$(\mathbf{e}_n \cdot \mathbf{J}) dS = \left(\frac{\partial J_1}{\partial x_1} + \frac{\partial J_2}{\partial x_2} + \frac{\partial J_3}{\partial x_3} \right) dV = (\nabla \cdot \mathbf{J}) dV,$$

Fig. F.1 Fluxes crossing the outer surfaces of an elementary volume

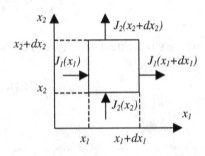

where $\nabla \cdot \mathbf{J}$ is the divergence of the vector field $\mathbf{J} = (J_1, J_2, J_3)$. Then:

$$\oint_S (\mathbf{e}_n \cdot \mathbf{J}) dS = \int_V (\nabla \cdot \mathbf{J}) dV. \qquad (F.3.4)$$

This is the *divergence theorem*, also called *Gauss-Ostrogradsky's theorem*, or *Green's theorem*. In 1D, it gives trivially,

$$\int_a^b \frac{df}{dx} dx = f(b) - f(a).$$

Finally, substituting Eq. (F.3.4) into (F.3.1) we obtain:

$$\int_V \left[\frac{\partial(\rho f)}{\partial t} + \nabla \cdot \mathbf{J} \right] dV = 0 \quad \Rightarrow \quad \frac{\partial(\rho f)}{\partial t} + \nabla \cdot \mathbf{J} = 0, \qquad (F.3.5)$$

In tensor notation, the divergence theorem can be expressed as follows,

$$\oint_S (\mathbf{e}_n \cdot \mathbf{T}) dS = \int_V (\nabla \cdot \mathbf{T}) dV, \quad \text{that is,} \quad \oint_S (e_i T_{ij}) dS = \int_V \left(\frac{\partial T_{ij}}{\partial x_i} \right) dV,$$

or, when $\mathbf{T} = f\mathbf{I}$,

$$\oint_S (\mathbf{e}_n f) dS = \int_V (\nabla f) dV, \quad \text{that is,} \quad \oint_S (e_i f) dS = \int_V \left(\frac{\partial f}{\partial x_i} \right) dV.$$

Leibnitz' formula and Reynolds' transport theorem
Consider a volume element $V(t)$ moving (and deforming) together with a material fluid element. Proceeding as in the previous Section, we can determine the integral balance equation of any physical quantity F, obtaining Eq. (F.3.1),

$$-\frac{d}{dt} \int_{V(t)} (\rho f) dV = \oint_{S(t)} J_m dS = \oint_{S(t)} (\mathbf{e}_n \cdot \mathbf{J}_r) dS, \qquad (F.3.6)$$

where we have applied the Cauchy lemma (F.3.2). Here, though, the volume and its bounding surface, are moving in time, so that now $\mathbf{J}_r$ in a relative flux, expressing the *net* amount of F crossing the outer surface dS. This implies that, if F moves with an absolute velocity $\mathbf{v}$ and dS moves with velocity $\mathbf{w}$, the velocity with which F crosses dS will be $\mathbf{v} - \mathbf{w}$ and therefore $\mathbf{J}_r = \mathbf{J} - \rho f \mathbf{w}$. For example, in the case of mass transport, $f = 1$, $\mathbf{J} = \rho\mathbf{v}$ and so the relative flux is $\rho(\mathbf{v} - \mathbf{w})$. Now, the RHS of Eq. (F.3.6) becomes,

$$\oint_{S(t)} (\mathbf{e}_n \cdot \mathbf{J}_r)dS = \oint_{S(t)} (\mathbf{e}_n \cdot \mathbf{J})dS - \oint_{S(t)} [\mathbf{e}_n \cdot (\rho f\mathbf{w})]dS$$

$$= \oint_{V(t)} (\nabla \cdot \mathbf{J})dV - \oint_{S(t)} [\mathbf{e}_n \cdot (\rho f\mathbf{w})]dS$$

and applying the balance equation (F.3.5) we obtain:

$$\frac{d}{dt} \int_{V(t)} (\rho f)dV = \oint_{V(t)} \frac{\partial(\rho f)}{\partial t} dS + \oint_{S(t)} [\mathbf{e}_n \cdot (\rho f\mathbf{w})]dS. \qquad (F.3.7)$$

This is the *Leibnitz formula*, indicating the rule of differentiation under integral sign. It generalized the 1D relation,

$$\frac{d}{dt} \int_0^{a(t)} f(x,t)dx = \int_0^{a(t)} \frac{\partial f}{\partial t}dx + wf(a(t),t),$$

where $w = da/dt$. Naturally, Eq. (F.3.7) can also be derived directly, considering a moving infinitesimal volume, as we did to prove the divergence theorem.

When the volume $V(t)$ is a *material volume* $V_m(t)$, and therefore the velocity $\mathbf{w}$ coincides with the fluid velocity $\mathbf{v}$, then, by definition, the temporal derivative d/dt appearing in Eq. (F.3.7) coincides with the material derivative D/Dt. Therefore, applying the divergence theorem, Eq. (F.3.7) reduces to the so-called *Reynolds transport theorem*:

$$\frac{D}{Dt} \int_{V_m(t)} (\rho f)dV = \int_{V_m(t)} \left[\frac{\partial(\rho f)}{\partial t} + \nabla \cdot (\rho f\mathbf{v})\right]dV. \qquad (F.3.8)$$

Note that, substituting the continuity equation (6.2.2), this equation becomes:

$$\frac{D}{Dt} \int_{V_m(t)} (\rho f)dV = \int_{V_m(t)} \rho\frac{Df}{Dt}dV. \qquad (F.3.9)$$

Solutions of the Problems

Chapter 1

Problem 1.1 Denote with $M = \rho L^3$ the mass of the turkey, where L is a linear dimension. Now, $L^2 \approx 2\alpha t$, where t is the characteristic heat diffusion time and α is the thermal diffusivity; assuming that the cooking time τ is the time that takes the temperature at the center of the turkey to reach 95 % of its equilibrium value (i.e., the oven temperature), we see that $\tau = 10\,t$ and therefore we obtain:

$$\tau \Big/ M^{2/3} \approx 1 \Big/ \left(20\rho^{2/3}\alpha \right) = C.$$

Evaluating C as indicated in the table below, we see that C does not change very much.

M (kg)	τ (s)	C (s kg$^{-2/3}$)
2.5	55 min/kg × 60 s/min × 2.5 kg = 8250	4480
4.5	45 min/kg × 60 s/min × 4.5 kg = 12150	4460
7.5	40 min/kg × 60 s/min × 7.5 kg = 18000	4690
11.5	35 min/kg × 60 s/min × 11.5 kg = 24150	4740

We can also estimate the thermal diffusivity of the turkey as:

$$\alpha \approx 1 \Big/ \left(20\rho^{2/3}C \right) = 10^{-3}\,\text{cm}^2/\text{s}.$$

Therefore, we see that α is about the same as the thermal diffusivity of water. That means that as a first-order approximation, a turkey (and us, as well) can be modeled as a ball full of water.

Chapter 2

Problem 2.1 The pressure depends only on the height z, i.e., $p = p_0 + \rho g z$, where we have assumed that $z = 0$ corresponds to the water free surface, while $z = 2L$ at

© Springer International Publishing Switzerland 2015
R. Mauri, *Transport Phenomena in Multiphase Flows*,
Fluid Mechanics and Its Applications 112, DOI 10.1007/978-3-319-15793-1

the bottom. Therefore, at mid height, when $z = L = 10$ cm, we find the same pressure in all three cases,

$$\Delta p = \rho g L = 10^3 \text{ Kg m}^{-3} \times 9.8 \text{ m s}^{-2} \times 10 \text{ cm} = 980 \text{ Pa};$$
$$p \cong 1.01 \times 10^5 \text{ Pa}.$$

The force acting on the lateral walls is due only to the excess hydrostatic pressure Δp, since the atmospheric pressure acts on the walls from the outside as well. Therefore, considering that the elementary surface area is $\pi L dz$, we obtain:

$$F = \left| \int_0^{2L} \Delta p (\pi L dz) \right| = \rho g \pi L \left(\frac{4L^2}{2} \right) = \rho g L (2\pi L^2) = 61.58 \text{ N}.$$

Since $2\pi L^2$ is the area of the lateral surface, we see that F can be obtained assuming that Δp is constant and equal to the pressure drop at mid height, as one would expect, due to the linear dependence of the pressure drop on the height. Conversely, the torque calculated with respect to the center of mass is:

$$\Gamma = \left| L \int_0^{2L} \Delta p (L - z) dz \right| = \frac{2}{3} \rho g L^4 = 10.45 \text{ Nm}.$$

In the other cases with conical containers, the pressure is the same, but the surfaces are larger in the lower part, i.e., when $z > L$, so that the force is larger.

Problem 2.2 The force acting on the infinitesimal volume indicated in Fig. 2.2P is $dF = 2\pi \rho L \omega^2 r^2 dr$. At steady state this force balances the pressure drop, i.e., $dp = dF / 2\pi r L = \rho \omega^2 r dr$. Note that this relation reduces to Eq. (2.1.1), when the centrifugal acceleration $(\omega^2 r)$ is replaced by gravity $(-g)$. Integrating we find:

$$p_2 - p_1 = \int_{r_1}^{r_2} \omega^2 \rho r dr = \frac{1}{2} \omega^2 \rho (r_2^2 - r_1^2).$$

Problem 2.3 At any point at a distance r from the axis and height z (measured from the bottom of the cylinder) the pressure is $p = p_0 - \rho g z + \rho \omega^2 r^2 / 2$, (Note: p decreases at increasing z and decreasing r) where $p_0 = p(z = 0, r = 0)$. Now, imposing that at the free surface, $z = h$, the pressure is constant and equal to the atmospheric pressure, p_a, we obtain:

$$h(r) = h_0 + \omega^2 r^2/2g.$$

Here, $h_0 = (p_0 - p_a)/\rho g$ is the height of the free surface at $r = 0$ and can be determined by imposing a fixed total volume of the fluid, V_f, i.e., $V_f = \int_0^R h(r)2\pi r dr$, obtaining:

$$h_0 = V_f/\pi R^2 - \omega^2 R^2/4g.$$

Note that the profile of the free surface depends on gravity, but it is independent of the atmospheric pressure.

Problem 2.4 Substituting (2.1.3) into (2.1.1) we obtain: $dp/p = -(M_w g/RT)dz$. Now, if we assume that T is constant, we obtain the barometric equation (2.1.4). Instead, when T is a linear decreasing function of the height, i.e., $T = T_0 - \alpha z$, we obtain:

$$\frac{dp}{p} = -\left(\frac{M_w g}{RT_o}\right)\frac{dz}{1 - \alpha z/T_o}.$$

Integrating between $z = 0$ (where $p = p_o$) and z, we find:

$$\int_{p_0}^p \frac{dp}{p} = \ln\left(\frac{p}{p_o}\right) = -\left(\frac{M_w g}{RT_o}\right)\int_0^z \frac{dz}{1 - \alpha z/T_o} = \left(\frac{M_w g}{R\alpha}\right)\ln\left(1 - \frac{\alpha}{T_o}z\right).$$

Hence:

$$p = p_o(1 - (\alpha/T_o)z)^{\frac{M_w g}{R\alpha}}, \quad \text{so that } z = \frac{T_o}{\alpha}\left[1 - (p/p_o)^{\frac{R\alpha}{M_w g}}\right].$$

Here, $T_o = 288$ K; $\alpha = 0.005$ K/m; $M_w = 29$ Kmol/Kg; $g = 9.8$ m/s^2; $R = 8314$ J/Kmol $\times$ K; $p_o/p = 2$. Therefore, we find: $z = 5553$ m.

Problem 2.5 The air mean density can be evaluated using Eq. (2.1.3), obtaining:

$$\rho = \frac{M_w p}{RT} = \frac{29 \frac{\text{Kg}}{\text{Kmol}} \times 570\,\text{mmHg} \times \frac{10^5\,\text{N/m}^2}{760\,\text{mmHg}}}{8314 \frac{\text{J}}{\text{KmolK}} \times 273\,\text{K}} = 0.96 \frac{\text{Kg}}{\text{m}^3}.$$

Now, using Eq. (2.1.2) we find:

$$z = \frac{\Delta p}{\rho g} = \frac{380 \, \text{mmHg} \times \dfrac{10^5 \, \text{N/m}^2}{760 \, \text{mmHg}}}{0.96 \, \frac{\text{Kg}}{\text{m}^3} \times 9.8 \, \frac{\text{m}}{\text{s}^2}} = 5324 \, \text{m}.$$

Note that this result is quite close to that of the previous problem.

Problem 2.6 Applying Eq. (2.2.2) we obtain:

$$\Delta p = (\rho_A - \rho_B) g \Delta z = (13.6 - 1.6) 10^3 \, \text{Kg/m}^3 \times 9.8 \, \text{m/s}^2 \times 0.2 \, \text{m} = 23.52 \, \text{kPa}$$

Problem 2.7 After finding $D_d = 0.35$ cm, applying Eq. (2.3.6) we obtain:

$$\sigma = 0.244 \, \rho g D_d^3 / D_c = 76.4 \, \text{dyn/cm} = 0.076 \, \text{N/m}.$$

Although this is larger than the tabulated value, the result is satisfactory, considering that it is based on an approximate coefficient (i.e., 1.6) and therefore a 10 % error is quite normal.

Problem 2.8 Considering that $\dot{V} = V_d f$, where $V_d = \pi D_d^3 / 6$ is the volume of a drop with diameter D_d, applying Eq. (2.3.5) we find: $V_d = 0.68 \, (\pi \sigma D_c) / (\rho g)$, hence: $X / \dot{V} = 1.46$, in good agreement with the experimental findings.

Problem 2.9 The Weber number, $We = (\rho v^2) / (\sigma / D)$, expresses the ratio between inertial and capillary forces, where $v = \dot{V} / (\pi D^2 / 4)$ is the liquid velocity within the capillary. Imposing $We = 1$, we find:

$$\dot{V} = (\pi^2 D^3 \sigma / 16 \rho)^{1/2}.$$

Then, since $\dot{V} = V_d f$, where $V_d = \pi D_d^3 / 6$ is the volume of a drop with diameter D_d, applying Eq. (2.3.4) we obtain:

$$f(D/g)^{1/2} = 0.366 \, Bo^{1/2}. \quad \text{In our case}: \, f = 13 \, \text{drop/s}.$$

Chapter 3

Problem 3.1 We want to determine a dimensionless number N, such that $N = L^a \mu^b \rho^c V^d$, where a, b, c, and d are constant to be determined. Indicating length by L, mass by M, and time by T, the units of the various quantities (indicated by the square brackets) are:

$$[L] = L; \quad [\mu] = ML^{-1}T^{-1}; \quad [\rho] = ML^{-3}; \quad [V] = LT^{-1}.$$

Therefore:

$$[N] = L^{a-b-3c+d}M^{b+c}T^{-b-d},$$

and then, considering that, $[N] = L^0M^0T^0$, we obtain:

$$a - b - 3c + d = 0; \quad b + c = 0; \quad b + d = 0, \quad \text{that is, } a = -b = c = d.$$

Thus:

$$N = (\rho VL/\mu)^a.$$

showing that, using the four indicated quantities we can form only one dimensionless number, namely the Reynolds number, $Re = \rho VL/\mu$. Obviously, any power of Re is also a dimensionless number.

Problem 3.2 The boundary layer thickness grows with the distance L from the tube inlet as $\delta \approx \sqrt{(Lv/V)}$. Imposing $\delta = R$ we find:

$$R = \sqrt{\frac{Lv}{V}}, \quad \text{hence}: \quad L = R\left(\frac{VR}{v}\right), \quad \text{i.e., } \frac{L}{R} = \frac{1}{2}Re, \quad \text{where } Re = \frac{VD}{v}; \quad D = 2R.$$

This calculation overestimates the length L that is needed to reach the fully developed flow regime, since here we have assumed that the fluid moves with velocity V outside the boundary layer. In reality, we obtain, approximately:

$$\frac{L}{R} = 0.1 \, Re.$$

The same result can be obtained considering that momentum disturbances diffuse from the wall to the center of the tube within a time $t \cong R^2/2v$. In fact, assuming that during that time momentum is convected longitudinally with velocity V (leading to an overestimate of the result, as noted above), we find:
$$L \approx Vt = VR^2/v = \frac{1}{2}R\,Re.$$

Chapter 4

Problem 4.1 The kinetic energy flux, J_E, is equal to the kinetic energy, E_K, contained within a cylinder of volume $Sv\Delta t$, divided by S and by Δt. Considering that $E_K = (\rho v^2/2)(Sv\Delta t)$, we obtain:

$$J_{cE} = \frac{E_K}{S\Delta t} = \frac{1}{2}\rho v^3.$$

Problem 4.2 From Fig. 4.1P it appears that, since the total force applied to the system is null, the excess of force, $p\Delta S$, exerted on the tube inlet and outlet sections, must be balanced by a force that is exerted on the pipe walls.

Problem 4.3 In this case, the force exerted on the pipe walls is $\Delta S(p + \Delta p/2)$. Neglecting the $\Delta S\Delta p$ term, this coincides with the static result.

Problem 4.4 The correction factor is defined as $\alpha = \langle v^3\rangle/\langle v\rangle^3$. In our case, $v = V_{max}(1 - \xi^2)$, where $\xi = y/L$, with $0 < \xi < 1$. Thus:

$$\langle v\rangle = V_{max}\int_0^1 (1 - \xi^2)d\xi = V_{mqx}\left(1 - \frac{1}{3}\right) = \frac{2}{3}V_{max}.$$

$$\langle v^3\rangle = V_{max}^3\int_0^1 (1 - \xi^2)^3 d\xi = V_{max}^3\int_0^1 (1 - 3\xi^2 + 3\xi^4 - \xi^6)d\xi$$

$$= V_{max}^3\left(1 - \frac{3}{3} + \frac{3}{5} - \frac{1}{7}\right) = \frac{16}{35}V_{max}^3.$$

Therefore: $\alpha = (16 \times 3^3)/(35 \times 2^3) = 54/35 = 1.54$. This result is the same for any 2D parabolic profile (e.g., when $-1 < \xi < 1$).

For a 3D parabolic profile in a circular pipe, where $v = V_{max}(1 - \xi^2)$, with $\xi = y/R$, we obtain:

$$\langle v\rangle = \frac{1}{\pi R^2}V_{max}\int_0^R \left(1 - \frac{r^2}{R^2}\right)2\pi r dr = 2V_{max}\int_0^1 (1 - \xi^2)\xi d\xi = 2V_{mqx}\left(\frac{1}{2} - \frac{1}{4}\right) = \frac{1}{2}V_{max}.$$

$$\langle v^3\rangle = 2V_{max}^3\int_0^1 (1 - \xi^2)^3\xi d\xi = 2V_{max}^3\int_0^1 (\xi - 3\xi^3 + 3\xi^5 - \xi^7)d\xi$$

$$= 2V_{max}^3\left(\frac{1}{2} - \frac{3}{4} + \frac{3}{6} - \frac{1}{8}\right) = \frac{1}{4}V_{max}^3.$$

Therefore : $\alpha = 2^3/4 = 2$.

Problem 4.5

$$\rho_1 = \frac{p_1 M_W}{RT} = \frac{7 \times 10^5 \times 29}{8314 \times 300} = \frac{54}{35} = 8.14 \text{ kg/m}^3; \quad \rho_2 = \rho_1/7 = 1.16 \text{ kg/m}^3.$$

Mass balance: $\rho_1 v_1 S_1 = \rho_2 v_2 S_2$, hence

$$v_1 = (\rho_2/\rho_1)v_2 = (1.16/8.14) \times 30 = 4.3 \text{ m/s}.$$

Momentum balance:

$$\mathbf{F} = \left(\rho v^2 + \tilde{p}\right)_1 \mathbf{S}_1 - \left(\rho v^2 + \tilde{p}\right)_2 \mathbf{S}_2,$$

hence:

$$F = \left(\frac{\pi}{4}0.1^2\right)\left[8.14 \times 4.3^2 + \left(7 \times 10^5 - 10^5\right) - 1.16 \times 30^2 - \left(10^5 - 10^5\right)\right]$$
$$= 1.18 + 4715 - 8.2 - 0 = 4708 \text{ N}$$

where we have considered that $\tilde{p}$ is the relative pressure and that $\mathbf{S}_1$ and $\mathbf{S}_2$ are both directed along the same direction.

Problem 4.6 Here the mass balance is trivially: $v_1 = v_2$, while, compared to the previous problem, here the applied force in the momentum balance has $\mathbf{S}_2 = -\mathbf{S}_1$ so that the momentum rate of changes at the inlet and at the outlet of the tube will sum, instead of subtracting from each other, i.e.,

$$F = \left(\frac{\pi}{4}0.1^2\right)\left[2 \times 1000 \times 20^2 + 2 \times 10^5 + 1.6 \times 10^5\right] = 6283 + 1571 + 1257$$
$$= 9110 \text{ N}$$

Problem 4.7 Denoting by 1 the section on the free surface and by 2 the section at the exit, the Bernoulli equation gives,

$$\frac{1}{2}\left(v_1^2 - v_2^2\right) + gH = 0,$$

where H is the height of the free surface. In addition, mass balance gives $S_1 v_1 = S_2 v_2$. Therefore, assuming that $S_1 \gg S_2$, so that $v_2 \gg v_1$, we obtain

$$v_2 = \sqrt{2gH}.$$

This is the Torricelli equation, which is valid when all friction losses are negligible. Note that v_2 is the velocity of a body falling without frictions from a height H.

Problem 4.8 From a mass balance we obtain:

$$dM/dt = -\rho v_2 L_2^2$$

where v_2 is the outlet velocity of the liquid, L_2^2 is the area of the outlet cross section, while $M = \rho H L_1^2$ is the mass of the liquid (H is the height of the free surface, L_1^2 is the area of the reservoir cross section. Considering the Torricelli equation (see the previous Problem), we obtain,

$$dH/dt = -(L_2/L_1)^2(2g\,H)^{1/2}.$$

Integrating from time $t = 0$, with $H(t = 0) = H_0$, to time $t = \tau$, with $H(t = \tau) = 0$, we find:

$$\tau = \sqrt{(2H_0/g)}\,(L_1/L_2)^2.$$

Note that we have substituted into a time-dependent mass balance the Torricelli equation, which has been derived in stationary conditions, i.e., with constant H. This is called the Quasi Steady State (QSS) approximation, which can be applied when the relaxation time characterizing the variation of H, i.e., τ in this case, is much longer than the time needed to reach steady state, τ_{ss}. Here, if H is changed abruptly, the steady state Torricelli velocity at the exit is reached within a time $\tau_{ss} \cong L_2/v_2$, so that the QSS approximation is valid when

$$\frac{L_2}{\sqrt{gH_0}} \ll \left(\frac{H_0}{g}\right)^{1/2}\left(\frac{L_1}{L_2}\right)^2 \quad \Rightarrow \quad \frac{L_2^3}{L_1^2 H_0} \ll 1.$$

Considering typical values, such as $H_0 = 1$ m, $L_1 = 1$ m and $L_2 = 1$ cm, we see that this relation is satisfied.

Problem 4.9 Applying the Bernoulli equation, we find: $\Delta p/L = \rho h_f/L$, where

$$\frac{h_f}{L} = \frac{2fv^2}{D} = \frac{32f\dot{V}^2}{\pi^2 D^5}, \quad \text{with } f = f(Re) \text{ and } Re = \frac{vD}{\nu} = \frac{4\dot{V}}{\nu\pi D}.$$

Find: $v = 5.1$ m/s and $Re = 2.5 \times 10^4$, hence $f = 0.0063$.
Thus $h_f/L = 6.6\,(\text{m}^2/\text{s}^2)/\text{m} = 6.6\,(\text{J/m})/\text{kg}(\text{m}^2)$
The friction losses are: $\Delta p/L = 6.6 \times 10^3 = 6.6$ kPa/m $= 0.066$ bar/m.
The energy "dissipated" by friction is $\dot{Q}/L = vA\Delta p/L = \dot{V}\Delta p/L = 0.066$ kW/m.
As an alternative, it can be evaluated as follows: $\dot{Q}/L = \dot{m}h_f/L$.

Problem 4.10 From $\Delta p/L = \rho h_f/L = (2/D)f\rho v^2$, we find the following iteration scheme:

$$v^{(n+1)} = C/\sqrt{f^{(n)}}, \quad \text{where } C = \sqrt{\frac{\Delta p}{L}\frac{D}{2\rho}} = 0.41 \text{ m/s}, \text{ while } f^{(n)}$$

$$= 0.0791/\left[Re^{(n)}\right]^{1/4}.$$

Assume (iteration 0) that the fluid velocity is $v^{(0)} \cong 0.5$ m/s (note: we know from the previous problem that this velocity is 10 times smaller than its exact value), therefore $Re^{(0)} = 2500$, and $f^{(0)} = 0.011$. Next (iteration 1), we find $v^{(1)} = 3.88$ m/s, and $Re^{(1)} = 1.9 \times 10^4$, hence $f^{(1)} = 0.0067$ and therefore (iteration 2) $v^{(2)} = 5.01$ m/s. Note that, even if Re has increased by a factor 10, f has only decreased by a factor 2. Continuing, we see that the result converges to $v = 5.1$ m/s, so that the volumetric flow rate is 0.01 m³/s.

As an alternative, from $\frac{\Delta p}{L} = \frac{2}{D} \times 0.0791 \left(\frac{v}{vD}\right)^{1/4} \rho v^2$, we obtain: $v = \left(\frac{\Delta p}{L} \frac{D^{5/4}}{2\rho v^{1/4}}\right)^{4/7}$ $= 5.1$ m/s.

Problem 4.11 Applying the Bernoulli equation, the maximum pumping power is

$$h_f = \frac{\dot{W}}{\dot{m}} = 2f \frac{L}{D} \frac{\dot{V}^2}{(\pi D^2/4)^2} = 10^3 \frac{m^2}{s^2}.$$

Hence,

$$D^5 = \frac{32 L \dot{V}^2}{\pi^2 h_f} f = 1.6210^{-3} f \quad \Rightarrow \quad D = 0.277 f^{0.2},$$

where f depends on $Re = vD/v = 1274/D$.

Iterative solution: assume (iteration 0) $f^{(0)} = 0.006$, hence (iteration 1) $D^{(1)} = 0.099$ m and $Re^{(1)} = 12795$, so that $f^{(1)} = 0.00725$. Then (iteration 2), $D^{(2)} = 0.103$ m etc. The iteration converges to $D = 10.4$ cm.

Alternative solution: $f = 0.0791/Re^{1/4} = 0.01322 D^{1/4}$, i.e., $D^{4.75} = 2.14 \times 10^{-5}$, or $D = 0.104$ m.

Problem 4.12 Applying the mass and momentum balances between the cross section at the exit of the faucet (0) and the cross section at a distance L (1), we obtain:

$$v_0 D_0^2 = v_L D_L^2;$$
$$(v_L^2 - v_0^2)/2 - gL = 0.$$

Thus we obtain: $v_0^2 [(D_0 / D_L)^4 - 1] = 2gL$, and so: $(D_L/ D_0) = (1 + 2gL/v_0^2)^{-1/4}$.

Problem 4.13 Applying the Bernoulli equation between the free surfaces of the two reservoirs we obtain:

$$(v_1^2 - v_2^2)/2 + g(z_1 - z_2) + (p_1 - p_2)/\rho + \eta w_p = h_f.$$

Since $v_1 = v_2$ and $p_1 = p_2$, we find $\eta w_p = g\Delta z + h_f$. Here, $g\Delta z = 382$ m²/s², while:

$$\eta w_p = \eta \dot{W}_p/\dot{m} = (0.7 \times 15010^3)/(10^3 \dot{V}) = 105/\dot{V} \ m^2/s^2.$$

In addition, $h_f = 2 f (L/D) v^2$, where $f = 0.0791/Re^{1/4}$. Expressing v and Re as functions of the volumetric flow rate, $v = 4\dot{V}/(\pi D^2) = 5.1 \dot{V}$ and $Re = 4\dot{V}/(\pi D v) = 2.5 \times 10^6 \dot{V}$, we obtain: $h_f = 202 \dot{V}^{1.75}$, and so:

$$\frac{105}{\dot{V}} = \underbrace{382}_{pot.en.} + \underbrace{202 \dot{V}^{1.75}}_{friction} .$$

Note that we have neglected all concentrated pressure drops, due to contraction, expansion and bending of the tube. In fact, their contribution to h_f is of order $0.1 \, v^2$, while the distributed pressure drop are of order $10 \, v^2$ (to see that, assume a reasonable value $f = 0.004$).

Solving by iteration, assuming an initial value corresponding to no friction at all, we obtain: $\dot{V}^{(0)} = 0.27 \text{ m}^3/\text{s}$. Then, the iterations soon converge to $\dot{V} = 0.26 \text{ m}^3/\text{s}$. This corresponds to a fluid velocity $v = 1.4$ m/s, showing that the friction losses constitute only 5 % of the total losses (equal to the pumping power); the remaining 95 % of the pressure losses are due to the gravitational term.

Problem 4.14 First, convert the cross section areas into SI units, finding $S_1 = 0.032 \text{ m}^2$ and $S_2 = 0.008 \text{ m}^2$. Applying a momentum balance we have:

$$F = \left(\rho v_1^2 + \tilde{p}_1\right)S_1 - \left(\rho v_2^2 + \tilde{p}_2\right)S_2$$

Note that $\tilde{p}$ is referred to the outside pressure p_2, so that $\tilde{p}_2 = 0$. Now, we must determine v_1 and $\tilde{p}_1$. The former is obtained from mass conservation, i.e., $v_1 = v_2 S_2/S_1 = 5.5$ m/s. $\tilde{p}_1$ can be determined from the Bernoulli equation, neglecting all frictions, i.e.,

$$\frac{1}{\rho}(p_1 - p_2) = \frac{1}{\rho}\tilde{p}_1 = \frac{1}{2}(v_2^2 - v_1^2) = 226 \text{ m}^2/\text{s}^2.$$

Finally:

$$F = \underbrace{968 - 3872}_{recoil} + \underbrace{7232 - 0}_{pressure} = 4328 \text{ N}$$

Problem 4.15

(a) Applying a momentum balance between the vertical sections 1 and 2 (of heights H_1 and H_2, and width L_W), located just before and just after the barrier, respectively, we obtain:

$$F = \left(\rho v_1^2 + \tilde{p}_1\right)S_1 - \left(\rho v_2^2 + \tilde{p}_2\right)S_2,$$

where $v_2 = \dot{V}/(H_2 L_W) = 7$ m/s and $v_1 = \dot{V}/(H_1 L_W) = 1.17$ m/s are the water mean velocities. The relative pressures, $\tilde{p}_1$ and $\tilde{p}_2$, are referred to the atmospheric pressure and are the mean hydrostatic pressures acting on the cross sections 1 and 2, i.e.,

$$\tilde{p}_1 S_1 = \rho g \frac{H_1}{2}(H_1 L_W); \quad \tilde{p}_2 S_2 = \rho g \frac{H_2}{2}(H_2 L_W).$$

Finally we obtain:

$$\frac{F}{L_W} = 4083 + 44100 - 24500 - 1225 = 22.5 \times 10^3 \text{ N/m}.$$

The positive sign indicates that **F** has the same direction as the flux, as indicated in the figure. In fact, the dominant term in the expression of F is $\tilde{p}_1$, i.e., the hydrostatic pressure, pushing forward.

(b) Assuming turbulent flow, so that the velocity profile is approximately flat, the non-uniformity of the forces exerted on the barrier is due to the hydrostatic pressure linearly varying with the height. Denoting by $L = H_1 - H_2 = 2.5$ m the height of the barrier, and preceeding as in Problem 2.1, we find a pressure force $P = \rho g\, L_w L^2/2$ and a torque $\Gamma = \rho g\, L_w L^3/12$.

(c) Imposing $F = 0$, we obtain:

$$0 = 48183 - 12500/H_2 - 4900/H_2^2, \quad \text{hence } H_2 = 0.26 \text{ m}.$$

In fact, in the expression of F the kinetic term, v_2^2, must balance the hydrostatic term, $\tilde{p}_1$, and therefore we must decrease the section S_2.

Problem 4.16 From a mass balance we obtain: $\bar{v}_0 D_0^2 = \bar{v}_1 D_1^2$. Now, if we apply the Bernoulli equation, considering that both pressures and heights are constant, we obtain: $\bar{v}_0^2 = \bar{v}_1^2$, and therefore we conclude that $D_1 = D_0$. However, we should consider that in the energy conservation equation that eventually leads to the Bernoulli equation we have:

$$\frac{1}{2}\rho\langle v^3 S\rangle_0 - \frac{1}{2}\rho\langle v^3 S\rangle_1 = 0,$$

where the bracket indicates an average over the cross section. Therefore, defining the correction factor α as $\langle v^3\rangle = \alpha \bar{v}^3$, we see that $\alpha_0 = 2$ and $\alpha_1 = 1$, since the velocity profile is parabolic at the exit of the capillary, while it is flat in the jet, as the air resistance is negligible. Finally we obtain:

$$2\bar{v}_0^3 D_0^2 = \bar{v}_1^3 D_1^2 \Rightarrow \frac{D_1}{D_0} = \left(\frac{1}{2}\right)^{1/4} = 0.84.$$

The discrepancy with respect to the experimental value, 0.87, is due to the fact that the parabolic velocity profile induces internal energy dissipations that are not included within the Bernoulli equation. So, the correct result could be obtained applying a momentum balance: considering that there are no external forces and that there are no pressure drops, we find:

$$\rho\langle v^2 S\rangle_0 - \rho\langle v^2 S\rangle_1 = 0.$$

Now, considering that $\langle v^2 \rangle = \beta \bar{v}^2$, where $\beta_0 = 4/3$ for a parabolic profile, while $\beta_1 = 1$, we find:

$$\frac{4}{3}\bar{v}_0^2 D_0^2 = \bar{v}_1^2 D_1^2 \quad \Rightarrow \quad \frac{D_1}{D_0} = \left(\frac{3}{4}\right)^{1/2} = 0.86 \cong 0.87,$$

in agreement with the experimental data.

Problem 4.17 Considering that $\dot{V}_{12} = \dot{V}_{23} = \dot{V}_{35} = \frac{1}{3}\dot{V}$ and $\dot{V}_{13} = \frac{2}{3}\dot{V}$, where $\dot{V}$ is the inlet volumetric flow rate, the following equations are obtained

$$p_1 - p_2 = C/D_1^4$$
$$p_1 - p_3 = 2C/D^4$$
$$p_3 - p_4 = C/D_2^4$$
$$p_3 - p_6 = 2C/D^4$$

where $C = (128\mu L\dot{V})/(3\pi)$. Knowing $\dot{V}$, we have 4 unknowns: p_1, p_3, D_1 and D_2. Therefore, the problem is well posed. Considering that $p_2 = p_4 = p_6 = p_a$, summing the second and forth equations yield: $p_1 - p_a = 4C/D^4$. Therefore, from the first and third equation we obtain:

$$D_1/D = (1/4)^{1/4} = 0.71; \quad D_2/D = (1/2)^{1/4} = 0.84$$

Problem 4.18 First of all, note that, since $Re \approx 10$, the flow regime is laminar. In addition, the hydrostatic pressure losses, $\rho g L \approx 10^4$ Pa, are much smaller than the losses due to friction. In fact, the latters are, typically, $2(L/D)f\rho v^2$, with $f = 16/Re$, where $Re = v_{12}D/\nu$ and $v_{12} = (4\dot{V})/(\pi D^2)$, and so they are about equal to 3×10^5 Pa.

Applying the Bernoulli equation between 1 and 2, 2 and 4, and 2 and 5, we obtain:

$$p_1 - p_2 = Cv_{12}; \quad C = 32\,L\,\mu/D^2$$
$$p_2 - p_4 = Cv_{24} + \rho g L$$
$$p_2 - p_5 = 2Cv_{25} + \rho g L$$
$$v_{12} = v_{24} + v_{25}.$$

(the last equation expresses mass conservation at point 2) Considering that $p_4 = p_5 = p_{atm}$ and that v_{12} is known, we have 4 unknowns (p_1, p_2, v_{24} and v_{25}) and 4 equations, so that the problem is well posed.

(a) From the last 3 equations we obtain:

$$v_{24} = 2\,v_{25} = (2/3)v_{12}, \text{ where } v_{12} = (4V)/(\pi D^2) = 1.27 \text{ m/s}.$$

(b) From the first two equations we have, neglecting the hydrostatic pressure:

$$p_1 = (5/3)\,C\,v_{12} = (160\,L\mu v_{12}/3\,D^2) = 6.8 \times 10^5 \text{ Pa}.$$

(c) Now the Reynolds number is about 1000 times larger, and therefore the flow regime is turbulent. In addition, the friction losses are of the same magnitude as the hydrostatic pressure drops, since f is about 100 times smaller than in the previous case. Therefore, the equations are now:

$$p_1 - p_2 = Cv_{12}^{7/4}; \quad C = 0.158\,\rho\,v^{1/4}\,L/D^{5/4}$$
$$p_2 - p_4 = Cv_{24}^{7/4} + \rho g L$$
$$p_2 - p_5 = 2Cv_{25}^{7/4} + \rho g L$$
$$v_{12} = v_{24} + v_{25}.$$

From the last three equations we obtain:

$$v_{24}^{7/4} = 2\,v_{25}^{7/4}, \text{ cioè } v_{24} = 1.48\,v_{25}.$$

Problem 4.19 Applying the Bernoulli equations at the two horizontal conduits we obtain:

$$w_p - g\Delta H = h_f,$$
$$g\Delta H = h_f$$

where $h_f = 2\,(L/D)\,f\,v^2$, with $f = 0.0791/Re^{1/4}$; $Re = vD/v$ and $v = (4\,V)/(\pi D^2)$.

Find: $v = 8$ m/s, $Re = 3.2 \times 10^5$, $f = 0.00333$; $h_f = 106$ m²/s².

(a) From the second equation we obtain: $\Delta H = h_f/g = 10.9$ m.
(b) From the first equation we obtain: $w_p = 2\,g\Delta H = 207{,}8$ m²/s², and therefore the pumping power is $W_p = \rho\, V\, w_p = 2.1$ kW.

Problem 4.20 Applying the Bernoulli equation we find: $\frac{1}{2}\,(v_1^2 - v_2^2) + gH = h_f = \frac{1}{2}\,v_2^2\,0.45\,(1-\beta)$. Since by continuity $v_1 = \beta v_2$, we find:

$$v_2^2[1 - \beta^2 + 0.45(1 - \beta)] = 2gH.$$

Thus, even when $\beta = 0$ the Torricelli equation should be modified to account for the pressure drops at the orifice, obtaining: $v_2 = \sqrt{(1.38gH)}$.

Problem 4.21 Applying the Bernoulli equation we find:

$$-\eta w_p = (v_1^2 - v_2^2)/2 + g(z_1 - z_2) + (p_1 - p_2)/\rho - h_f.$$

Here $v_1 = 0$ and $p_1 = p_2$, hence, $\eta w_p = v_2^2/2 - g(z_1 - z_2) + h_f$, where, $h_f = 2f(L/D)\,v^2$, with $f = 0.0791/Re^{1/4}$. Now, expressing v_2 and Re in terms of the volumetric flow rate as:

$v_2 = 4\dot{V}/(\pi D^2) = 9.7$ m/s and $Re = 4\dot{V}/(\pi D v) = 10^6$, we obtain: $f = 0.003$, and therefore,

$$\eta w_p = (9.7)^2/2 - 9.8(160 - 151) + 2 \times 0.003 \times (9.7)^2(91.5/0.12)$$
$$= 47.3 - 89.6 + 431 = 389 \text{ m}^2/\text{s}^2.$$

Thus, $w_p > 0$, which means that a pump is needed to support the given volumetric flow rate. Note that here the pressure drops are due predominantly to friction.

Problem 4.22 Consider the Bernoulli equation

$$(v_1^2 - v_2^2)/2 + g(z_1 - z_2) + (p_1 - p_2)/\rho = h_f.$$

Choose the two control sections 1 and 2 as the water free surface and the exit of the capillary, respectively. Then, $z_1 - z_2 = 0.3$ m, $v_1 = 0$ and $v_3 = 30$ m/s, $\rho = 10^3$ kg/m³, while $p_1 - p_2 = \Delta p$ is the relative pressure of the air in the reservoir. In addition, h_f is the sum of two distributed terms (in the larger tube and in the capillary), the concentrated term at the elbow of the larger tube, and the two contractions (from the reservoir to the larger tube and from the larger tube to the capillary). These latter have a contraction ratio $\beta \approx 0$ and $\beta = D_3^2/D_2^2 = 0.09$. In addition, $v_1 = v_2 = v_3 D_3^2/D_2^2 = 2.7$ m/s, hence $Re_1 = Re_2 = 2.7 \times 10^4$ and $f_1 = f_2 = 0.006$, while $Re_3 = 9 \times 10^4$ and $f_3 = 0.0045$. Summarizing,

$$h_f = (2L_1/D_1)f_1\,v_1^2 + (2L_2/D_2)f_2v_2^2 + (2L_3/D_3)\,f_3\,v_3^2 + k_gv_1^2/2 + 0.45v_1^2/2$$
$$+\,0.45(1 - 0.09)\,v_3^2/2 = 1.74 + 1.74 + 163 + 0.6 + 1.6 + 180 = 349.$$

Thus, we see that most of the friction losses take place within the capillary (both distributed and concentrated). Finally, the Bernoulli equation gives:

$$(\Delta p/\rho) = \tfrac{1}{2}(30)^2 - (9.8)(0.3) + 349 = 796\ \text{m}^2/\text{s}^2.$$

thus: $\Delta p = (1000)\,(796) = 8 \times 10^5$ Pa.

Problem 4.23 Generally, for low Re, we have: $P_2 - P_1 = \rho g(z_1 - z_2) + C\,V_{12}L/D^4$, where V is the volumetric flux, and $C = 128\,\mu/\pi$. Applying the Bernoulli equation, neglecting the inertial contribution, we find:

$$P_2 - P_a = 2\rho gL - CV_{12}L/D^4 = C\,V_{23}L/D_1^4 = -\rho gL + 2CV_{25}L/D^4.$$

Since $V_{23} = V_{25} = V_{12}/2 = V$, from the first and third relation we obtain: $CVL/D^4 = (3/4)\rho gL$, and therefore: $P_2 - P_a = (1/2)\rho gL$. Now, from the second relation we have:

$$(1/2) = (3/4)\,(D/D_1)^4 \quad \text{and therefore}: \quad D_1/D = (3/2)^{1/4}.$$

When the pipe 2–5 is horizontal, we have:

$$C\,V_{23}L/D_1^4 = 2CV_{25}L/D^4, \quad \text{and therefore}: \quad D_1/D = (1/2)^{1/4}.$$

Problem 4.24

(a) Taking as reference sections the free surface and the tube outlet, applying the Bernoilli equation we find:

$$g(h + H) = (4fH/d + k + 1)u^2/2,$$

where $k = 0.45(1-\beta)$, and β is the ratio between the area of the tube cross section and that of the reservoir.

(b) In our case, $f = 16/Re$, with $Re = ud/v$ and $fH/d \gg 1$. Therefore,

$$gH = 4(16v/ud)(H/d)u^2/2, \quad \text{hence}: \quad u = gd^2/32v.$$

(c) From a mass balance we obtain: $dV/dt = S\,dh/dt = -0.2\,u(\pi\,d^2/4)$, with h $(t = 0) = h_0$.

Therefore: $t_s = (160/\pi)\,(vSh_0/gd^4)$.

If the hypothesis $h \ll H$ is removed, we find: $u = gd^2/32v\,(H + h)/H$, and therefore:

$$dh/(H + h) = -0.2\,gd^2\ldots etc.\ldots$$

Chapter 5

Problem 5.1 The geometry of the problem corresponds to one half of a 2D channel (see Fig. 5.1P). Applying a force balance on the elementary volume (of thickness b) indicated in the Figure ann considering that the shear stress on the centerline is null, we obtain:

$$(by)p - (by)(p - dp) - (bdz)t = 0, \quad \text{i.e.,}$$
$$dp/dz = \tau/y = \tau_w/H.$$

Note that the relation $\tau = \tau_w y/H$ is identical to Eq. (5.1.2). Then, proceeding as in Sect. 5.1, we find:

$$v = \frac{\tau_w}{2\mu H}(H^2 - y^2) = v_{max}\left[1 - \left(\frac{y}{H}\right)^2\right], \quad \text{with } v_{max} = \frac{\tau_w H}{2\mu} = \frac{dp}{dz}\frac{H^2}{2\mu}.$$

Now, since the mean velocity equals 2/3 of the maximum velocity (see Problem 4.4), we find: $\bar{v} = 2v_{max}/3 = (H/3\mu)\tau_w$, and therefore, since $\tau_w = H(dp/dz)$, we obtain:

$$\frac{dp}{dz} = \frac{3\mu\bar{v}}{H^2}.$$

Problem 5.2 This is very similar to Problem 5.1. Here, the mass flow rate is $\dot{m} = \rho\bar{v}(Hb)$, and the mass flow rate per unit width is $\Gamma = \dot{m}/b = \rho\bar{v}H$. In addition,

Fig. 5.1P Sketch of a channel flow

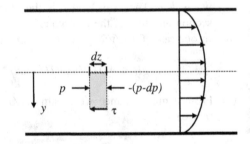

the pressure drop per unit length is $dp/dz = \rho g \cos \beta$. Therefore, since also $dp/dz = 3\mu\bar{v}/H^2$ (see above), we find:

$\rho g \cos \beta = 3\mu(\Gamma/\rho H)/H^2$ and from here the final result.

Problem 5.3 As the flow regime is laminar at low Re, we can neglect the pressure drop associated with the elbow. Here, $\bar{v} = \dot{V}/(4HW)$, where we have considered that the volumetric flow rate in each of the two channels $\dot{V}/2$. In addition, in Problem 5.1 we found: $dp/dz = 3\mu\bar{v}/H^2$, so:

$$\frac{dp}{dz} = \frac{3\mu\bar{v}}{H^2} = \frac{3\mu\dot{V}}{4H^3W}; \quad p(z) = p_a + \frac{3\mu\dot{V}}{4H^3W}z,$$

where p_a is the atmospheric pressure at the channel outlet. Finally:

$$\Delta p = (3\mu L\dot{V})/(4WH^3) \quad \text{and, vice versa,} \quad \dot{V} = (4WH^3\Delta p)/(3\mu L).$$

Problem 5.4 In general, la force exerted by the fluid on an elementary surface Wdz of the upper slab is $pWdz$. This force must balance the weight Mg plus the effect of the atmospheric pressure. So:

$$Mg = 2W \int_0^L \tilde{p}(z)dz = 2WL\bar{\tilde{p}} = WL\Delta p,$$

where Δp has been determined in Problem 5.3. Here, we have considered that the pressure varies linearly from the atmospheric pressure, p_a, to a maximum value, $p_a + \Delta p$.

(a) In laminar flow regime, the pressure drops have been determined in Problem 5.3, so that $C_1 = 3$ and we obtain:

$$Mg = \frac{3\mu\dot{V}L^2}{4H^3} \quad \Rightarrow \quad H = \left[\frac{3\mu\dot{V}L^2}{4Mg}\right]^{1/3}.$$

(b) In turbulent flow regime, we find:

$$Mg = C_2\frac{\rho v^2 WL}{H} = C_2\frac{\rho\dot{V}^2L}{16H^3W} \quad \Rightarrow \quad H = \left[\frac{C_2\rho\dot{V}^2L}{16MgW}\right]^{1/3}.$$

Problem 5.5 First of all, we see that out of symmetry the velocity field is unidirectional along the azimuthal direction, depending only on r, i.e., $v = v_\theta(r)$. In addition, since the fluid trajectories are closed, there cannot be any pressure drop. Therefore, from a momentum balance on the fluid volume indicated in Fig. 5.4P, denoting by L its height and by τ_w the unknown shear stress at the wall, we have: $\tau_w R_i \theta$

$L = \tau r \theta L; \Rightarrow \tau = \tau_w R_i / r$. Now, since $\tau = -\mu dv/dr$, after integration we find: $v(r) = -(\tau_w/\mu)R_i \ln r + C$. Finally, imposing that $v(R_i) = \omega R_i$ and $v(R_o) = 0$, we obtain:

$$\tau_w = \mu\omega \frac{1}{\ln(R_o/R_i)}; \quad v(r) = \omega R_i \frac{\ln(R_o/r)}{\ln(R_o/R_i)}.$$

Naturally, the torque is

$$\Gamma = \int_0^{2\pi} R_i \tau_w (R_i L d\theta) = 2\pi R_i^2 L \tau_w = \mu\omega \frac{2\pi R_i^2 L}{\ln(R_o/R_i)}.$$

In the small gap limit case, i.e., when $R_i/R_o = 1 - \varepsilon$, with $\varepsilon \ll 1$, considering that $\ln(1 + \varepsilon) \cong \varepsilon$, we find:

$$\tau_w = \mu \frac{\omega R_i}{\delta}; \quad v(\tilde{y}) = \omega R_i \tilde{y},$$

where $\delta = R_o - R_i = \varepsilon R_o$, and $\tilde{y} = (R_0 - r)/(R_o - R_i)$ is the dimensionless distance from the outer cylinder.

Problem 5.6 Proceeding as in Sect. 5.2, we find: $\tau = -K(dv/dr)^2 = \tau_w r/R$; Taking the square root and integrating we obtain:

$$\int_0^v dv = \sqrt{\frac{\tau_w}{RK}} \int_R^r \sqrt{r} dr, \text{ hence } v = \frac{2}{3}\sqrt{\frac{\tau_w}{RK}}\left(R^{3/2} - r^{3/2}\right)$$

$$= v_{max}\left[1 - \left(\frac{r}{R}\right)^{3/2}\right], \text{ with } v_{max} = \frac{2R}{3}\sqrt{\frac{\tau_w}{K}}.$$

Now we want to find the pressure drops in terms of the mass flow rate. To do that, we determine the mean velocity, finding:

$$\bar{v} = \frac{1}{\pi R^2}\int_0^R v(2\pi r)dr = 2v_{max}\int_0^1 \left(1 - y^{3/2}\right)y dy = \frac{3}{7}v_{max} = \frac{2R}{7}\sqrt{\frac{\tau_w}{K}}.$$

Then, since from Eq. (4.5.2) the pressure drops are proportional to sehar stresses at the wall, we obtain:

$$\frac{\Delta p}{L} = \frac{2\tau_w}{R} = \frac{2}{R}\left(\frac{49K\bar{v}^2}{4R^2}\right) = \frac{49}{2}K\frac{\bar{v}^2}{R^3}.$$

These results coincide with those of Sect. 5.3 when $n = 2$.

Problem 5.7 The equivalent diameter of long cylindrical particles of radius R is $D_p = 6V_p/S_p = 3R$. Therefore, considering the Blake-Kozeny Eq. (5.4.8) in terms of Φ and α, we obtain:

$$\Phi = \frac{25\alpha}{6(1-\alpha)^3}.$$

Comparing these results with the experimental findings, we can see that the agreement is good when $\alpha > 0.1$ and it breaks down only when the solid fraction is low, i.e., for relatively empty filters, a condition that is encountered only very rarely. In reality, one would expect a good agreement only when viscous forces are dominant, that is when α is not far from unity. Therefore, it is surprising that the agreement is good also at relatively low solid fractions, probably due to a series of cancellation of errors not uncommon when dealing with the Blake-Kozeny equation.

Problem 5.8

(a) From a mass balance on an elementary surface of width W and length dx, we obtain, with $q = \dot{m}/pW$ denoting the volumetric flow rate per unit width,

$$\dot{m}(x) - \dot{m}(x + dx) = pW[q(x) - q(x + dx)] = \frac{dM}{dt} = pv_eWdx + pvWdx,$$

where v is the penetration velocity in the soil, which satisfied the Darcy equation,

$$v = \frac{\kappa}{\mu}\frac{d}{dy}(pgz) = \frac{\kappa}{\mu}pg\cos\phi.$$

Here x indicates the longitudinal coordinate, y the transversal coordinate and z the vertical coordinate. In the limit $dx \to 0$ we obtain,

$$\frac{dq}{dx} = -\left(v_e + \frac{\kappa}{\mu}pg\cos\phi\right) \quad \Rightarrow \quad q(x) = q_0 - \left(v_e + \frac{\kappa}{\mu}pg\cos\phi\right)x.$$

(b) The flux is vanishes at a distance L. Thus,

$$L = \frac{q_0}{v_e + \frac{\kappa}{\mu} \rho g \cos \phi} = \frac{10^{-2} \, \text{m}^2/\text{s}}{10^{-2} \, \text{m/s} + 10^{-11} \, \text{m}^2 \times 9.8 \, \text{m/s}^2 \times 9.8/10^{-6} \, \text{m}^2/\text{s}^2}$$
$$= 100 \, \text{m}.$$

Note that evaporation does not play any role.

(c) As $v = dy/dt$, approximating $\cos \phi \approx 1$ we obtain:

$$\frac{dy}{dt} \approx \frac{dz}{dt} = \frac{\kappa}{v} g \quad \Rightarrow \quad z(t) = \frac{\kappa}{v} gt = \left(9.8 \times 10^{-3} t\right) \, \text{cm},$$

where t is expressed in seconds. Therefore, after one hour, the penetration depth is 35 cm, which is adequate to wet the roots of the seasonal plants.

Problem 5.9 The porosity of the material, both volumetric and superficial, is $\varepsilon = m r^2/R^2$. Since the interstitial velocity is equal to the mean fluid velocity inside the tubes, we obtain:

$$v_i = \frac{r^2}{8\mu} \frac{\Delta p}{L} \quad \Rightarrow \quad v_s = \varepsilon v_i = \frac{1}{8\mu} \frac{m r^4}{R^2} \frac{\Delta p}{L}.$$

Now, applying the Darcy law, we find the permeability; $\kappa = \varepsilon r^2/8$.

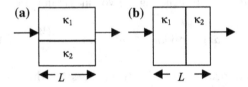

Problem 5.10 In the case (a) we obtain:

$$\frac{\Delta p}{L} = \frac{\mu}{\kappa_1} v_1 = \frac{\mu}{\kappa_2} v_2 = \frac{\mu}{\kappa} v_s,$$

where v_s is the superficial velocity, while we have defined the effective perme-ability, $\bar{\kappa}$. The total volumetric flow rate is the sum of two parts, pertaining the two phases, that is $S v_s = S_1 v_1 + S_2 v_2$, where S is the total cross section, while S_1 and S_2 are the cross sections of the two phases. Therefore, defining $\varepsilon_1 = S_1/S$ and $\varepsilon_2 = S_2/S$ the volume fraction of the two phases, we obtain:

$$\frac{\Delta p}{L} = \frac{\mu}{\kappa_1} v_s = \frac{\mu}{\kappa}(\varepsilon_1 v_1 + \varepsilon_2 v_2) = \frac{\mu}{\kappa}\left(\varepsilon_1 + \frac{\varepsilon_2 v_2}{\kappa_1}\right) v_1 = \frac{\mu}{\kappa_1} v_1,$$

hence,

$$\bar{\kappa} = \varepsilon_1 \kappa_1 + \varepsilon_2 \kappa_2.$$

This case corresponds to two resistances in parallel, where the conductance are summed.

In the case (b), instead, we obtain:

$$\mu v = \bar{\kappa}_1 \frac{\Delta p_1}{L_1} = \bar{\kappa}_2 \frac{\Delta p_2}{L_2} = \bar{\kappa} \frac{\Delta p}{L},$$

where $\Delta p = \Delta p_1 + \Delta p_2$ and $L = L_1 + L_2$, and we have defined the effective permeability, $\bar{k}$. Now, proceeding as for the previous case, we find:

$$\bar{\kappa} = \left(\frac{\varepsilon_1}{\kappa_1} + \frac{\varepsilon_2}{\kappa_2}\right)^{-1},$$

where $\varepsilon_1 = L_1/L$ and $\varepsilon_2 = L_2/L$. This case corresponds to two resistances in parallel, where the resistances are summed.

Problem 5.11 Assuming QSS conditions and laminar flow conditions, we obtain (see Problem 5.1),

$$\dot{V} = 2HW\bar{v} = \frac{2WH^3 \Delta p}{3\mu L},$$

where L is the penetration depth of the drop. Here, the pressure drop is given by the Young-Laplace equation, where one of the curvature radii is $H/\cos\theta$, while the other is infinite, so that $\Delta p = \sigma \cos\theta /H$. This result could also be obtained considering that the pressure drop is the ratio between the driving force $F = 2W \sigma \cos\theta$ (the factor 2 accounts for the fact that there are 2 walls) and the cross section area, $S = 2WH$. Finally, considering that

$$\dot{V} = \frac{dV}{dt} = 2HW\frac{dL}{dt},$$

We obtain: $L(dL/dt) = (H\sigma\cos\theta)/(3\mu)$, hence:

$$L = \sqrt{\frac{2\cos\theta\, H\sigma}{3}\frac{1}{\mu}t}.$$

Vice versa, the time t_g needed for the whole drop to enter the conduit, with $L_g = V_g/(2WH)$, is then

$$t_g = \frac{3\mu V_g^2}{8H^3 W^2 \sigma \cos\theta}.$$

Chapter 7

Problem 7.1 The force, directed upward, exerted by the fluid on a circular crown of thickness dr at a distance r from the center is $P(r)2\pi r dr$. The total upward force is opposed by the sum, directed downward, of the weight Mg and the effect of the atmospheric pressure, p_a. So we obtain:

$$Mg = 2\pi \int_{R_0}^{R} \tilde{P}(r) r dr,$$

where $\tilde{P} = P - p_\alpha$ is the relative pressure. Hence, substituting Eq. (7.3.8), we obtain:

$$Mg = \left(\frac{3\mu \dot{V} R^2}{2H^3}\right) I; \quad \text{where}$$

$$I = -\int_{R_0/R}^{1} \xi \log \xi \, d\xi \cong -\left[\frac{1}{2}\xi^2 \log \xi d\xi\right]_0^1 + \int_0^1 \frac{1}{2}\xi d\xi = \frac{1}{4},$$

where we have assumed that $R_0/R \ll 1$. Therefore:

$$H = \left(\frac{3\mu \dot{V} R^2}{8Mg}\right)^{1/3}.$$

Problem 7.2 From a force balance on the elementary volume of Fig. 7.1P we obtain: $\tau = (\Delta p/L)y$, that is the same result as for the single phase case. Therefore, the velocity profile is composed of two parabolic pieces. In particular, $\tau_w = (\Delta p/L)$ H, and therefore, $\int_0^{v_{H/2}} dv = -\left(\frac{\Delta p}{\mu L}\right) \int_H^{H/2} y \, dy$, hence:

$$v_{H/2} = \frac{3}{8}\left(\frac{\Delta p H^2}{L\mu}\right). \text{ Also, } \int_{v_H}^{v_{H/2}} dv = -\left(\frac{\Delta p}{10\mu L}\right) \int_{H/2}^{0} y \, dy, \text{ hence:}$$

$$v_H = v_{H/2} + \frac{1}{80}\left(\frac{\Delta p H^2}{L\mu}\right) = \frac{31}{80}\left(\frac{\Delta p H^2}{L\mu}\right).$$

We could also solve the Navier-Stokes equation,

$$dp/dx = 10\mu d^2 v/dz^2, \quad \text{when } 0 < z < H/2;$$
$$dp/dx = \mu d^2 v/dz^2, \qquad \text{when } H/2 < z < H,$$

with $dp/dz = -\Delta p/L = \cos t.$ with the following boundary conditions,

$$dv/dz(0) = 0; 10\mu dv/dz(H/2^-) = \mu dv/dz(H/2^+); v(H/2^-) = v(H/2^+); v(H)$$
$$= 0,$$

where we have considered that at the interface $y = H/2$ both velocity and shear stresses are continuous.

Problem 7.3 Start from the Navier-Stokes equation,

$$dp/dx = \mu d^2 v/dz^2, \quad \text{per } 0 < z < H$$

with $dp/dz = -\Delta p/L = \cos t..$, to be solved with boundary conditions,

$$dv/dz(0) = 0; \quad a|dv/dz(H)| = v(H).$$

Integrating once we obtain: $-(\Delta p/L)z = \mu dv/dz + C_1$, where $C_1 = 0$ as dv/dz $(0) = 0$.

Integrating again we find: $-(\Delta p/L)(z^2/2) = \mu v + C_2$, where, applying the other boundary condition,

$$C_2 = -(\Delta p/L)(H^2/2) - \mu v_w = -(\Delta p/L)(H^2/2) - \mu a|dv/dz|_w$$
$$= -(\Delta p/L)(H^2/2) - a(\Delta pH/L).$$

(pay attention to the sign!!!) Therefore, we obtain:

$$v(z) = (\Delta pH^2/2\mu L)\left[1 + 2(a/H) - (z/H)^2\right]$$

Obviously, when $a = 0$, we obtain the usual Poiseuille flow field.

Note: the velocity profile remains parabolic, with the same curvature as when $a = 0$. The difference is that the slip velocity at the wall introduces a constant offset, so that the velocity profile appears translated by a constant quantity, proportional to a/H.

Chapter 8

Problem 8.1 Comparing a station wagon with its sedan counterpart, we see that the lack of a tail in the station wagon means that the boundary layer will detach earlier, decreasing its adherence. This is why there are no Ferrari station wagons.

Problem 8.2 Imposing $f(\eta) = \eta$ into Eq. (8.4.11), we obtain:

$$\alpha_1 = 1/2, \; \alpha_2 = 1/6 \, e \, \beta = 1, \quad \text{hence}: \; \delta(x) = 2\sqrt{3}(vx/U)^{1/2}.$$

Therefore: $\tau_w = \mu U/\delta = f_F \, (\tfrac{1}{2}\rho U^2)$ with $f_F = (1/\sqrt{3}) \, Re_x^{-1/2} = 0.58 \, Re_x^{-1/2}$.
We could find this result also applying Eq. (8.4.18), with $f_F = \delta_2/x = \alpha_2\delta/x$.
Also, note that near the wall we have: $u/U = y/\delta = 0.29 \, y/(vx/U)^{1/2}$.
Finally, $\delta_1 = \delta/2 = \sqrt{3}(vx/U)^{1/2} = 1.73(vx/U)^{1/2}$.
Comparing this result with the exact solution, we see that, despite its simplicity, it is not too far off.

Problem 8.3 The velocity profile $f(\eta) = \sin(\pi\eta/2)$ satisfies the conditions: $f(0) = 0$, $f(1) = 1 \, e \, f'(1) = 0$. We obtain: $\alpha_1 = 1-2/\pi = 0.363$, $\alpha_2 = 1/2(4/\pi-1) = 0.137$ and $\beta = \pi/2 = 1.571$, hence: $\delta(x) = 4.79 \, (vx/U)^{1/2}$.
Therefore, $\delta_1 = \alpha_1\delta = 1.74 \, (vx/U)^{1/2}$, while from Eq. (8.4.18) we find:

$$f_F = \delta_2/x = \alpha_2\delta/x = 0.656 \, Re_x^{-1/2},$$

in very good agreement with the exact solution.

Chapter 9

Problem 9.1 The heat leaving the room is $Q = \Delta T/\Sigma R_{th}$, where R_{th} are the following thermal resistances:

$$R_1 = L/(Ak_v) = 6.67 \times 10^{-3} \, \text{K/W}; \quad R_2 = L/(2Ak_v) = 3.33 \times 10^{-3} \, \text{K/W} = R_1/2;$$
$$R_3 = l/(Ak_a) = 4.17 \times 10^{-2} \, \text{K/W}; \quad R_4 = 1/(Ah) = 0.1 \, \text{K/W}.$$

We see that: (1) the thermal resistances of the glasses are small; (2) the largest resistance is that of the outside atmosphere, that is subjected to large variations, depending on the weather. In the two cases (with single or double glasses) we find:

$$R_I = R_1 + R_3 = 0.107 \, \text{K/W}; \quad R_{II} = R_1 + R_3 + R_4 = 0.148 \, \text{K/W}.$$

Therefore, with double glasses we save about 40 % of the heat losses.
In the first case, we find $Q = 140$ W, and so in one year, with 8760 h, we consume 1230 kWh. In the second case, $Q = 102$ W with a 894 kWh heat consumption.
It is instructive to calculate the temperature T_L of the outer surface of the window. With a single glass, imposing,

$$J = Q/A = k_v(T_1 - T_L)/L = h(T_l - T_0), \text{ we find}: T_L = 24\,^\circ\text{C}.$$

Therefore, we see that the temperature profile in the glass is almost flat, as the maximum temperature drop occurs on the outer surface of the window. So, if the heat transfer coefficient h increases, the heat losses increase almost proportionally.

In the case of a double glass, instead, we find, with obvious notation:

$$J = Q/A = k_v(T_1 - T_{12})/L = k_a(T_{12} - T_{13})/l = k_v(T_{13} - T_L)/l = h(T_l - T_0),$$
hence $T_L = 20.21\,^\circ\text{C}.$

(also, $T_{12} = 24.67\,^\circ\text{C}$, and $T_{13} = 20.53\,^\circ\text{C}$). Thus, we see that part of the temperature excursion is absorbed by the air gap, so that the total thermal resistance depends less on the weather.

Problem 9.2 The total thermal resistance is: $AR_{th} = (1/h) + (L_2/k_1) + (L_2/k_2) + (1/h)$. In the first case, without coating, i.e., when $L_2 = 0$, we obtain:

$$AR_{th} = (2/100) + (0.1/0.1) \cong 1\ \text{m}^2\,\text{K/W}.$$

Therefore, the total hear flux is

$$Q = \Delta T/R_{th} = 2\,\text{KW}.$$

In the second case, we find: $R_{th} = 4\ \text{m}^2\text{K/W}$, hence: $Q = \Delta T/R_{th} = 0.5\ \text{KW}$.

Problem 9.3 Assuming $A = 1\ \text{m}^2$ we find:

$$R_{th} = (1/A)[(1/h_1) + (L_1/k_1) + (L_2/k_2) + (1/h_2)]$$
$$= (0.040 + 0.333 + 0.139 + 0.125)\text{K/W} = 0.637\ \text{K/W}.$$
$$J_Q = Q/A = (T_1 - T_0)/(AR_{th}) = 20/0.637 = 31.4\ \text{W/m}^2.$$

It may be instructive to calculate the temperatures in the wall, T_{1w} and T_{2w}, and that at the interface, T_i. Considering that $J_Q = h_1(T_1 - T_{1w}) = h_2(T_{2w} - T_2) = k_1(T_{1w} - T_i)/L_1$, we find:

$$T_{1w} = 43.7\,^\circ\text{C and } T_{2w} = 28.9\,^\circ\text{C and } T_i = 33.2\,^\circ\text{C}.$$

Therefore, most of the temperature excursion takes place across the insulator.

Problem 9.4 Consider a reference tube length $L = 1$ m. The total thermal resistance is:

$$R_{th} = (1/2\pi L)[1/(h_i R_1) + \ln(R_2/R_1)/k_1 + \ln((R_2 + d)/R_2)/k_2 + 1/(h_o(R_2 + d))]$$

Therefore: $Q/L = (T_i - T_o)/(R_{th}A) = .. 61.2$ W/m

Problem 9.5 When the external radius R_o increases, at constant R_i, two competing effects can be observed. On one hand, the thickness of the insulating layer grows, and so the conductive thermal resistance also increases; on the other hand, the convective resistance at the outer wall decreases, as the heat exchange surface increases. In fact,

$$Q = 2\pi L(\Delta T)/[\ln(R_o/R_i)/k + 1/(R_o h)]$$
$$dQ/dR_o = 0 \Rightarrow \ldots R_o = R_c = k/h. \quad d^2Q/dR_o^2 < 0$$

This is a point of maximum. Therefore, when $R_o < R_c$, heat transfer increases as R_o increases, while when $R_o > R_c$, heat transfer decreases as R_o increases.

Naturally, when $R_i > R_c$, it is always true that, as expected, the tube becomes better insulated as we add insulation.

Problem 9.6 Let us calculate the Biot number. At 300 K thermal conductivity of cupper is $k = 401$ W/m K and therefore $Bi = hR/k \cong 0.03 < 0.1$. Therefore, we may safely assume that temperature is uniform within the sphere. Applying Eq. (9.4.4) we can then find T_f. Knowing the data: $\rho = 8933$ kg/m^3; $c_p = 385$ J/kg K, and with $T_0 = 800$ °C, $T_a = 20$ °C, $V = \frac{4}{3}\pi R^3$ and $A = 4\pi R^2$, we obtain:

$$\tau = \rho c_p R/3h = 63.7 \text{ s}.$$

Thus, with $t = 45$ s, we have: $\Theta = (T - T_a)/(T_0 - T_a) = 0.49$ and $T = 405$ °C.

Problem 9.7 Assume that $Bi = hL/k \ll 1$, where L is a characteristic length of the metallic object. That means that the temperature T_1 of the object is uniform. From an overall heat balance at steady state we have: $V_1\rho_1c_1(T_{10} - T_f) = V_2\rho_2c_2(T_f - T_{20})$, so we find T_f. In addition, applying two separate heat balances for the object and for the liquid, we obtain:

$$V_1\rho_1c_1 dT_1/dt = -hS\,(T_1 - T_2) \quad \Rightarrow dT_1/dt = -A_1(T_1 - T_2),$$
$$V_2\rho_2c_2 dT_2/dt = hS\,(T_1 - T_2) \quad \Rightarrow dT_2/dt = A_2(T_1 - T_2),$$

where $A_1 = hS/(V_1\rho_1c_1)$ and $A_2 = hS/(V_2\rho_2c_2)$ are the inverses of the characteristic times. Subtracting these two equations from one another find:

$$d\Delta T / dt = -(A_1 + A_2)\Delta T, \quad \text{where } \Delta T = (T_1 - T_2).$$

Now integrate with $\Delta T(t = 0) = \Delta T_0$ obtaining: $\Delta T(t) = \Delta T_0 \exp[-(A_1 + A_2)t]$. Finally, from this result and from $dT_1/dt = -A_1\Delta T$, we find:

$$dT_1/dt = -A_1 \Delta T_0 \exp[-(A_1 + A_2)t].$$

Integrating again, we obtain:

$$T_1 - T_{10} = -[A_1/(A_1 - A_2)]\Delta T_0 \{1 - \exp[-(A_1 + A_2)t]\}.$$

Problem 9.8 The problem is complex, since the boundary condition at $r = R_i$ changes with time. Therefore, suppose that we can apply the Quasi Steady State approximation, assuming that the characteristic time τ with which the boundary conditions change (that is, the characteristic relaxation time of the fluid temperature inside the sphere) is much smaller than the relaxation time τ_{ss} with which the temperature profile within the spherical crown reaches its steady state value.

Applying a heat balance on the fluid inside the sphere we obtain:

$$\left(\frac{4}{3}\pi R_i^3\right) pc \frac{dT}{dt} = -\dot{Q} = -\left(4\pi R_i^2\right) J_i,$$

where J_i is the heat flux leaving the liquid (and therefore entering the spherical crown), while ρ and c are the fluid density and specific heat, respectively. Now, if the QSS approximation is valid, J_i is given by Eq. (9.2.13) and, denoting $\Delta T = T - T_0$ and $L = R_o - R_i$, we obtain:

$$\frac{d\Delta T}{\Delta T} = \frac{dt}{\tau}; \quad \tau = \frac{pcLR_i^2}{3k_sR_o}\frac{1 + Bi'}{Bi'}; \quad Bi' = \frac{hLR_o}{k_s R_i}.$$

Then, solving with initial condition $\Delta T(t = 0) = \Delta T_i = T_i - T_0$, we obtain: $\Delta T = \Delta T_i e^{-t/\tau}$, showing that τ is the above-mentioned relaxation time. Finally, let us check when the QSS condition can be applied, considering that $\tau_{ss} = L^2/a_s$, where $a_s = k_s/\rho_s c_s$ is the thermal diffusivity of the solid. Imposing that $\tau \gg \tau_{ss}$ we obtain:

$$\frac{L}{R_i} \ll \left(\frac{pc}{\rho_s c_s}\right)\left(\frac{R_i}{R_o}\right)\left(\frac{1 + Bi'}{Bi'}\right),$$

which, basically, is equivalent to Eq. (9.4.12).

Problem 9.9 Case (a):

$$\dot{Q} = J_1 S_1 + J_2 S_2 = (k_1 S_1 + k_2 S_2)\Delta T/L,$$

where J_1 and J_2 are the thermal fluxes in the two regions, of sections S_1 and S_2. Considering that $\varepsilon_1 = S_1/S = V_1/V$ and $\varepsilon_2 = S_2/S = V_2/V$ are the volumetric fractions of the two phases, we obtain:

$$k_{eff} = \varepsilon_1 k_1 + \varepsilon_2 k_2.$$

This case corresponds to having two resistance in parallel, requiring that the two conductances are summed.

Case (b):

$$\Delta T = \Delta T_1 + \Delta T_2 = J\left(\frac{L_1}{k_1} + \frac{L_2}{k_2}\right), \quad \text{as } J = k_1\frac{\Delta T_1}{L_1} = k_2\frac{\Delta T_2}{L_2}, \quad \text{where } L = L_1 + L_2.$$

Finally, proceeding as in the previous case (a), we obtain:

$$\frac{1}{k_{eff}} = \frac{\varepsilon_1}{k_1} + \frac{\varepsilon_2}{k_2},$$

where $\varepsilon_1 = L_1/L$ and $\varepsilon_2 = L_2/L$. This case corresponds to having two resistances in series.

Problem 9.10 Since, at steady state, each section is crosses by the same the heat flux, we find:

$$J_U = \frac{k_1}{L_1}(T_1 - T_{12}) = \frac{k_2}{L_2}(T_{12} - T_2), \quad \text{hence,} \quad T_{12} = \left(\frac{k_1 T_1}{L_1} + \frac{k_2 T_2}{L_2}\right)\bigg/\left(\frac{k_1}{L_1} + \frac{k_2}{L_2}\right).$$

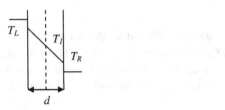

Problem 9.11 The temperature profile at the dividing wall is represented in the figure aside. Out of symmetry, the temperature at the center is constant and equal to $T_1 = (T_R + T_L)/2$. Therefore, applying Eq. (9.2.5) with $L = d/2$, we obtain:

$$J_U = k_{eff} \frac{(T_1 - T_R)}{d/2} = k_{eff} \frac{(T_1 - T_R)}{d}; \quad k_{eff} = k_s \frac{Bi}{1 + Bi}; \quad Bi = \frac{hd}{2k}.$$

At the end, we obtain again Eqs. (9.4.10)–(9.4.11), with

$$\tau = \frac{pcLd}{2k_{eff}} = \left(\frac{pcL}{p_s c_s d}\right)\left(\frac{p_s c_s d}{h} + \frac{1}{2}\frac{p_s c_s d^2}{k_s}\right) = \left(\frac{Mc}{M_s c_s}\right)(\tau_c + \tau_d),$$

where $\tau_c = p_s c_s d/h$ and $\tau_d = d^2/2a_s$, with $a_s = k_s/p_s c_s$, are the characteristic times of the convection at the wall surface and of the diffusion inside, while $M = pLA$ and $M_s = p_s dA$ are the mass of the liquid and that of the dividing wall. Now, we know that the relaxation time needed to reach steady state (i.e., how long it takes the temperature profile to reach a new steady state after a sudden chance in its boundary conditions) is $\tau_{ss} = \tau_c + \tau_d$ and therefore the QSS approximation, $\tau \gg \tau_{ss}$, is valid when $Mc \gg M_s c_s$, which coincides with Eq. (9.4.13).

Problem 9.12 This case is identical to Problem 9.11, when $d/2$ indicates the wall thickness.

Chapter 10

Problem 10.1 Since the problem is linear, the superposition principle can be applied. So, first we solve the problem without heat source, i.e. with $\dot{q} = 0$, and with the given boundary conditions, i.e., with $T(0) = T_0$ and $T(L) = T_L$, obtaining the linear profile, $T(x) = T_0 + (T_L - T_0)x/L$. Then, we solve the problem with the heat source and homogeneous B.C., i.e., $T(0) = T(L) = 0$, obtaining $T(x) = \dot{q}x(L - x)/2k$. Now, the complete solution is the sum of the two individual solutions, i.e.,

$$T(x) = T_0 + (T_L - T_0)x/L + \dot{q}x(L - x)/2k.$$

Problem 10.2 Proceeding as in Sect. 10.1.2, we derive Eq. (10.1.10), with $C_1 = 0$, i.e., $J = \dot{q}r/2$. Then, integrating using the Fourier equation, we obtain: $T(r) = -\dot{q}r^2/(4k) + C$, where C is a constant that can be determined by applying the B.C.: $J(R) = h[T(R) - T_0]$. Finally we obtain:

$$T = T_0 + (\dot{q}/4k)(R^2 - r^2) + \dot{q}R/(2h).$$

In particular, here we obtain: $T_{max} = T(0) = 570\ °C; T(R) = 153\ °C$.

Problem 10.3 This case is identical to Problem 10.2. Using dimensionless coordinates, we again obtain the following temperature profile:

$$\Theta(\xi) = \frac{1}{4}\left(1 + \frac{2}{Bi} - \xi^2\right), \quad \text{where} \quad \xi = r/R, \quad \Theta = \frac{T - T_0}{\dot{q}R^2/k}, \quad Bi = \frac{hR}{k}.$$

Thus, we obtain the relation between $\dot{q}_{max}$ and $T_{max} = T(0)$, that is,

$$\dot{q}_{max} = \frac{4k(T_{max} - T_0)}{R^2(1 + 2k/hR)}.$$

Since T cannot exceed 500 °C, we find: $\dot{q} < \dot{q}_{max} = 6000$ MW/m^3.

Problem 10.4

$$\text{Plane case:} \quad \bar{\Theta} = \frac{1}{2}\int_{-1}^{1} \Theta(\xi)d\xi = \ldots = \frac{1}{3}.$$

$$\text{Cylindrical case:} \quad \bar{\Theta} = \frac{1}{\pi}\int_{0}^{1} \Theta(\xi)2\pi\xi d\xi = \ldots\frac{1}{8}.$$

$$\text{Spherical case:} \quad \bar{\Theta} = \frac{1}{(4\pi/3)}\int_{0}^{1} \Theta(\xi)4\pi\xi^2 d\xi = \ldots\frac{1}{15}.$$

Problem 10.5 Proceeding as in Sect. 10.1.3, we find in the two regions, I $(0 < r < R)$ and II $(R, r < R + d)$:

$$J^I = \dot{q}\,r/2; \quad \text{and} \quad J^{II} = C/r.$$

Since $J^I(R) = J^{II}(R)$, we find: $C = \dot{q}R^2/2$. Now, applying the Fourier law, integrating J^{II} and imposing the boundary condition $T(R + d) = T_0$, we obtain:

$$T^{II}(r) - T_0 = (\dot{q}R^2)/(2k)\ln[(R + d)/r].$$

Then, integrating J^I, imposing that $T^I(R) = T^{II}(R)$, we obtain:

$$T^I(r) - T_0 = (\dot{q}R^2)/(2k)\{1/2[1 - ((r/R)^2] + \ln(R + d)/R].$$

Problem 10.6 Proceeding as in Sect. 10.1.3, we find the heat flux $J = \dot{q}r/3$ and the temperature distribution $\Theta(\xi) = -\xi^2/6 + C$, where Θ and ξ are defined in (10.3.2), while C is a constant. From the B.C. $J(R) = h[T(R) - T_0]$, we obtain $\Theta(\xi) = (1/6)[1 - \xi^2 + 2/Bi]$, where $Bi = hR/k$ is the Biot number. Thus, the maximum temperature is

$$T_{max} = T(0) = T_0 + (\dot{q}R^2)/(6k)[1 + 2/Bi].$$

Note that when $Bi \gg 1$, $T_{max} - T_0 = (\dot{q}R^2)/(6k)$ and therefore it is independent of h, while it is proportional to R^2. On the contrary, when $Bi \ll 1$, $T_{max} - T_0 = (SR)/(3\,h)$ and therefore it is independent of k, while it is proportional to R. The heat flux is $J_Q = -k(dT/dr)_R = \dot{q}R/3$ and it is independent of both h and k. This is easily explained considering that, at steady state, the product of the flux J_Q by the area $4\pi R^2$ equals the total heat produced, $\dot{q}\,(4\pi R^3/3)$.

Problem 10.7 The velocity profile is: $v(x) = v_{max}[1-(x/B)^2]$, where v_{max}, depending on the flow rate, is proportional to the pressure drop as $v_{max} = (\Delta p/L)(B^2/2\mu)$. In addition, the viscous dissipation power per unit volume is: $\dot{q} = \mu(dv/dx)^2 = (4\mu v_{max}^2/B^4)x^2$. Therefore, the heat equation is

$$-\kappa \frac{d^2T}{dx^2} = \dot{q} = \frac{4\mu v_{max}^2}{B^4}x^2.$$

Integrating between 0 and B, with $dT/dx(x = 0) = 0$ and $T(x = B) = T_0$, we obtain:

$$T - T_0 = \frac{\mu v_{max}^2}{3\kappa}\left[1 - \left(\frac{x}{B}\right)^4\right].$$

The heat flux is:

$$J_Q = -\kappa \left(\frac{dT}{dx}\right)_{x=\pm L} = \pm\frac{4\mu v_{max}^2}{3B},$$

i.e., the flux is positive at $x = B$ and negative at $x = -B$. Note that the heat flux is independent of the thermal conductivity k, as one can also see from:

$$J_Q = \int_0^B \dot{q}dx = \frac{4\mu v_{max}^2}{3B}.$$

Problem 10.8 The steady state heat equation is:

$$-\frac{k}{r}\frac{d}{dr}\left(r\frac{dT}{dr}\right) = \dot{q}; \quad \text{with} \quad \text{B.C.}: \ k\frac{dT}{dr}(R) = h[T(R) - T_0] \ \text{and} \ \frac{dT}{dr}(2R) = 0.$$

(a) Using dimensionless variables we obtain:

$$-\frac{1}{\zeta}\frac{d}{d\zeta}\left(\zeta\frac{d\Theta}{d\zeta}\right) = -1; \quad \text{with B.C.} \quad \frac{d\Theta}{d\zeta}(1) = Bi\Theta(1) \quad \text{and} \quad \frac{d\Theta}{d\zeta}(2) = 0,$$

where $\Theta = (T - T_0)/(\dot{q}R^2/k)$, $\zeta = r/R$ and $Bi = hR/k$.

(b) When $Bi \gg 1$, the first B.C. reduces to: $T(R) = T_0$, i.e., $\Theta(1) = 0$, and the solution is:
$d\Theta/dr = -\zeta/2 + A/\zeta$ and $\Theta = -\zeta^2/4 + A\,log\zeta + B$. Imposing that the B.C. are satisfied, we find:

$$A = 2 \text{ and } B = {}^1\!/_4. \quad \text{Therefore}: T(2R) = T_0 + (\dot{q}R^2/k)[-1 + 2\,log2 + {}^1\!/_4].$$

Problem 10.9 From an energy balance we obtain: $McdT/dt = \dot{Q} - h(T - T_0)A$, with $\Delta T = T - T_0$, where $M = \rho\frac{4}{3}\pi R^3$ is the mass of the water, while $A = 4\pi R^2$ is the exchange area of the heater. This equation can be written as:

$$\frac{d\Delta T}{dt} = -\frac{1}{\tau}(\Delta T - \Delta T_f), \quad \text{where } \tau = \frac{Mc}{hA} \text{ and } \Delta T_f = \frac{\dot{Q}}{hA}.$$

Here τ is a characteristic relaxation time, while ΔT_f is the steady state temperature difference, i.e. the solution as $t \to \infty$. The general solution is: $T(t) = T_0 + \Delta T_f(1 - e^{-t/\tau})$.

Note that for short times, i.e., when $\Delta T = \Delta T_f$, the temperature changes linearly with time, i.e., $T(t) = T_0 + (\Delta T_f/\tau)t = T_0 + (\dot{Q}/Mc)t$.

Problem   **10.10** $\Delta T_{max} = \frac{\dot{q}R^2}{4k}$ (see   Sect.   10.1.2). $R^2 = \frac{4k\Delta T}{\dot{q}} =$

$$\frac{4 \times 1 \text{ W/mK} \times 25 \text{ K}}{10^6 \text{ W/m}^3} = 10^{-4} \text{ m}^2 = 1 \text{ cm}^2.$$

Therefore, the maximum radius is 1 cm.

Chapter 11

Problem 11.1 We easily find $N = 5$ (note that N is a sort of square root of the Biot number) and therefore the fin efficiency is $\eta = TanhN/N = 0.2$ ($TanhN \cong 1$ when $N > 2$). Thus: $Q = \eta hA(T_w - T_a) = 1.5$ kW, as $A = 2$ m^2 is the fin surface area (above and below), while the temperature difference is 75 °C.

Problem 11.2 First, let us check whether the Biot number is small. Here, we find: $Bi = hR/k = 0.008 < 0.1$, since $k = 43$ W/mK. Therefore, the temperature profile can be assumed to be uniform, and we can use Eq. (9.4.4). Thus, considering that $\rho = 7840$ kg/m^3; $c_p = 460$ J/kgK, $V = \pi R^2 L$ and $A = 2\pi RL$, where L is the length of the steel bar, we obtain: $\tau = \rho\, c_p V/hA = 2885$ s.

The time interval is: $t = -\tau \ln[(T - T_a)/(T_0 - T_a)] = 6494$ s, i.e., $t \cong 1$ h and 48 min.

Problem 11.3 Data: $\rho = 7840$ kg/m^3; $c_p = 460$ J/kg K; $k = 52$ W/m K.

Therefore $Bi = hL/k = 3.6 \times 10^{-4} \ll 0.1$, so that the temperature profile can be assumed to be uniform, and we can use Eq. (9.4.4). We obtain: $\tau = \rho \, c_{pL}/h = 343$ s.

The time interval is: $t = -\tau \ln[(T_0 - T_a)/(T_i - T_a)] = 16.1$ min.

Problem 11.4 Since $\dot{Q} = m_T c_T (T_{Tin} - T_{Tout}) = 1.25 \times 10^4$ W $= m_S c_S (T_{Sout} - T_{Sin})$, we find: $T_{Sout} = 49.9$ °C.

Co-current case: $\Delta T_{In} = 46.0$ °C and therefore $S = \dot{Q}/(h_{tot}\Delta T_{In}) = 0.61$ m^2.

Counter-curernt case: $\Delta T_{In} = 54.6$ °C and therefore $S = \dot{Q}/(h_{tot}\Delta T_{In}) = 0.51$ m^2.

In general, counter-current heat exchangers are more efficient than co-current ones, i.e. they require smaller exchange areas.

Problem 11.5 We have: $T_{s,in} = 80$ °C; $T_{s,out} = 48$ °C; $T_{t,in} = 20$ °C; $T_{t,out} = 34$ °C; $\dot{m}_t = 40$ kg/s.

(a) $\dot{Q} = m_t c_t (T_{t,in} - T_{t,out}) = 1.3410^6$ W.

(b) $\dot{m}_t = \dot{Q}/[c_s(T_{s,in} - T_{s,out})] = 10.0$ Kg/s.

(c) Denoting by $N = 75$ the number of tubes and L the tube length, the exchange area (referred to the outer diameter) is $S = N\pi DL$ and $\dot{Q} = Sh_{tot}F\Delta T_{In}$. Here,

$$(\Delta T)_{lm} = \frac{\Delta T_1 - \Delta T_2}{\ln(\Delta T_1/\Delta T_2)} = \frac{T_{S,in} - T_{t,out}}{\ln[(T_{s,in} - T_{t,out})/(T_{s,out} - T_{t,in})]} = 36.3 \, \text{K}.$$

In addition, $R = (T_{s,in} - T_{s,out})/(T_{t,out} - T_{t,in}) = 2.29$; $P = (T_{t,out} - T_{t,in})/(T_{s,in} - T_{t,out}) = 0.23$. Therefore, from Fig. 11.3.4 we see that $F = 0.94$.

Now we must evaluate h_{tot} from Eq. (9.3.9), neglecting the thermal resistance of the metal. Thus,

$$h_{hot} = \left[\frac{R_o}{R_i h_i} + \frac{R_o \ln(R_o/R_i)}{k} + \frac{1}{h_o}\right]^{-1}.$$

Knowing that $h_o = 300$ W/m^2 K, we see that the thermal resistance of the metal can be neglected. Now, in order to calculate h_i, we need Re for the glycerol flow. Considering that the flow rate for each tube is $\dot{m}_t/N$, we obtain: $Re = 4\dot{m}_t/(\pi DN\mu) = 47.9$. Therefore, the regime is laminar and, since the heat flux is about constant along the heat exchanger (this is true for both counter-current and cross-current heat exchangers), we obtain: $Nu = 4.36$, hence, $h_i = Nu \, k/D = 73.5$ W/m^2 K.

Finally: $h_{tot} = 59$ W/m^2 K and we find: $L = \dot{Q}/(N\pi Dh_{tot}F\Delta T_{In}) = 171$ m. Then, since each tube presents 8 passages, the length of the heat exchanger is approximately $L/8 = 21.4$ m.

Chapter 12

Problem 12.1 Data for iron: at 0 °C.: $\rho = 7870$ kg/m^3; $c_p = 447$ J/kg K; $k = 80.3$ W/m K.

Hence : $\alpha = k/\rho c_p = 2.28 \times 10^{-5}$ m^2/s. $y/\sqrt{(4\alpha t)} = 0.4056$; $erf(0.4056) = 0.4337$.

Therefore from Eq. (12.2.3) we have: $(T-T_w)/(T_0-T_w) = 0.4337$ and $T = 34$ °C.

Problem 12.2 Data for nickel at 20 °C.: $\rho = 8900$ kg/m^3; $c_p = 444$ J/kg K; $\alpha = 2.30$ 10^{-5} m^2/s. From Eq. (12.3.3), when $y = 0$ we have: $T - T_0 = U/[\rho \, c_p\sqrt{(\pi\alpha t)}] = 160$ °C and thus, $T = 180$ °C.

Problem 12.3 Data: $k = 0.52$ W/m K; $\alpha = 1.33 \times 10^{-7}$ m^2/s. From $\Theta = (T_0 - T_a)/(T - {_i}T_a) = 0.4$ $Bi = hR/k = 1$, in Fig. 12.4.1 we see: $Fo = \alpha t/R^2 = 0.46$, and thus, $t = 1.4$ h.

This result could also be obtained approximately, knowing that $t \approx R^2/2\alpha \approx 1.7$ h.

Problem 12.4 Defining $\Theta(\tilde{x},\tilde{t})$, with $\Theta = \frac{T-T_0}{T_1}$, $\tilde{x} = x/L$ and $\tilde{t} = t/(L^2/\alpha)$, we obtain: $\frac{\partial\Theta}{\partial \tilde{t}} = \frac{\partial^2\Theta}{\partial \tilde{t}}$; $\Theta(0,\tilde{t}) = \Theta(1,\tilde{t}) = 0$; $\Theta(\tilde{x},0) = 1$.

Now, assume $\Theta(\tilde{x},\tilde{t}) = X(\tilde{x})T(\tilde{t})$ (i.e., separation of variables). Proceeding as in Sect. 12.4, we obtain two equations for T and X. The former leads to: $T(\tilde{t}) = Ae^{-\lambda^2 t}$; the latter is:

$$\frac{d^2X}{d\tilde{x}^2} + \lambda^2 X = 0,$$

to be solved with B.C. $X(0) = X(1) = 0$. This problem admits an infinite number of solutions, of the type $\sin(\lambda\tilde{x})$, satisfying the first B.C. Imposing that also the second B.C. is satisfied, we obtain: $\sin\lambda = 0$, hence: $\lambda = n\pi$, with n integer, and therefore $X(\tilde{x}) = \sin(n\pi\tilde{x})$. So, any function $XT = \sin(n\pi\tilde{x})\exp(-n^2\pi^2 t)$ is a solution of the problem and therefore, by superposition, we conclude that the genera solution is

$$\Theta(\tilde{x},\tilde{t}) = \sum_{m=0}^{\infty} c_m\sin[m\pi\tilde{x}]e^{-m^2\pi^2\tilde{t}}$$

In order to determine the constants c_m, we impose that the initial condition is satisfied, i.e.,

$$\Theta(\tilde{x},0) = 1 = \sum_{m=0}^{\infty} c_m \sin[m\pi\tilde{x}]$$

Now, consider that

$$\int_0^1 \sin[m\pi\tilde{x}] \sin[n\pi\tilde{x}] d\tilde{x} = \begin{array}{ll} 0 & \text{per } m \neq n \\ 1/2 & \text{per } m = n \end{array}$$

Thus, multiplying the above equation by $\sin[n\pi\tilde{x}]$ and integrating between 0 and 1, we obtain:

$$\int_0^1 \Theta(\tilde{x},0) \sin[n\pi\tilde{x}] d\tilde{x} = \sum_{m=0}^{\infty} c_m \int_0^1 \sin[m\pi\tilde{x}] \sin[n\pi\tilde{x}] d\tilde{x},$$

hence:

$$c_n = \frac{2}{n\pi}[1 - \cos(n\pi)] = \begin{array}{ll} 0 & \text{per } n \text{ pari} \\ \frac{4}{n\pi} & \text{per } n \text{ dispari} \end{array}$$

The final result is,

$$\Theta(\tilde{x},\tilde{t}) = \sum_{n \text{ dispari}} \frac{4}{n\pi} \sin(n\pi\tilde{x}) e^{-n^2\pi^2\tilde{t}}.$$

In particular, at $\tilde{x} = 1/2$, we obtain:

$$\Theta\left(\frac{1}{2},\tilde{t}\right) = \frac{4}{\pi} e^{-\pi^2\tilde{t}/4} - \frac{4}{3\pi} e^{-9\pi^2\tilde{t}/4} + \cdots,$$

which coincides with Eq. (12.4.6a) at $\tilde{x} = 0$.

Problem 12.5 Letting $\Theta = \frac{T-T_a}{T_i-T_a}$, the problem is identical to Eq. (12.4.2), as one would expect out of physical considerations. Therefore, at the wall $x = 0$ we find:

$$\Theta(0,\tilde{t}) = \frac{4}{\pi} e^{-\pi^2\tilde{t}/4} - \frac{4}{3\pi} e^{-9\pi^2\tilde{t}/4} + \cdots.$$

Chapter 13

Problem 13.1 First, we must know the values of α, κ and ν for air. Choose an average temperature, $T = 60$ °C, finding (Perry, Table 2–368): $Pr = 0.7$; $\kappa = 0.03$ W/m K and $\nu = 19 \times 10^{-6}$ m^2/s.

Thus, $Re = UL/\nu \cong 3 \times 10^5$, indicating that the flow regime is turbulent, so that the mean heat transfer coefficient is:

$$h = (2/3)(k/L)\,Re^{1/2}Pr^{1/3} \cong 10 \ \text{W/m}^2 \ \text{K}.$$

Hence, the heat exchanged (above and below the flat plate) is $Q = 2hS\Delta T \cong 6$ KW.

Problem 13.2 The problem depends only on the x coordinate, with $0 < x < L$, with

$$V\frac{dT}{dx} = \alpha\frac{d^2T}{dx^2}; \quad T(0) = T_0; \quad T(L) = T(L),$$

while the heat flux, $J_U(L)$, is the following:

$$J_U(L) = \rho cV[T(L) - T_0] - \kappa\frac{dT}{dx}(L).$$

In dimensionless terms we have:

$$\Theta = (T - T_0)/(T_L - T_0), \quad \xi = x/L \ \text{e} \ Pe = VL/\alpha,$$

obtaining:

$$Pe\frac{d\Theta}{d\xi} = \frac{d^2\Theta}{d\xi^2}; \quad \Theta(0) = \Theta_0; \quad \Theta(0) = 1.$$

It is convenient to express the heat flux in terms of the Nusselt number, $Nu = J_U/J_{U,cond}$, where $J_{U,cond}$ is the heat flux due to conduction only. In our case, $J_{U,cond} = -\kappa(T_L - T_0)/L$ and

$$Nu_L = \frac{J_U(L)}{J_{U,cond}} = \frac{d\Theta}{d\xi}(1) - Pe.$$

From this point on, the problem is identical to the one solved in Sect. 17.1.

Problem 13.3 Since $Pe \gg 1$ and $Re > Pe$, there is a thermal boundary layer, of thickness δ_T, external to the momentum boundary layer. At the edge $y = \delta_T$, equating convection and diffusion in the heat equation, we obtain:

$$(U/R) \approx (\alpha/\delta_T^2), \quad \text{hence:} \quad \delta_T/R \approx Pe^{-1/2}.$$

Now, since the heat flux is $J = -k(dT/dr)_{r=R}$, defining $Nu = J/(k\Delta T/R)$, we obtain:

$$Nu = R/\delta_T \approx Pe^{1/2} \approx Re^{1/2} Pr^{1/2}.$$

Chapter 15

Problem 15.1 For an ideal gas, $c = p/(RT) = 1$ atm/(82 cm^3 atm gmole^{-1} K^{-1}) (298 K) $= 4 \times 10^{-5}$ gmole/cm^3.

At the interface with solid naphthalene, $x_{A1} = p_{vap}/p = 9.6 \times 10^{-5} \ll 1$. Thus, in this dilute case, as convection is negligible, the concentration profile is linear and $J_A = cD dx_A/dz = cDx_{A1}/\Delta z$.

Therefore, the molar mean velocity is

$$v^* = x_A v_A = J_A/c = D x_{A1}/Dz = \dots 7.7 \times 10^{-7} \text{cm/s}$$

and is uniform along the tube. The same holds for the mass mean velocity, considering $v_B = 0$, i.e.,

$$v = \frac{\rho_A}{\rho_A + \rho_B} v_A = \frac{x_A M_{WA}}{x_A M_{WA} + x_B M_{WB}} v_A \cong \frac{M_{WA}}{M_{WB}} x_A v_A = \frac{M_{WA}}{M_{WB}} v^*$$
$$= 3.5 \times 10^{-6} \text{ cm/s}$$

Note that for non-dilute mixtures the molar mean velocity is still uniform, while the mass average velocity changes along the tube. Instead, v_A varies with z, as $v_A = J_A/c_A = v^*/x_A$. Thus, as x_A decreases with z, v_A increases. For example, at $z = 5$ cm, we obtain: $x_A = \frac{1}{2} x_{A1} = 4.8 \times 10^{-5}$ gmole/cm^3 and $v_A = v^*/x_A = 7.7 \times 10^{-7}$ cm/s $/4.8 \times 10^{-5}$ gmole/cm^3 $= 1.6 \times 10^{-2}$ cm/s. Thus, we see that the velocity of naphthalene is much larger than the mean velocity, as one would expect, due to its small molar fraction.

Problem 15.2 From a molar balance in a spherical crown of thickness dr we obtain:

$$J_A(r)(4\pi r^2) = J_A(r + dr)(4\pi(r + dr)^2) \quad \Rightarrow \quad J_A(r)r^2 = constant = K_1.$$
$$J_B(r)(4\pi r^2) = J_B(r + dr)(4\pi(r + dr)^2) \quad \Rightarrow \quad J_B(r)r^2 = constant = J_B(R)R^2 = 0,$$

where we have considered that air is insoluble in water. Now, proceeding as in Sect. 15.1, we obtain:

$$J_A = -\frac{cD}{1-x_A}\frac{dx_A}{dr} = \frac{K_1}{r^2}.$$

Integrating between $r_1 = R$ and $r_2 = R + \delta$, we find:

$$J_{A1} = \frac{K_1}{R^2} = cD\frac{\ln(x_{B2}/x_{B1})}{\delta}\left(1+\frac{\delta}{R}\right).$$

This result can be written i terms of a molar heat exchange coefficient,

$$J_{A1} = k(c_{A1} - c_{A2}), \quad \text{with}, \quad k = \frac{D}{\delta(x_B)_{\ln}}\left(1+\frac{\delta}{R}\right).$$

Repeating the integration between R and r we obtain:

$$\left(\frac{1-x_A}{1-x_{A1}}\right) = \left(\frac{1-x_A}{1-x_{A1}}\right)^{\frac{r-r_1}{r_2-r_1}\frac{r_2}{r}}.$$

When $\delta \ll R$, it is easy to verify that we obtain again the plane case that has been studied in Sect. 15.1, with $k = D/(\delta(x_B)_{\ln})$. In the dilute case, $x_B \cong 1$ and so: $J_{A1} = \frac{K_1}{R^2} = \frac{cD}{\delta}\left(1+\frac{\delta}{R}\right)$, that is $k = \frac{K_1}{R^2} = \frac{D}{\delta}\left(1+\frac{\delta}{R}\right)$. In particular, when $\delta \to \infty$, $J_{A1} = \frac{K_1}{R^2} = \frac{D}{R}$, i.e., $k = D/R$.

Problem 15.3 Imposing that the total flux crossing any cylindrical surface is constant, we see that $J_A = K_1/r$. Proceeding as in Sect. 15.2, we obtain: $J_A = -\frac{cD}{1-\beta x_A}\frac{dx_A}{dr} = \frac{K_1}{r}$. Integrating between R and $R + \delta$ with $x_A(R) = 0$ and $x_A(R + \delta) = x_{A0}$, we find the flux of A at the particle surface:

$$J_A = \frac{cD\ln(1-\beta x_{A0})}{\beta R \ln(1+\delta/R)}.$$

This results reduces to the plane case when $\delta \ll R$.

In the dilute case, $v^* = v_B = 0$ and thus we obtain:

$$J_A = J_{A,d} = -cD\frac{dx_A}{dr} = \frac{K_1}{r}; \quad \text{hence: } x_A(r) = A_1 + A_2\log(r),$$

where the constant A_1 and A_2 can be determined imposing the B.C. (15.2.12) obtaining:

$$x_A(r) = x_{A0}\frac{\log(r/R)}{\log(1 + \delta/R)}, \quad \text{hence: } J_{A,d}(r = R) = -\frac{cDx_{A0}}{R\ln(1 + \delta/R)}.$$

Therefore, the increment of the molar flux due to convection, namely $Sh = J_A/(J_A)_{diff}$, is again Eq. (15.2.9), that is the same as for the plane and spherical geometries.

Problem 15.4 Solve the following problem:

$$\frac{1}{\xi}\frac{d}{d\xi}\left(\xi\frac{dy}{d\xi}\right) - Day = 0; \quad y(1) = 1; \quad \frac{dy}{d\xi}(0) = 0,$$

with $y = x/x_0$, $\xi = r/R$ e $Da = kR^2/D$. We want to find $Sh = |dy/d\xi|_{\xi=1}$. Note that, even if the flux is negative, as it enters the particle, we consider its modulus, as Sh is a positive-definite dimensionless quantity. However, Si noti che il flusso è negativo perché entrante nella particella.

(a) $O(1)$:

$\frac{1}{\xi}\frac{d}{d\xi}\left(\xi\frac{dy}{d\xi}\right) = 0;$ $y_0(1) = 1;$ $\frac{dy_0}{d\xi}(0) = 0$, Thus, y_0 $(\xi) = 1$ and $Sh_0 = 0$

$O(Da)$:

$\frac{1}{\xi}\frac{d}{d\xi}\left(\xi\frac{dy_1}{d\xi}\right) - y_0 = 0;$ $y_1(1) = 0;$ $\frac{dy_1}{d\xi}(0) = 0$. Find: $y_1 = \left(\xi^2 - 1\right)/4$ thus $Sh_1 = 1/2$.

Concluding: $Sh = Da/2 + O(Da^2)$.

(b) The boundary layer is identical to the plane case, as curvature effects can be neglected. Thus,

    (i) Define a stretching coordinate, $\zeta = (1 - \xi)\sqrt{Da}$;
    (ii) The governing equation becomes $d^2y/d\zeta^2 = y$;
    (iii) Obtain: $y = e^{-\zeta} = \exp\left(-(1 - \xi)\sqrt{Da}\right)$;
    (iv) Hence: $Nu = \sqrt{Da}$.

Problem 15.5 At steady state, J_A and J_B are constant; in addition, for each mole of A reaching the catalyst surface there is one mole of B leaving it, so that $J_A = -J_B$, showing that the molar transport is diffusive. Consequently, $J_A = -cD\, dx_A/dz = \cos t$, with $x_A(z = 0) = 1$ and $x_A(z = L) = 0$ (note that the reaction on the catalyst surface is instantaneous). Therefore, x_A is linear, $N_A = cD/L$, and $c = P/RT$, so that we may conclude that: $D = LN_ART/P$.

Problem 15.6 In this case, $N_B = -2N_A$ and therefore (see Eq. (15.2.3) with $\beta = -1$):

$N_A = -[cD/(1 + x_A)]\, dx_A/dz = \cos t.$, with: $x_A(z = 0) = 1$ and $x_A(z = L) = x_{AL}$, where x_{AL} scan be determined from the equilibrium condition shown below. Integrating between $z = 0$ and $z = l$, we find:

$$\int_0^L \frac{N_A}{cD}\, dz = \int_1^{x_{AL}} \frac{1}{1 + x_A}\, dx_A \quad \Rightarrow \quad N_A = \frac{cD}{L}\ln\frac{2}{1 + x_{AL}}.$$

Now, let us determine x_{AL}. From:

$$K = c_B^2(L)/c_A(L) = c(1 - x_{AL})^2/x_{AL},$$

we see that x_{AL} satisfies the algebraic equation:

$x_{AL}^2 - 2kx_{AL} + 1 = 0$, where $k = 1 + K/2c > 1$ is a known constant. Thus: $x_{AL} = k - \sqrt{(k^2 - 1)}$

(note that we have discarded the other solution, as it is larger than 1).

Problem 15.7 At steady state, the ammonia flux reaching the dorn surface at $r = R_i$ equals the amount of ammonia that is consumed, that is $F = D(dc_A/dr)_{di}$ $(2\pi R_i L) = S(\pi R_{max}^2 L)$. The ammonia concentration, c_A, is equal to 0 at $r = R_i$ and $\alpha c_{A,}$ $_{atm} = 0.031$ mol/pa^3 at $r = R_{max}$. Applying Eq. (9.2.8), we find: $F = 2\pi LD\alpha c_{A,atm}/\ln$ (R_{max}/R_i). Thus:

$$R_{max}^2\ln\frac{R_{max}}{R_i} = \frac{2D}{S}\alpha c_{A,atm} = 177.8re^2, \quad \text{hence}: \ R_{max} = 8re.$$

Chapter 16

Problem 16.1 See Appendix F, Eqs. (F.3.6)–(F.3.9).

Problem 16.2 The dimensionless problem is as follows:

$$-\frac{1}{\xi^2}\frac{d}{d\xi}\left(\xi^2\frac{d\Theta}{d\xi}\right) = 0, \quad \text{with B.C. } \Theta(1) = 1 \quad \text{and} \quad \Theta(\infty) = 0,$$

where $\Theta = x/x_0$, $\xi = r/R$. The solution is $\Theta = 1/\xi$, leading to the flux $J = -Dc(dx/dr)_R = Dcx_0/R$ (positive because it leaves the drop). From a material balance on the drop find:

$$c_L dV/dt = -JS,$$

where $V = 4/3\pi R^3$, $S = 4\pi R^2$ and c_L is the molar concentration (mole/volume) of water in the drop. Note that in the material balance J should be the relative flux, but applying the QSS approximation we set it equal to the absolute flux. Therefore, $c_L dR/dt = -N = -Dcx_0/R$, hence:

$$RdR = dR^2/2 = -(Dcx_0/c_L)dt \quad e \quad R_i^2 - R^2 = 2Dt(cx_0/c_L).$$

The evaporation time is thus: $\tau = 1/2 \ (R_i^2/D) \ (c_L/ cx_0)$.

The QSS approximation is valid when the evaporation time $\tau \approx (R_i^2/D)(c_L/cx_0)$, is much larger than the relaxation time that is needed to reach steady state, i.e., $\tau_{ss} \approx (R_i^2/D)$. Therefore, we find: $c_L \gg cx_0$, which is obviously satisfied.

Problem 16.3 Consider the reaction $nA + B \rightarrow P$. The concentration fields of A and B during the processes of diffusion and reaction satisfy the following equations:

$$\frac{\partial c_A}{\partial t} - D_{AS}\frac{\partial^2 c_A}{\partial z^2} = -nkc_A c_B \quad \text{and} \quad \frac{\partial c_B}{\partial t} - D_{BS}\frac{\partial^2 c_B}{\partial z^2} = -kc_A c_B,$$

for $0 < z < \infty$. Note that in the dilute approximation both A and B diffuse in an atmosphere composed of pure S, with diffusion coefficients D_{AS} and D_{BS} that, in general, are different from one another. Now, if the reaction is very fast (see discussion at the end of Sect. 15.4), there will be a reaction front, advancing as methane burns the oxygen. So the equations become:

$$\frac{\partial c_A}{\partial t} = D_{AS}\frac{\partial^2 c_A}{\partial z^2} = 0 \quad \text{at} \quad 0 < z < z_f; \quad \text{with } c_A(z=0) = c_{A0} \text{ and } c_A(z=z_f) = 0,$$

$$\frac{\partial c_B}{\partial t} = D_{BS}\frac{\partial^2 c_B}{\partial z^2} = 0 \quad \text{at} \quad z_f < z < \infty; \quad \text{with } c_B(z=\infty) = c_{B0} \text{ and } c_B(z=z_f) = 0,$$

to be solved with initial conditions: $c_A(t = 0) = 0$ and $c_B(t = 0) = c_{B0}$. Now the two problems are uncoupled and can be solved separately, obtaining two error functions,

$$\frac{c_A(z,t)}{c_{A0}} = f_A(\eta) = 1 - \frac{erf(\eta)}{erf(\eta_f)}$$

$$\frac{c_B(z,t)}{c_{A0}} = f_B(\eta) = 1 - \frac{1 - erf(\sqrt{\lambda}\eta)}{1 - erf(\sqrt{\lambda}\eta_f)}$$

where $\eta = z/(4D_{Ast})^{1/2}$, while $\lambda = D_{AS}/D_{BS}$. Now we must determine the position η_f of the reaction front. To do that, we use the condition,

$$nJ_A(z_f) = -J_B(z_f) \quad \Rightarrow \quad nD_{AS}\frac{\partial c_A}{\partial z} - D_{BS}\frac{\partial c_B}{\partial z},$$

obtaining the following implicit equation,

$$\frac{c_{A0}}{c_{B0}}\sqrt{\lambda} = \frac{erf(\eta_f)}{1 - erf(\sqrt{\lambda}\eta_f)}e^{(1-\lambda)\eta_f^2}.$$

Note that, when $n = \lambda = 1$ and $c_{A0} = c_{B0}$, we find $erf(\eta_f) = \frac{1}{2}$. Finally, we obtain the flux of A at the wall, $z = 0$,

$$J_{A0} = -D_{AS}\left(\frac{\partial c_A}{\partial z}\right)_{z=0} = \frac{c_{A0}}{erf(\eta_f)}\sqrt{\frac{D_{AS}}{\pi t}},$$

corresponding to a diffusion length,

$$N_{A0} = -\frac{D_{AS}(c_{A0} - 0)}{\delta_{diff}}; \quad \delta_{diff} = erf(\eta_f)\sqrt{\pi D_{AS} t}.$$

Problem 16.4 The concentration c_A depends on r and t and satisfies the following problem:

$$\frac{\partial c_A}{\partial t} = D\frac{1}{r^2}\frac{\partial}{\partial r}\left(r^2\frac{\partial c_A}{\partial r}\right) \quad c_A(r = R) = 0; \quad c_A(t = 0) = c_A(r = \infty) = c_{A0}$$

Defining: $y(r', t) = r[c_{A0} - c_A(r, t)]/Rc_{A0}$, with $r' = r - R$, we obtain:

$$\frac{\partial y}{\partial t} = D\frac{\partial^2 y}{\partial r'^2} \quad y(r' = 0) = 1; \quad c_A(t = 0) = c_A(r' = \infty) = 0$$

hence: $y(r', t) = erfc(\eta)$, where $\eta\frac{r'}{\sqrt{4Dt}} = \frac{r-R}{\sqrt{4Dt}}$. So, finally, we obtain the solution:

$$c_A(r, t) = c_{A0}\left[1 - \frac{R}{r}erfc(\eta)\right].$$

Chapter 17

Problem 17.1 This problem is analogous to the one in Sect. 17.1. Here, however, the effects of diffusion and convection sum each other. At the end, defining $\Theta = c/c_0$, we obtain:

$$\Theta(\xi) = \frac{e^{pe}(1 - e^{-Pe(1-\xi)})}{e^{pe} - 1} \quad and \quad Sh = Pe\frac{e^{Pe}}{e^{Pe} - 1}.$$

Therefore, while when $Pe \ll 1$ we find $Sh = 1 + 1/2Pe + O(Pe^2)$, when $Pe \gg 1$, we find $Sh = Pe$.

Problem 17.2 The velocity profile is $v = v_{mx}(1 - r^2/R^2)$, where v_{mx} is the maximum velocity, $v_{mx} = 2\bar{v}(2\dot{V})/\pi R^2$ and $\dot{V}$ is the volume flow rate. Thus, we obtain: $\frac{dv}{dr}\big|_w = \gamma \frac{4}{\pi R^3} \dot{V}$. The mass flow rate absorbed at the wall, Eq. (17.3.10) is:

$$\dot{m} = 0.808 c_\infty (\gamma D^2)^{1/3} WL^{2/3} = 5.50 c_\infty (\dot{V}D^2)^{1/3} L^{2/3},$$

where $W = 2\pi R$ is the tube perimeter. This analysis is valid when the velocity profile is linear inside the boundary layer. Therefore, the thickness δ_M of the mass boundary layer at $z = L$ must be much smaller than R, that is:

$$\delta_M(L) \approx g(L) \approx \sqrt[3]{DL/\gamma} \ll R \quad \Rightarrow \quad L \ll \dot{V}/D.$$

Problem 17.3 A material balance on the water drop yields: $c_A^L dR/dt = -N_A = -k_M c_A^V$, where c_A^L and c_A^V are, respectively, the water concentration in the drop and the saturation concentration of water in air. Assuming to be in QSS conditions, it is reasonable to set $Re \gg 1$ and $Sc \cong 1$ (i.e., $v \cong D$), so that:

$$k_M = \frac{D}{R} Sh = a \frac{D}{R} \left(\frac{VR}{D}\right)^{1/2} = aD^{1/2} g^{1/4} \frac{1}{R^{1/4}},$$

where a is an $O(1)$ dimensionless constant. Finally, we obtain:

$$R^{1/4} dR = aD^{1/2} g^{1/3} \frac{c_A^V}{c_A^L} dt$$

And integrating between $t = 0$, when $R = R_0$ and $t = \tau$, when $R = 0$, we find:

$$\tau = \frac{4}{5a} \frac{R_0^{5/4}}{D^{1/2} g^{1/4}} \frac{c_A^L}{c_A^V}.$$

The QSS condition is satisfied when τ is much larger than $\tau_{ss} \approx \delta_M^2/D \approx \delta_M/Sh$, where δ_M is the thickness of material boundary layer, i.e., $\delta_M \approx R/Sh$. Therefore:

$$\tau \approx \frac{R_0}{k_M} \frac{c_A^L}{c_A^V} \gg \tau_{ss} \approx \frac{\delta_M}{k_M} \quad \Rightarrow \quad \frac{1}{Sh} \frac{c_A^V}{c_A^L} \ll 1.$$

Problem 17.4 A material balance on the candy bar gives: $c_A^S d(\pi R^2 L)/dt = J_A(2\pi RL)$, and therefore we obtain: $c_A^S dR/dt = J_A = -k_M \Delta c$. Here, c_A^S is the sugar concentration in the bar, while Δc is the sugar concentration difference across the material boundary layer, $\Delta c = c_A^L - c_{A\infty}$, where c_A^L is the sugar concentration at the surface of the candy (in general, the ratio c_A^L/c_A^S depends on the solubility of sugar in water), while $c_{A\infty}$ is the unperturbed sugar concentration, and therefore $c_{A\infty} = 0$. Since $Re < 1$ and $Pe \gg 1$ $Pe \gg 1$ we have:

$$k_M = \frac{D}{R} Sh = a \frac{D}{R} \left(\frac{VR}{D} \right)^{1/3} = a \frac{D^{2/3} V^{1/3}}{R^{2/3}},$$

where a is an $O(1)$ dimensionless constant. Finally, we obtain: $R^{2/3} dR = a D^{2/3} V^{1/3} \frac{c_A^L}{c_A^S} dt$, and integrating between $t = 0$, when $R = R_0$, and $t = \tau$, when $R = 0$, find:

$$\tau = \frac{3}{5a} \frac{R_0^{5/3}}{D^{2/3} V^{1/3}} \frac{c_A^S}{c_A^L}.$$

The conditions of applicability of the QSS approximation is identical to the case of Sect. 17.5, leading to Eq. (17.5.4), i.e.,

$$\tau \approx \frac{R_0 \, c_A^S}{k_M \, c_A^L} \gg \tau_{ss} \approx \frac{\delta_M}{k_M} \quad \Rightarrow \quad \frac{1}{Sh} \frac{c_A^L}{c_A^S} \ll 1.$$

Problem 17.5 A material balance on the nucleus of B yields, $c_B^S dR/dt = -J_B = J_A = D\Delta c_A/R$, where J_A is the diffusive flux of A leaving the particle, while $\Delta c_A = c_A^R - c_A^\infty = c_B^S \delta/R$ is the driving force of this flux. Therefore, we obtain: $R^2 dR = D\delta \, dt$ and thus $R^3 = 3D\delta t$.

The QSS condition is valid when $\tau_{ss} = R^2/D \ll \tau \approx R^3/D\delta$, that is, when $\delta \ll R$.

Chapter 19

Problem 19.1 Repeating the calculations of Sect. 19.2 using a generic T_0, we obtain:

$$\tilde{v}(\eta) = \frac{1}{12} Gr(\eta - \eta^3) - \frac{h^3}{v^2 \rho} \left(\frac{dP}{dz} - \rho g \beta (T_m - T_0) \right) (1 - \eta^2),$$

hence,

$$\langle \tilde{v} \rangle = \frac{1}{2} \int_{-1}^{1} \tilde{v}(\eta) d\eta = -\frac{h^3}{3v^2 \rho} \left(\frac{dP}{dz} - \rho g \beta (T_m - T_0) \right). \tag{19.2.5}$$

In the absence of a net flux, that is when $\langle v \rangle = 0$, we find: $dP/dz = \rho g \beta \, (T_m - T_0)$ and therefore it is reasonable to assume $T_m = T_0$.

Problem 19.2 In this case Eq. (19.2.6a) gives: $v_{max} = 0.032 \, (v/h) \, Gr$.

In the case of water at 50 °C, we have $v = 0.55 \times 10^{-6} \, m^2/s$ and $\beta = 3.7 \times 10^{-4} \, K^{-1}$; therefore, we obtain: $\gamma = g\beta/v^2 = 1.2 \times 10^{10} \, m^{-3} \, K^{-1}$ and $Gr = \gamma \, h^3 \Delta T = 6 \times 10^5$. Finally, $v_{max} = 1.06 \, m/s$.

In the case of air, we have $v = 1.8 \times 10^{-5} \, m^2/s$, $\gamma = 1.0 \times 10^8 \, m^{-3} K^{-1}$ and $Gr = 5 \times 10^3$, so that we find: $v_{max} = 0.29 \, m/s$.

Note that, although water is much more viscous than air, the natural circulation velocity is larger for water than for air, due to a much larger value of γ.

Problem 19.3 At the mean temperature $T_m = 70 \, °C = 373 \, K$, using the equation of state of ideal gases, we find:

$$\rho = PM_W/RT = (10^5 \, Nm^{-2} \times 29 \, kg \, kmoli^{-1})/(8314 \, J \, kmoli^{-1} \, K^{-1} \times 343 \, K) = 1 \, kg/m^3.$$
$$\beta = 1/T_m = 1/(343 \, K) = 2.9 \times 10^{-3} \, K^{-1}.$$

In addition, $\mu = 2 \times 10^{-5} \, kg \, m^{-1} \, s^{-1}$; $k = 0.03 \, W \, m^{-1} \, K^{-1}$. Thus: $Pr = 0.7$.

$$Gr = d^3 g \beta \Delta T/v^2 = (0.015^3 \times 9.8 \times 2.9 \times 10^{-3} 60)/(2 \times 10^{-5}/1)^2 = 1.44 \times 10^4;$$
$$Ra = Gr \, Pr = 10^4.$$

Now, applying the experimental correlation reported in Sect. 19.5, after checking that all the validity conditions are satisfied, we obtain:

$$Nu = hd/k = 0.2 \, Ra^{0.25} (L/d)^{-0.11} = 1.35, \quad \text{hence,} \quad J_U = h\Delta T = k \, Nu \Delta T/d = 162 \, W/m^2.$$

Based on the physical meaning of the Nusselt number, we see that natural convection increases purely conductive heat trasfer by about 35 %.

Problem 19.4 In Eq. (19.3.8) we should consider that, at the edge of the thermal boundary layer, the viscous term dominates the convective term, balancing the $O(1)$ driving force. Therefore, we have:

$$\frac{1}{Gr^{1/2}} \tilde{\nabla}^2 \tilde{v}_x \approx 1 \quad \Rightarrow \quad \frac{1}{Gr^{1/2}} \frac{\tilde{v}_x}{\tilde{\delta}_T^2} \approx Pr^{2a-b} \approx 1 \quad \Rightarrow \quad 2a - b = 0.$$

Note that, in the case of forced convection, the viscous force inside the boundary layer does not balance any other force and therefore $\partial^2 v_x/\partial y^2 = 0$, indicating that v_x is linear in y and so $a = b$.

An analogous analysis performed on Eq. (19.3.9) gives the following result:

$$\tilde{\mathbf{v}} \cdot \tilde{\nabla}\tilde{T} \approx \tilde{v}_x \approx Pr^{-b} = \frac{1}{Pr\,Gr^{1/2}}\tilde{\nabla}^2\tilde{T} \approx \frac{1}{Pr\,Gr^{1/2}}\frac{1}{\tilde{\delta}_T^2} \approx Pr^{-1+2a} \quad \Rightarrow$$

$$2a + b = 1.$$

This relation is valid in forced convection as well. Therefore, from $2a - b = 0$ and $2a + b = 1$, we obtain:

$$a = \frac{1}{4}; \quad b = \frac{1}{2}.$$

On the other hand, for forced convection, from $a = b$ and $2a + b = 1$, we obtain: $a = b = 1/3$.

Finally, we may conclude that, when $Gr \gg 1$ and $Pr \gg 1$ we obtain:

$$\frac{\delta_T}{L} \approx \frac{\delta}{L}Pr^{-1/4} \simeq Gr^{-1/4}\,Pr^{-1/4} = Ra^{-1/4},$$

From the above analysis we see that $v_x/V_c \approx (\delta_T/\delta)^2$, showing that the velocity profile at the wall is parabolic, instead of being linear, as in forced convection. This is easily explained considering that within the boundary layer there is a pressure gradient causing the formation of a Poiseuille-like velocity profile.

Chapter 20

Problem 20.1 The net flux can be evaluated from Eq. (19.2.13) with $F_{12} = 1$ and $A_1/A_2 = 0$, obtaining,

$Q_{net,1-2} = A_1\sigma e_1(T_1^4 - T_2^4)$, where T_1 is the silicon temperature and $T_2 = T_e = 300$ K is the temperature of the room walls. Therefore, the heat balance coincides with Eq. (20.3.5), leading to Eq. (20.3.7), where $Bi = e_1\sigma T_e^3 L/k = 0.61$ and $\Theta_0 = 2.67$. Finally, we obtain:

$$t = 240\,\text{s} = 4\,\text{min}.$$

Problem 20.2 In general, we have:

$$ds = \left(\frac{\partial S}{\partial T}\right)_v dT + \left(\frac{\partial S}{\partial V}\right)_T dV = \frac{1}{T}C_v dT + \left(\frac{\partial p}{\partial T}\right)_v dV,$$

where we have considered that $C_V = (\partial U/\partial T)_V = T(\partial S/\partial T)_V$ and we have applied the third of the Maxwell equations. Now, since for a black body $p = u/3$ and $u = bT^4$, i.e., $p = (b/3)\,T^4$ and therefore $(\partial p/\partial T)_V = dp/dT = (4/3)bT^3$. In addition, as

$U = Vu = VbT^4$, we finally obtain: $C_V = (\partial U/\partial T)_V = 4VbT^3$. So, concluding, we have:

$$dS = 4VbT^2 dT + \frac{4}{3}bT^4 dV.$$

Problem 20.3 Consider a radiation in equilibrium with the cavity walls. If we expand the volume of the cavity, keeping the wall temperature constant, we have to provide the cavity the energy of the radiation that fills the additional volume. If the process is reversible and isothermal, the relation that has been found in Problem 20.2 becomes: $Q = T \Delta S = (4/3) bT^4 (V_{fin} - V_{iniz})$.

For an adiabatic transformation, instead, imposing $dS = 0$ we obtain: $dV/V = -3$ dT/T, and therefore, $VT^3 = const$.

Background Reading

(1) R. B. Bird, W.E. Stewart and E.N. Lightfoot, *Transport Phenomena*, Wiley (1960). This is our gold standard, the textbook that has taught transport phenomena to generations of students.

(2) S. Middleman, *An Introduction to Mass and Heat Transfer*; and *An Introduction to Fluid Mechanics*, Wiley (1998). A beautiful textbook, with an impressive number of intelligent problems.

(3) W.M. Denn, *Analysis of Transport Phenomena*, Oxford University Press (1998). An advanced textbook, particularly clear in the dimensional analysis.

(4) L.G. Leal, *Laminar Flow and Convective Transport Processes*, Butterworth-Heineman (1992). An advanced textbook, based on the mythical graduate lectures taught by Prof. Andreas Acrivos.

(5) G.K. Batchelor, *An Introduction to Fluid Mechanics*, Cambridge University Press (1967). This is the reference text on fluid mechanics, written by one of the greatest fluid dynamicists of the 20th century.

© Springer International Publishing Switzerland 2015
R. Mauri, *Transport Phenomena in Multiphase Flows*,
Fluid Mechanics and Its Applications 112, DOI 10.1007/978-3-319-15793-1

Index

© Springer International Publishing Switzerland 2015   457
R. Mauri, *Transport Phenomena in Multiphase Flows*,
Fluid Mechanics and Its Applications 112, DOI 10.1007/978-3-319-15793-1

Printed in the United States
By Bookmasters